1975 Yearbook of Science and the Future

Encyclopædia Britannica, Inc.

Chicago Toronto London
Geneva Sydney Tokyo Manila
Johannesburg Seoul

1975 Yearbook of Science and the Future

MANAGING EDITOR
Lawrence K. Lustig

EDITOR
Dave Calhoun

EDITORIAL CONSULTANT
Howard J. Lewis, Director, Office of Information,
National Academy of Sciences—National Academy
of Engineering—National Research Council

STAFF EDITORS
Daphne Daume, Arthur Latham,
Mary Alice Molloy

ART DIRECTOR
Cynthia Peterson

DESIGN SUPERVISOR
Ron Villani

SENIOR PICTURE EDITOR
Catherine Judge

PICTURE EDITOR
Jeannine Deubel

LAYOUT ARTIST
Richard Batchelor

ART PRODUCTION
Richard Heinke

CARTOGRAPHER
Chris Leszczynski

EDITORIAL PRODUCTION MANAGER
J. Thomas Beatty

PRODUCTION COORDINATOR
Ruth Passin

PRODUCTION STAFF
Sujata Banerjee, Necia Brown, Charles Cegielski,
Barbara Wescott Hurd, Marilyn Klein,
Lawrence Kowalski, Winifred Laws, Susan Recknagel,
Julian Ronning, Harry Sharp, Susan Stucklen

COMPUTER SERVICES
Patricia Wier, Robert Dehmer

COPY CONTROL
Mary Srodon, Supervisor; Mary K. Finley

INDEX
Frances E. Latham, Supervisor;
Rosa E. Casas, Mary Neumann, Mary Reynolds

SECRETARY
Fleury Nolta

DIRECTOR, YEARBOOKS
Margaret Sutton

THE UNIVERSITY OF CHICAGO
The Yearbook of Science and the Future
is published with the editorial advice of the faculties of
the University of Chicago

Encyclopædia Britannica, Inc.

CHAIRMAN OF THE BOARD
Robert P. Gwinn

PRESIDENT
Charles E. Swanson

VICE-PRESIDENT, EDITORIAL
Charles Van Doren

Science and Shortages

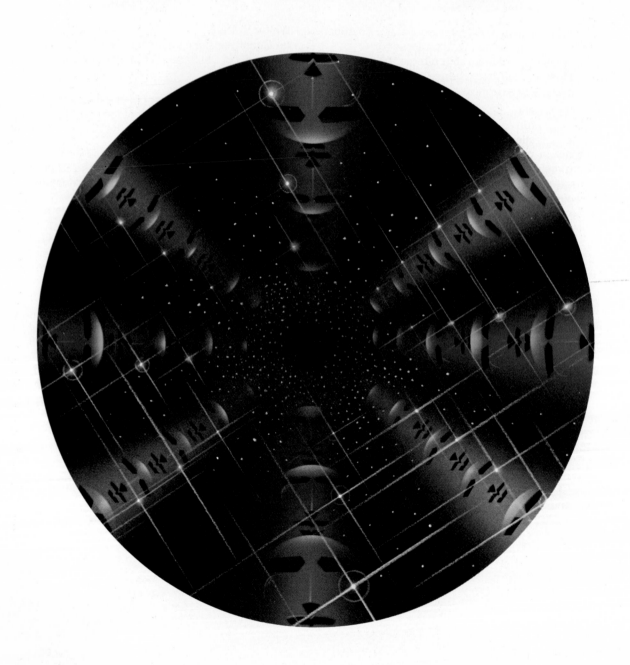

An all-too-pervasive theme during the past year was that of shortages: fuel and power, food and water, and paper and plastics were among the most important in a long and growing list. At times, these shortages became sufficiently acute to generate crisis conditions. In the United States and Western Europe the greatest impact was felt in the area of energy, but in parts of the less developed world food and water were in dangerously short supply and famine seemed close at hand.

This situation presents a major challenge to science, a challenge very different in kind but certainly not in complexity or difficulty from that of landing men on the moon. Can scientists find the means to prevent, or perhaps circumvent, ever worsening shortages? In a world with finite natural resources and a rapidly expanding population—predicted to double by 2000—the achievement of such a goal will require great ingenuity.

Scientists today are addressing themselves to the problem, and their efforts are reflected in this *1975 Yearbook of Science and the Future*. For example, dependence on oil and coal for energy may end if researchers are successful in developing alternate sources of power. The enormous potential of the sun as such a source is described in "Power from the Sun," and an ancient method of harnessing energy is brought up to date in "Power from the Wind." The steam, hot water, and molten rock that lie beneath the earth's surface might also some day contribute greatly to the world's power supply, as discussed in "Power from the Earth."

Our supply of food depends upon protecting crops from the ravages of insect pests, but at the same time such general poisons as DDT have proved harmful to the natural environment. The achievements of researchers in overcoming this conflict are described in "Growth Regulators: A New Approach to Insect Control." Food supply in a specific area, that part of Africa just south of the Sahara, is under serious threat because of recent severe drought conditions. The result is an apparent expansion of the desert, portrayed in "The Advancing Sahara." Also threatened are the world's tropical forests. In many places they are being cleared for agriculture and utilized for wood products, but, as discussed in "The Tropical Forests: Endangered Ecosystems?," many scientists suggest that these practices may cause considerable harm to the global environment.

Conservation of space and materials has been a major concern of scientists in the field of communications, and one result has been the design of a system based on hair-thin glass fibers. "Fiber Optics: Communications System of the Future?" tells of the nature and potential advantages of this new development.

Of course, not all scientists in 1974 were concerned with shortages. Prevention of disease has continued to be the subject of much research, and in "The Lymphatic System: Man's Biological Defense Network" we learn of the recent progress made in understanding and treating the disorders of this major component of the human body. Of immeasurable aid to research physicians, as well as to scientists in many other fields, are the giant electron microscopes capable of resolving images of atoms and molecules. These and other advances in microscopy are described in "Probing the Invisible World."

Man's past is a fertile field of exploration for many scholars. When and why our ancestors first began to sow and harvest food crops is the subject of "Origins of Agriculture," describing a research effort that also shed light on changes in climate in the ancient world. An outstanding collection of cultural objects from past civilizations is the subject of a pictorial essay on Mexico's renowned National Museum of Anthropology. And the threat of destruction, by water and pollution, of a city that is a living museum of past glories is vividly described in "Can Venice Be Saved?"

Full-color photographs and drawings abundantly illustrate all of these and other Feature articles in the *Yearbook*. As in the past, distinguished authorities have been selected to write these features and also the factual Year in Review treatments of the individual disciplines within the area of science and technology.

Scientists, indeed, face a challenge of great magnitude. How they have responded to date provides fascinating reading in this *Yearbook*, and their future efforts will undoubtedly shape the contents of the editions in the years ahead.

—THE EDITOR

Contents

Contributors to the Science Year in Review

Edward M. Arnett *Chemistry: Structural Chemistry.* Professor of Chemistry, University of Pittsburgh, Pittsburgh, Pa.

Joseph Ashbrook *Astronomy.* Editor, *Sky and Telescope,* Cambridge, Mass.

C. E. Ballou *Molecular Biology: Biochemistry.* Professor of Biochemistry, University of California, Berkeley.

Hyman Bass *Mathematics.* Professor of Mathematics, Columbia University, New York, N.Y.

Louis J. Battan *Atmospheric Sciences.* Professor of Atmospheric Sciences and Director of the Institute of Atmospheric Physics, University of Arizona, Tucson.

Harold Borko *Information Science and Technology.* Professor in the School of Library Service, University of California, Los Angeles.

George M. Briggs *Foods and Nutrition.* Professor of Nutrition, University of California, Berkeley.

D. Allan Bromley *Physics: Nuclear Physics.* Henry Ford II Professor and Chairman, Department of Physics, and Director, A. W. Wright Nuclear Structure Laboratory, Yale University, New Haven, Conn.

James J. Brophy *Electronics.* Academic Vice-President, Illinois Institute of Technology, Chicago.

Richard L. Davidson *Molecular Biology: Genetics.* Associate Professor of Microbiology and Molecular Genetics, Harvard Medical School and Children's Hospital Medical Center, Boston, Mass.

F. C. Durant III *Astronautics and Space Exploration: Earth Satellites.* Assistant Director (Astronautics), National Air and Space Museum, Smithsonian Institution, Washington, D.C.

Robert G. Eagon *Microbiology.* Professor of Microbiology, University of Georgia, Athens.

Lothar K. Engelmann *Photography.* Dean, College of Graphic Arts and Photography, Rochester Institute of Technology, Rochester, N.Y.

Joseph Gies *Architecture and Building Engineering.* Science and technology writer, Chicago, Ill.

Robert Haselkorn *Molecular Biology: Biophysics.* F. L. Pritzker Professor and Chairman, Department of Biophysics and Theoretical Biology, University of Chicago.

L. A. Heindl *Earth Sciences: Hydrology.* Executive Secretary, U.S. National Committee for the International Hydrological Decade, National Academy of Sciences—National Research Council, Washington, D.C.

Richard S. Johnston *Astronautics and Space Exploration: Manned Space Exploration.* Director of Life Sciences, NASA Johnson Space Center, Houston, Tex.

Lawrence W. Jones *Physics: High-Energy Physics.* Professor of Physics, University of Michigan, Ann Arbor.

Lou Joseph *Medicine: Dentistry.* Assistant Director, Bureau of Public Information, American Dental Association, Chicago, Ill.

Walter S. Koski *Chemistry: Chemical Dynamics.* Professor of Chemistry, Johns Hopkins University, Baltimore, Md.

Ernest R. Kretzmer *Communications.* Director, Data Communications Technology & Applications Laboratory, Bell Telephone Laboratories, Holmdel, N.J.

John G. Lepp *Zoology.* President, Marion Technical College, Marion, Ohio.

Howard J. Lewis *Science, General.* Director, Office of Information, National Academy of Sciences—National Academy of Engineering—National Research Council, Washington, D.C.

Melvin H. Marx *Behavioral Sciences: Psychology.* Professor of Psychology, University of Missouri, Columbia.

Marcella M. Memolo *Agriculture.* Public Information Officer, Information Division, Agricultural Research Service, U.S. Department of Agriculture, Washington, D.C.

Raymond Lee Owens *Behavioral Sciences: Anthropology.* Assistant Professor of Anthropology, University of Texas, Austin.

George R. Pettit *Chemistry: Chemical Synthesis.* Professor of Chemistry, Arizona State University, Tempe.

Willard J. Pierson, Jr. *Marine Sciences.* Professor of Oceanography, University Institute of Oceanography, The City College of New York, New York, N.Y.

Froelich Rainey *Archaeology.* Professor of Anthropology and Director of the University Museum, University of Pennsylvania, Philadelphia.

John R. Rice *Computers.* Professor of Mathematics and Computer Science, Purdue University, West Lafayette, Ind.

Henry I. Russek *Medicine: Cardiology.* Research Professor in Cardiovascular Disease and Clinical Professor of Medicine, New York Medical College, New York, N.Y.

Byron T. Scott *Medicine: General Medicine.* Assistant Professor, School of Journalism, Ohio University, Athens.

Mitchell R. Sharpe *Astronautics and Space Exploration: Space Probes.* Science writer, author of *Satellites and Probes, Living in Space,* and *Dividends from Space,* Huntsville, Ala.

Albert Smith *Botany.* Associate Professor of Biology, Wheaton College, Wheaton, Ill.

Dorothy Plack Smith *Chemistry: Applied Chemistry.* Manager, News Service, American Chemical Society, Washington, D.C.

Frank A. Smith *Transportation.* Senior Vice-President-Research, Transportation Association of America, Washington, D.C.

J. F. Smithcors *Veterinary Medicine.* Editor, American Veterinary Publications, Santa Barbara, Calif.

William E. Spicer *Physics: Solid-State Physics.* Professor of Electrical Engineering and Materials Science, Stanford University, Stanford, Calif.

Gene H. Stollerman *Medicine: Rheumatic Diseases.* Professor and Chairman, Department of Medicine, University of Tennessee College of Medicine, Memphis.

Kenneth E. F. Watt *Environmental Sciences.* Professor of Zoology, University of California, Davis.

Robert L. Wesson *Earth Sciences: Geophysics.* Geophysicist, U.S. Geological Survey, Menlo Park, Calif.

James A. West *Energy.* Staff Assistant, Energy and Minerals, Office of the Secretary, Department of the Interior, Washington, D.C.

Walter H. Wheeler *Earth Sciences: Geology and Geochemistry.* Professor of Geology, University of North Carolina, Chapel Hill.

Richard C. York *Medicine: Crib Death.* Technical Editor, Environmental Division, Sargent and Lundy Engineers, Chicago, Ill.

A Pictorial Essay

The Museum as Work of Art

by Howard J. Lewis

Dedicated to the interpretation of Mexican culture
from its earliest beginnings, Mexico's National Museum
of Anthropology is itself an outstanding example
of the nation's artistic genius.

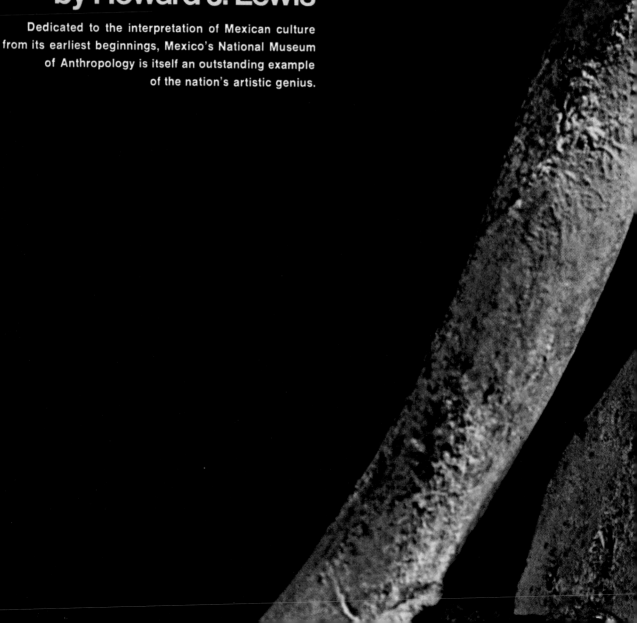

Dominating the central patio of Mexico's National Museum of Anthropology, a single bronze-clad column supports a ribbed aluminum canopy through which water cascades to the stone pavement below (opposite page). The structure combines modern engineering techniques with art forms suggesting the nation's Indian and Spanish roots, a theme explicitly set forth in exhibits devoted to the many cultures that have flourished on what is now Mexican soil. The jaguar head (above), part of a stone receptacle for the hearts of sacrificial victims, dates from the Late Postclassic Period, the last cultural phase before the Spanish conquest.

HOWARD J. LEWIS *is director of the Office of Information of the National Academy of Sciences, National Research Council, Washington, D.C.*

The National Museum of Anthropology in Mexico City enjoys the favor of museologists the world over, but it is even more distinguished by the apparently unanimous appreciation of its visitors. There are those who find fault with New York City's American Museum of Natural History, Chicago's Museum of Science and Industry, the Louvre, or even the eminently successful Deutsches Museum in Munich—but if there is a single soul who has strolled through the richly mounted rooms of the National Museum of Anthropology and has emerged less than enthralled, he has yet to make himself known.

After ten years as director of the museum, Ignacio Bernal is prepared to accept such immoderate praise, but he is also quick to point out that these comparisons are not fair to the other institutions. For one thing, the Museum of Anthropology has a single subject; the others attempt to be universal collections. For another, because many of the monumental objects shown at the Museum of Anthropology had been gathered before it was built, the architects were able to design the building around the holdings; in most other major museums, the curators have been required to accommodate new acquisitions to an existing edifice.

Bernal feels it is more appropriate to compare the Museum of Anthropology with other national museums having a similarly concentrated emphasis. Among these, the famous museum of Egyptian antiquities in Cairo and its counterpart in Athens were built long before the exquisite modern display techniques utilized by the Museum of Anthropology became available. For a true comparison, therefore, Bernal refers the global connoisseur to the newly built museums in Jerusalem and Teheran.

But if the splendid museum in Mexico City has benefited from its concentrated focus, it also bears cultural responsibilities far greater than many of its larger sister museums. It is designed to convey a complex and subtle message to its visitors—namely, the colossal nature of the feat whereby the early cultures created a civilization, without out the benefit of established cultures of still earlier peoples to build upon.

"We are trying to show that, contrary to what has so often been said, Mexican history did not begin with the arrival of the Spaniards." Bernal holds the view that, although this message is an important one to convey to foreign visitors (as well as to Mexicans of Spanish descent), it is even more significant to the Mexican Indians. He continues: "They live in areas outside the general flow of information and culture and frequently have the impression of having been set aside. When they visit our collections, especially in ethnography, they may see . . . the kind of dress that their own grandmother wore. . . . When they come across something familiar in a clean, well-lighted display cabinet, they say, 'These are things from our culture. We may be poor and small in number, but we have value.' " In this mission, the museum is aided greatly by the dedication of its multilingual guides, whose sympathies are made abundantly clear to visitors.

12

In explaining the role of his museum, Bernal takes great pains to point out that the Museum of Anthropology tells only part of the story of Mexico. "You cannot see this museum by itself. This museum shows Indian Mexico, but that is only part of the story. The museum at Tepoztlán shows Spanish or colonial Mexico and the Chapultepec Museum of History in Mexico City shows independent Mexico. The three together try to show the visitor the basic factors of Mexican culture and Mexican life—that it is a country that has mixed two cultures, two civilizations; and if you see only one or the other you are not understanding anything. You have to understand the combination, which is what Mexico is."

Even though the Museum of Anthropology has only one-third of the story to tell, the instructional task facing its director and his curatorial staff of eight is one of considerable magnitude. One may have known that many separate cultures developed in pre-Hispanic Mexico—the Aztec, Toltec, Mixtec, and Olmec, among others—but it is nonetheless startling to learn that when Hernán Cortés landed on the American continent, in 1519, 175 different languages were being spoken in Middle America. The great city of Teotihuacán—whose origins in the 2nd or 3rd century A.D. still baffle archaeologists—was probably inhabited by close to 150,000 people at its zenith. In both size and grandeur it eclipsed many major European cities of the first thousand years of the Christian era, none of which was aware of its existence.

The grand design

In their initial planning, the museum directorate made several basic design commitments. One was the museographic conception—in the words of Pedro Ramírez Vázquez, the museum's chief architect—of providing "a scientifically exact presentation which at the same time would be visually so effective that a museum visit might constitute a true dramatic spectacle and experience . . . to elicit some emotional response before the relic or work of art." Few would deny that this objective has been met.

A second decision was to withstand a common compulsion of museum curators to emphasize the taxonomic aspects of the subject mat-

Detail of a Late Classic stucco mask (opposite page) representing the Mayan sun god Kinich Ahau. More peaceful and benign than the warlike cultures that followed it, Mayan civilization first developed in the rain forests of Guatemala and achieved a remarkable renascence in the Yucatán Peninsula about the 9th century A.D. In the gardens of the Museum of Anthropology is the splendid example (above) of a Late Zapotec decorated wall from Mitla in the Valley of Mexico. The pieces forming the design were first cut so that they fitted together perfectly and then attached to the wall's surface. From the Veracruz region at approximately the same period, the clay figure (below, left) shows Xochipilli, the Aztec god of music and dance, adorned as for a festival.

ter—*i.e.*, those characteristics that tend to differentiate apparently similar objects. Instead, the aim was to show that there is a unity in the aspirations and cultural responses of man that transcends the differences arising among peoples of diverse backgrounds interacting with diverse environments.

Both Bernal and Vázquez are also aware of the limitations of human physiology. One of the requirements that Bernal imposed—and to which Vázquez quickly assented—was that the visitor could opt to visit only one exhibit hall if he wished. In this, they triumphantly overthrew one of the more sadistic traditions of the museographic profession: that the museum-goer must proceed through an interminable series of rooms in order to find what he wants to see.

In the Museum of Anthropology, it is impossible to move through three adjacent halls without going onto the patio. And having once set foot in that pleasant place, on a cool and sunny Mexican day, it is extraordinarily difficult not to lower oneself to a nearby bench, listen to the children, or be awed by the torrential downpour of water that flows continuously from the massive roof suspended from a single column over the lower portion of the patio. Or, at the other end of the patio, one can sit by a grass-spiked pool and be transported back in time by the melancholy sounds of a conch shell. Even on the second floor, it is necessary to step out onto the gallery that runs its full length. Vázquez notes that the break serves not only to relax the faculties of concentration but also to remind the visitor that he is in Chapultepec Park, "the most ancient and one of the most magnificent parks in the New World."

16

Spectacular museum exhibit (opposite page) recreates part of the Temple of Quetzalcóatl at Teotihuacán, the most important city of central Mexico until its destruction by the Toltecs about A.D. 650. Quetzalcóatl, the Feathered Serpent, was one of the oldest and most famous deities of the ancient Mexican pantheon. The statue at left, called "The Adolescent" (Huastec culture, c. A.D. 900–1250), represents a young priest of his cult or possibly the god himself as a priest. Low reliefs on the body include symbols of maize, said to be one of Quetzalcóatl's gifts to the human race.

Lee Boltin

Marc and Ev Bernheim from Woodfin Camp

*Clay figurines like those
of the ball players (above), buried
as votive offerings on the island
cemetery of Jaina, Yucatán,
illustrate the physical appearance
and clothing of the ancient Maya.
The cranial deformation was induced,
perhaps for cosmetic reasons,
but in many respects these
eight-inch statuettes resemble
the Maya of today.*

Serpentine and mosaic mask (far left), dating from the 5th century A.D., was found at Teotihuacán. At left is a votive stone hacha or ax of the Classic Period, from El Tajín, Veracruz, carved in the form of a man's head with a dolphin headdress.

A Brief Guide to the Museum

The subtle pedagogic efforts of the National Museum of Anthropology are most visible in the layout of its numerous halls. The first hall a visitor enters is not about Mexico but about anthropology. Eskimos, Lapps, and Hopis are shown, as well as the people of Middle America. There are objects from Mesopotamia and Egypt as well as from pre-Columbian America. A section on physical anthropology begins with the origin of man and extends through his cultural evolution. Prehistoric man is examined as a biological specimen.

A second section of the first hall deals with archaeology, how it is carried out and what it reveals about the relationship between prehistoric man and his environment, from the Lower Paleolithic era to the present. The study of linguistics occupies the central part of the hall, and the fourth section, dedicated to social ethnology, aims to show that "All men resolve the same needs with different resources and in distinct ways; all cultures are equally valuable."

The visitor moves from the Hall of Anthropology to the Hall of Mesoamerica, where he is introduced to the people of Middle America and their works. It is here that the guides introduce visitors to the incredible diversity of pre-Columbian cultures and to the characteristics they had in common. Their artifacts, religions, and intellectual achievements can be studied in detail or skimmed over by visitors eager to see the stunning monuments that wait in the adjoining halls.

The Hall of the Origins exhibits the routes by which the ancient wanderers arrived in Middle America. There are displays of settlements and activities of the Pleistocene era, including a diorama showing the killing of a mammoth that rivals any similar display in any natural history museum. From this, one moves sequentially through the halls that border the patio: Preclassic, Teotihuacán, Toltec, Mexica, Oaxaca, Gulf of Mexico, Maya, and the cultures of the North and West.

On the second level of the museum are the halls of ethnography, in which the products of contemporary Indian cultures are displayed. These include songs, toys, and games as well as clothing, shelters, and utensils. Again there is an introductory hall that permits—and encourages—a panoramic view of the throng of individual cultures. Eight adjoining halls are devoted to the principal cultural regions of indigenous Mexico, and a final hall is dedicated to the processes of social and cultural change that are taking place among the Indian peoples today.

*Colossal Olmec stone head (top), almost seven and one-half feet high, is displayed
in the museum gardens. In the workshops (below, left), exhibits are
prepared, such as that of a burial c. 1300–800 B.C. (below, right).*

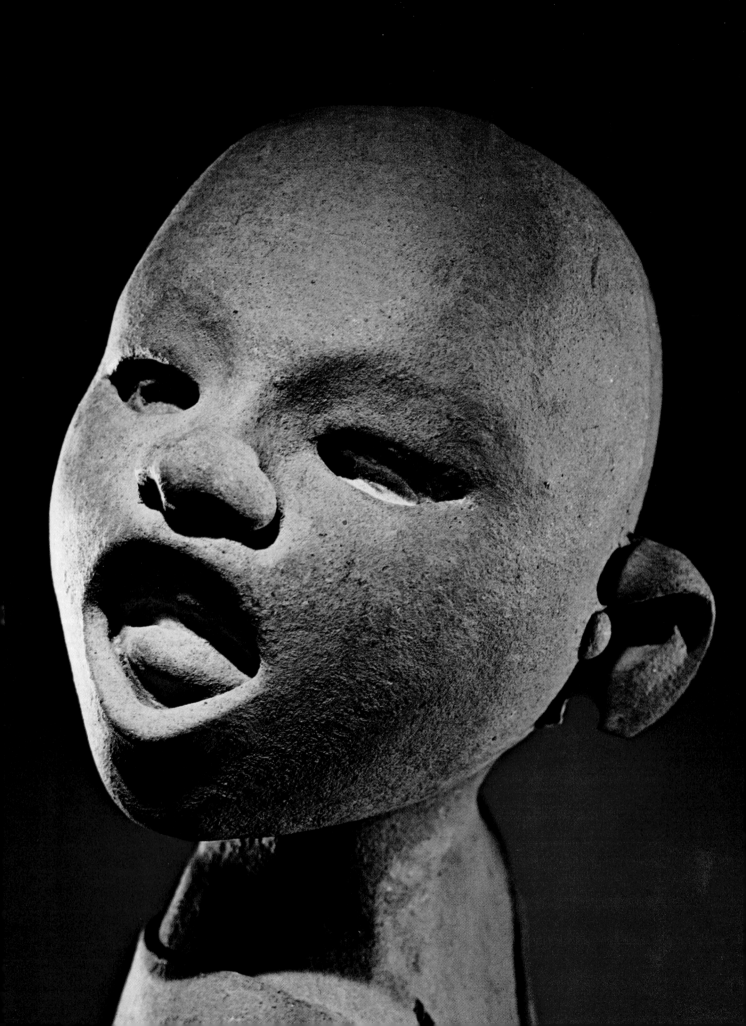

Historical background

The conceptual origins of the Museum of Anthropology can be traced back to the middle of the 16th century, when Bernardino de Sahagún (*c.* 1490–1590) and a handful of fellow Spaniards and Indians began to assemble what they could of the old myths and legends in an effort to preserve the vanquished culture. Unfortunately, their interest in the ancient civilizations was not shared by the political and religious leaders of the time. The fabulous gold ornaments were melted down into bullion; the great statues and temples were shattered in an effort to eradicate the symbols of paganism. Eventually even the scholarly efforts were abandoned. The invaluable and irreplaceable codices of ancient laws and institutions were transported to European vaults or burned on the spot.

Thus, as the golden age of Spanish colonialism dawned in Middle America, the dark ages began for the remnants of the ancient cultures. They did not end until the latter half of the 18th century, when a group of the new Mexicans, inspired by the ideas of the Enlightenment, awoke to the existence of the incalculable cultural treasure whose ruins lay underfoot. The renaissance was catalyzed by the fortuitous discoveries in 1790 of three immense monoliths—the Sun Stone, the statue of Coatlicue, and the monument to the victories of Tizoc.

These three pieces served as the nucleus of what came to be the forerunner of the present museum—established in 1825 and housed first in the University of Mexico and then, according to a decree signed by Maximilian on Dec. 5, 1865, in a new National Museum in an old palace on Moneda Street in Mexico City. In its early years the museum included both post-Hispanic objects and a natural history collection, but in 1940 it was exclusively dedicated to anthropological material. The present building was opened to the public in September 1964 by Pres. Adolfo Lopéz Mateos, whose remarkable and persistent devotion had made its construction possible.

The planning period

During the two years that preceded the construction of the museum, a group under the leadership of the director and his architect met at least once a week. Their concern was not only the architectural design of the building but also the relative proportion and distribution of the objects to be displayed. This was no mean task, for they had available to them one of the largest treasure troves of ancient artifacts ever discovered. For the zealotry of the Spaniards had been limited by one important circumstance—most of the items they had sought to pilfer or destroy were hidden from view, buried under thousands of cubic yards of Mexican soil. And this included not only small objects but entire cities.

There were fundamental decisions to be made regarding architectural design. The design group, under Vázquez, boldly reached back to the nearly forgotten traditions of pre-Hispanic architecture. "Once these constant elements of tradition were established," he said, "our

Mayan-Toltec stone statue from central Mexico (opposite page) represents a priest wearing the flayed skin of a young victim to appease Xipe Tétec, the god of spring. The grisly ceremony symbolized the taking on of a new spring mantle by the earth. Shown above is a Huastec vessel of the Postclassic Period, modeled from clay in the form of a creature resembling a monkey.

Photos, Lee Boltin

23

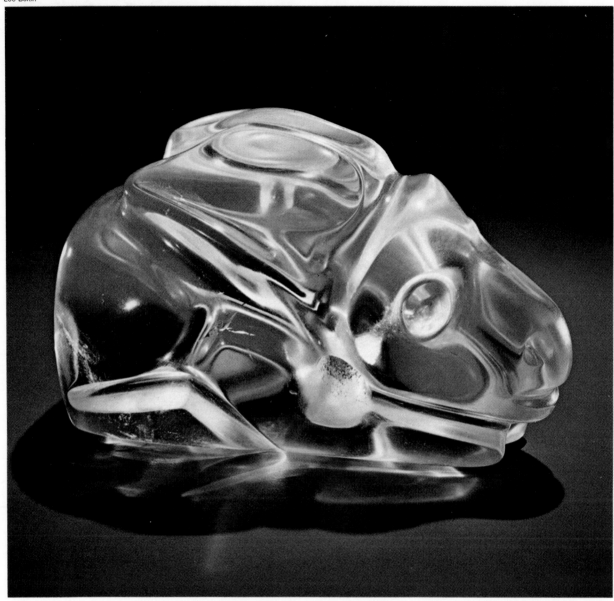

*Surprisingly modern jade mask (opposite page, top), made
in several parts, comes from the Mayan ceremonial center
of Palenque in what is now the state of Chiapas. The later Aztec
culture is represented by the stone sculpture of a coiled
rattlesnake (opposite page, bottom) and by the seven
and one-half-inch jadite frog (above).*

purpose became that of preserving and continuing them, even though their formal solution might not be so adopted. We do much the same thing when we respect and adhere to certain unchanging spiritual norms of our grandfathers without necessarily keeping their manner of dress.''

What were these norms? Several suggested themselves immediately: respect for and integration with the physical environment, generous use of space and sunlight, technical boldness, preservation of natural colors and textures, and—most important—''a plastic continuity perpetuated through the handiwork of artisans and modes of construction that are characterized by an ambition for permanence and boldness of design.''

Even the spiritual qualities of the ancient peoples were woven into the complex architectural fabric. Vázquez saw the Mexicans as drawing upon two religious traditions and inheriting from both a ''concept of architecture oriented toward eternity.

''Thus the Mexican love of the landscape, expressed in a striving for harmony between architecture and environment, began as part of an exalted conception of man, which was not limited to his physical and individual dimensions alone, but extended to encompass his dignity as a person in the grandeur of his collective expression.''

Installation

So much for the design of the museum. What about its contents? It was decided not to take existing scholarly conceptions for granted. A team of 40 scientific consultants was asked to prepare detailed monographs on all aspects of pre-Hispanic culture. Three thousand pages of original research resulted from this effort, a virtual encyclopaedia of Mexican anthropology.

The quest for new and more representative material resulted in the dispatch of 70 ethnographic expeditions throughout Mexico. They returned with information—including 15,000 photographs and hundreds of drawings—that permitted more accurate cataloging of the current holdings, and with thousands of additional secular and religious objects. But the expeditions produced more than objects and documentation. They produced people. Although it would have been possible for the museum staff to reconstruct the characteristic dwellings of the indigenous tribes for the ethnographic displays, the design team chose to involve the villagers themselves. Bringing their local materials with them, the villagers set to work within the shell of the museum. As the men recreated their village abodes according to traditional techniques, the women of the tribe made pottery and baskets. Even the children were invited to contribute, and many made toy animals for the displays.

Meanwhile, the gathering of objects continued. The total is now estimated at 60,000 and stretches back some 14,000 years. The earliest item in the collection is also one of the handsomest: a coyote head fashioned from a fossilized llama vertebra of 12000 B.C.

Moving the 170-ton statue of the rain god Tláloc (below) 30 mi from Coatlinchán to the museum required both engineering expertise and diplomatic skill in dealing with villagers who feared Tláloc would take the rains with him. Fearsome apparition depicted in stone (opposite page, top) is Coatlicue, "goddess of the serpent skirt," bride of the sun and mother of Quetzalcóatl. The statue, found in Mexico City on the site of the Aztec capital of Tenochtitlán, symbolizes the deity's attributes as mother goddess maintaining the cosmic balance of life and death. Museum exhibit shown on the opposite page, below, is a reconstruction of a Mayan burial.

Porterfield-Chickering from Robert Davis

26

development, but it does insure that the sperm does not contribute any of its genes.

If all the previous steps are carried out successfully, a small percentage of the recipient eggs with transplanted nuclei will develop into normal, viable animals, genetically identical to the animals that provided the donor nuclei. As outlined, this procedure provides only one or, at most, two animals identical to the donor.

By introducing a procedure called serial transplantation, scientists have found it possible to obtain a true "clonal" population from one successful transplant. This procedure circumvents the problem of low frequency of success (about 1 in 100) in transplantation experiments. As soon as it is clear that a transplant has been successful and normal development is ensuing, the embryo is dissociated much in the same way as the original donor tissue. The resulting nuclei of this embryonic dissociation are, themselves, used as the donor nuclei for subsequent transplants. Since all of these nuclei are identical, the second set of transplant animals will be identical to each other and to the original donor animal. Similarly, one of these first serial transplant embryos may be dissociated and its nuclei used for another set of transplants. In this way it becomes possible to produce a clonal population of animals limitless in number, all of which are genetically identical to each other and to the first donor animal.

Can mammals be cloned?

How feasible is it to obtain clonal organisms in higher animals, such as mammals? Although this is theoretically possible, it requires the solution of far greater technical problems than those encountered in the frog experiments. For example, the entire process of fertilization and embryonic development in frogs occurs externally to the animals and can be observed and manipulated in a simple culture dish. In mammals fertilization is internal, and subsequent embryonic development occurs inside the mother. The embryo is implanted at an early stage inside the uterus, where a complex association with the maternal blood and physiology develops. Any perturbation of the delicate balance between mother and fetus is likely to result in disaster.

Nuclear transplantation in mammals would still have to be performed in a culture dish. Suitable donor tissue (comparable to the frog intestinal tissue) could undoubtedly be found in a given donor animal. In fact, it is even possible to obtain viable mammalian eggs by flushing them from the oviduct of a female that has recently ovulated or by removing a piece of ovary tissue. These eggs would have to be kept viable in a culture medium and then enucleated. The enucleation would be difficult because the mammalian egg is much smaller and is organized very differently than the frog egg. It is possible, however, that ultraviolet radiation would be effective in achieving the enucleation. Transplantation of the donor nucleus would follow enucleation, and the egg would be reimplanted into the uterus of a suitable mother (probably the donor). Normal development would then occur.

Serial transplant procedure begins with the same first steps as in the process illustrated on page 35. During the blastula stage of the original transplant, the embryo is dissociated, in much the same way as was the original intestinal tissue. The genetically identical nuclei from the cells of the dissociated embryo are then transplanted into enucleated eggs. From these are produced a "first serial" clonal population. By dissociating one of the embryos of this population, a second serial group can be produced; this process can be continued indefinitely.

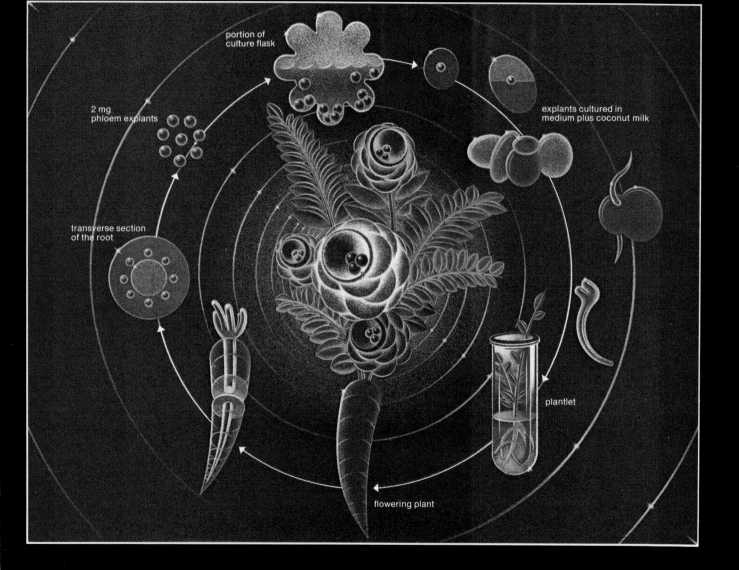

The following labels appear on the figure:

portion of
culture flask

2 mg
phloem explants

explants cultured in
medium plus coconut milk

transverse section
of the root

plantlet

flowering plant

A carrot is cloned (clockwise from bottom left) by cutting a transverse slice from a tap root and obtaining from it tissue called phloem. The phloem was put in a rotating culture dish, causing it to form growing clumps called calluses. Mechanical agitation detaches from these calluses cells that become multicellular nodules. When the nodules are placed in a solid growth medium, they eventually grow into normal flowering carrot plants.

All of the preceding steps are theoretically possible. Indeed, fertilization of mammalian eggs outside the mother's body and their reimplantation have been accomplished in mice and rabbits and, to a limited extent, in humans. However, the manipulations in the enucleation, transplantation, and reimplantation steps of the cloning process are likely to produce such trauma and changes in the transplant egg that normal development would not occur. Even if the transplant did "take," the resulting offspring would have a very high probability of being abnormal. The number of successes obtained would be so few in proportion to the amount of effort and expense that the entire endeavor would be impractical. However, the theoretical possibility of success does exist.

40

Plant cloning

The "vegetative propagation" of plants does not involve the complex association between mother and embryo that occurs in mammals. In fact, many plants have evolved a method of natural cloning, illustrated by the orchid *Malaxis*. In this plant, the cells along the edge of the mature leaf regain the property of cell division and subsequently develop into tiny, egg-shaped embryos. These clonally produced embryos look very much like the normal embryos that are formed in the ovary (ovule) of the plant during its regular reproductive cycle. The clonal embryos are surrounded by a sheathing jacket, and after they are detached from the leaf by the action of wind or rain, they develop into young plants that are genetically identical to the parent plant and to all the other plants derived from clonal embryos of that parent.

For plants that do not clone naturally, two classic research projects, one on carrots and one on tobacco plants, can be used to illustrate experimental clonability. The carrot experiments were performed initially at Cornell University by F. C. Steward. The first step in the procedure was to cut a piece of root from a donor plant and separate out a special class of tissue called phloem. This tissue normally functions in the transport of material throughout the plant. The phloem was next placed in a special rotating culture dish that contained a specific kind of growth medium, originally consisting mainly of coconut milk. Under these conditions, the tissue formed masses of growing clumps called calluses. Under constant mechanical agitation, single cells and groups of cells detached themselves from these calluses. Some of these isolated, suspended cells became multicellular nodules that subsequently developed roots. As long as these rooted nodules remained in suspension in the culture medium, no other organized structures were formed. However, when the nodules were transferred to a solid medium containing gelatinlike agar, they became anchored in a fixed position and many produced shoots. The shoots eventually formed complete plantlets that grew and developed all the normal structures of a carrot. Ultimately, some of these were transferred to soil, where they produced flowers and seeds and completed their life cycles. Thus, the result of the entire process was the production of clonal carrot plants identical to the original parent.

The early experiments of Steward were further confirmed and expanded upon by Vimla Vasil and A. C. Hildebrandt in 1965 in their research on tobacco plants. An important feature that distinguished these experiments from the carrot studies was that the resulting plants clearly originated from single cells. In the Steward experiments, it could only be surmised that the resulting clonal plant originated from one cell.

The tobacco culture procedures differed somewhat from those that were used with carrots. The initial plant tissue (in this case, pith) was excised and transferred directly to a solid agar culture medium. After a callus was formed there, tissue was transferred from it to a liquid culture medium containing mainly inorganic salts, sucrose, coconut milk,

Natural cloning in plants is illustrated by the orchid Malaxis. Cells along the edge of the mature leaf become capable of cell division and develop into tiny egg-shaped embryos (top). Each embryo is surrounded by a sheathing jacket (middle), and when it is detached from the leaf by wind or rain develops into a clone of the parent plant. At the bottom is a longitudinal section of a partially developed embryo, showing its relationship to the tissues of the leaf tip.

41

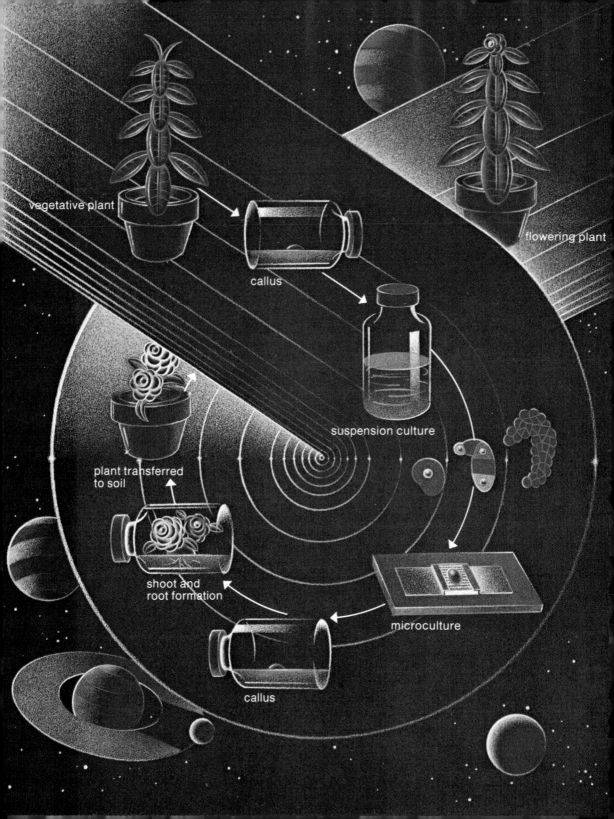

vegetative plant

flowering plant

callus

suspension culture

plant transferred
to soil

shoot and
root formation

callus

microculture

and vitamins. There, again through agitation, a suspension culture of single cells was formed. Individual single cells were then removed from this culture and placed on a microscope slide with a small amount of culture medium. In this microculture the cells could be directly observed and photographed as they progressively developed into small masses of tissue. These small tissue masses were again transferred to a solid agar medium in a bottle, after which they formed roots and shoots. The small plantlets were finally transferred to soil, resulting in the growth of mature clonal tobacco plants.

Though the procedure with tobacco plants was somewhat more complicated than the carrot studies in terms of the number of steps involved, the results clearly demonstrated the "totipotency" of individual differentiated cells. This was precisely the same conclusion drawn from the nuclear-transplant studies on frogs. Such comparisons reinforce the conclusion that the same basic laws of biology hold for plants as well as for animals.

The future of cloning

From the preceding studies, one can perceive that some day a high-yield plant with just the right combination of genes will have a small piece of tissue removed, and from it many cultures and subsequent clonal plants will be derived. Indeed, it is possible to envision literally thousands of plants identical to one high-yield parent becoming the foundation stock for clonally produced fields of crops.

It is less easy to envision the widespread application of cloning procedures to higher animals. However, many of the technical problems of fertilization outside the living mother and reimplantation are being intensively studied. In fact, an organization called "The Animal Bank," dedicated to the preservation of rare and nearly extinct animals, plans to freeze tissues of such animals and at a later time defrost and transplant the nuclei from these frozen samples to a host egg. They believe that the nucleus from the frozen tissue will direct the formation of a complete animal identical to the one from which the frozen tissue was originally derived. If this goal ever is achieved, man will have created a "bank" of animals from which he can obtain any rare or economically important species.

Cloning of tobacco plants begins when pith cells from a hybrid vegetative plant are transferred to establish a callus in a solid agar medium. Tissue from the callus is then transferred to a liquid medium, where a suspension culture of single cells is formed. When these single cells are removed to a microslide culture, they can be observed as they divide and form small masses of tissue. These masses are then placed in an agar medium, where they develop shoots and roots. These small plantlets are then transferred to soil and become mature clonal tobacco plants.

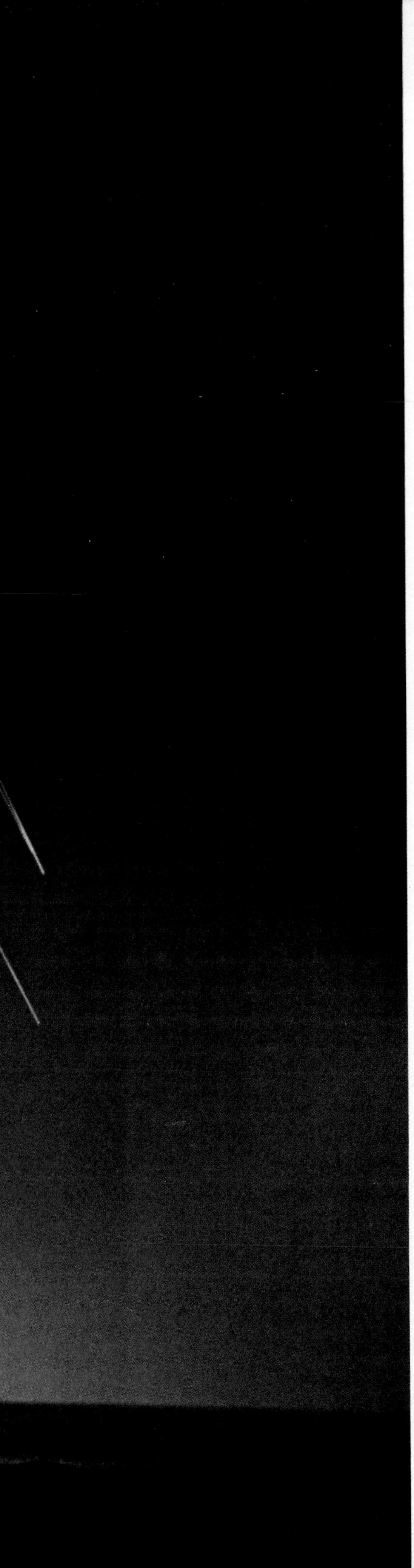

Fiber Optics: Communications System of the Future?

by Amnon Yariv

If the present intense research efforts are successful, many of the communications networks of tomorrow will be based on the flow of optical impulses through hair-thin glass fibers.

From the first verbal communication between individuals to the telephone and satellite communication systems of the present day the evolution of a civilization is mirrored in the amount of information exchanged between its people. In recent times the ever increasing complexity and sophistication of man's communications have led to an increase not only in the amount of information that is supplied daily to an average citizen but also in the average distance over which it is transmitted. It is now accepted as commonplace to be able to make an overseas telephone call or to witness in California a sports event at the time it actually takes place in West Germany.

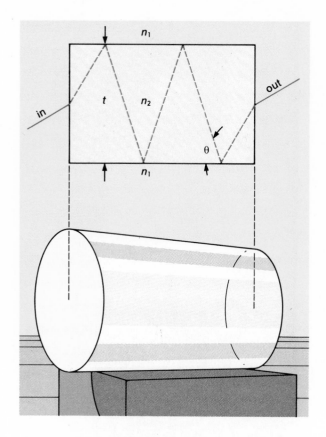 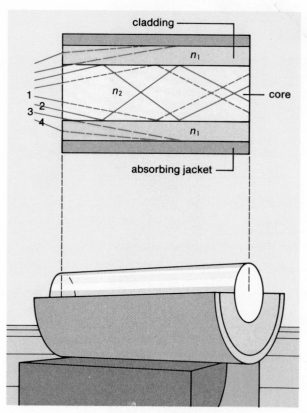

Figure 1 (left). An optical ray "zigzags" by total internal reflection inside a slab in which the velocity of light is less than that in the bounding medium. The conditions for such ray trapping are that θ be sufficiently small and that the optical index n_2 inside is larger than n_1 (n is the factor by which the velocity of light in a medium is reduced relative to that in a vacuum). In figure 2, right, rays 1 and 2 are trapped (propagating) modes, while 3 and 4 are not reflected at the core-cladding interface and are absorbed in the jacket.

AMNON YARIV *is professor of electrical engineering at the California Institute of Technology, Pasadena.*

The demands made on communications systems will almost certainly accelerate in the near and intermediate future. This is due in part to the ever increasing need and appetite for information that accompany an advancing civilization. Another important reason is the gradual depletion of conventional sources of raw materials. An increasing share of any future improvement in the standard of living will, therefore, have to come in the form of services that consume a minimum of material and energy. Consequently, the communications industry is attempting to design systems that will convey the maximum amount of information while using the least possible amount of material.

Up to the present time the mainstays of man's communications have been transmission systems based on copper. Coaxial cables or twisted wire pairs in 1974 still carried the vast majority of telephone conversations within the cities, and also interconnected computers within different business installations. However, a number of recent developments have made it clear that in the future an increasing amount of communications will flow in the form of optical impulses in hair-thin glass fibers.

A typical microwave copper coaxial cable has an outside diameter of 0.4 cm and is capable of handling a few million bits of information per second for distances of a few miles. By contrast, a bundle containing a few hundred optical fibers can be accommodated in the same diameter of 0.4 cm, and each individual fiber can carry over the same dis-

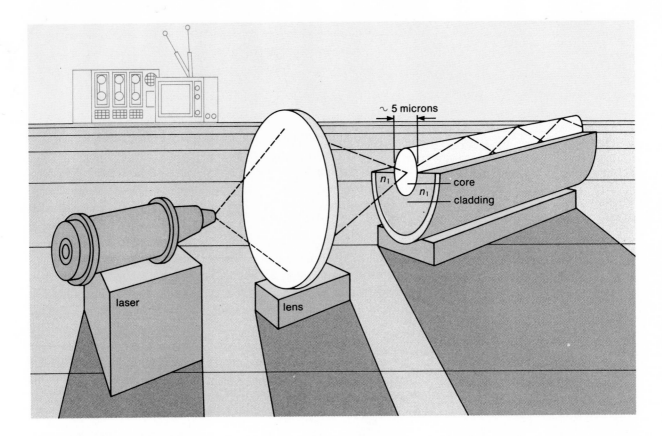

Figure 3. The highly directional output of a laser can be focused by a lens so that most of the power excites the (only) propagating mode of a single-mode fiber.

tance more information with less loss than the coaxial cable. In this way one of the main advantages of optical fiber systems, space economy, is clearly demonstrated. The same space presently occupied by copper cables can be used in optical systems to increase the channel capacity by a factor of a few hundred. Such an increase will almost certainly open up new areas of communications services. Such services may include, for example, banking and shopping using special fiber videotelephone links between private residences and commercial establishments.

By 1974 research and development in optical fiber transmission systems was taking place at an intense pace in a number of countries. Based on the amount of invested effort, it seemed reasonable to assume that such systems would be introduced by the late 1970s in West Germany, the United Kingdom, the United States, and Japan. The last two countries mentioned were supporting the largest efforts in this field.

The need for transmission, repeating, and receiving terminals for the fiber links has given rise to yet a new field, "integrated optics." In this area the effort is aimed at duplicating for light the feat that integrated electronics has already achieved for electricity, the ability to manipulate the flow of light in miniature, monolithic optical circuits. The goal is to incorporate a large number of optical devices, such as lasers, modulators, detectors, and filters, into one compact circuit.

47

Optical fiber waveguide

The first demonstration of light-guiding was that by British scientist John Tyndall in 1870. The principle involved in his experiment is that known in optics as "total internal reflection" and is demonstrated in figure 1. The first condition for trapping a light ray by total reflection is that the inner layer of the waveguide have a larger index of refraction, n, than the outer one. (A waveguide is a medium, often a metal pipe, for propagating electromagnetic waves of a given frequency.) This causes the ray to travel more slowly inside the layer than outside it. The second condition is that the bounce angle Θ be smaller than some maximum value. A rigorous electromagnetic treatment shows that for a given light wavelength and index difference, $n_2 - n_1$, only a discrete number of trapped angles Θ are allowed. By making t, the thickness of the slab, small enough this number can be reduced to one. Each of the rays that can be transmitted is referred to as a mode of the waveguide.

Light passes through an optical fiber in the same way as it does through a slab as in figure 1. The inner pencil-like glass core is surrounded by a glass cladding sheath of a smaller index of refraction (figure 2). The light is trapped in the inner dense core. Rays 1 and 2 correspond to trapped (propagating) modes. Rays 3 and 4 are too "steep" and are not reflected at the core-cladding interface but are instead absorbed in the jacket. The index difference, $n_2 - n_1$, is of the order of magnitude of 0.01.

Fused silica is used to fabricate a single-material fiber. The thick outer tube, 10 mm in diameter, contains a rod and a thin plate. When the tube is heated, the rod is fused to the plate and the plate to the tube. This assembly is then drawn out into a fiber with the diameter of a human hair.

Two main types of fibers appear to be contenders for optical communication. In the first of these, the single-mode fiber, the diameter of the core is small, about 5 microns (about 0.0002 in.) so that only one or a few modes exist. This type of fiber needs to be excited by a highly directional light wave so that it couples efficiently with the propagating mode, as shown in figure 3. Such waves are available only from lasers. The second major type, the multimode fiber, has a core diameter ranging from 75–100 microns. These fibers can support many rays (modes), and light sources such as light-emitting diodes can be used to generate the largely diverging incoherent output beams suitable for exciting the rays.

Sketches of a multimode and a single-mode fiber appear in figure 4. These fibers are produced by pulling a rod consisting of the core material through a sleeve made up of the cladding. The two are softened by heating to such an extent that their diameters contract to the requisite size.

Glass optical fibers have been used since the 1950s for specialized applications involving short-distance (less than a few yards) transmission. They were not considered as serious contenders for communications purposes because of the huge losses, of the order of 3,000 decibels per kilometer (dB/km), of the light propagating in them. A loss of that magnitude indicated that the light intensity was halved during every meter of propagation through the fiber.

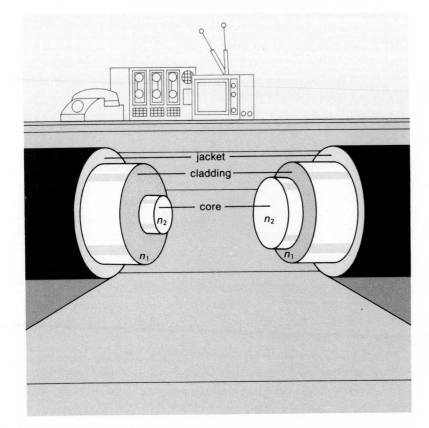

Figure 4. Optical fibers consist of a core and cladding, both of transparent glass, surrounded by a light-absorbing jacket. If n_2 is larger than n_1, the fiber will guide light, n being the refractive index of the material. The core at left is that of a single-mode fiber and measures only a few microns in diameter; the one at the right is of a multimode fiber and is several mils in diameter.

What changed this situation was the realization in 1968 by K. C. Kao and T. W. Davies of the Standard Telecommunications Laboratories in Great Britain that these high losses were not intrinsic to the fibers but were due to the impure material used in their preparation. The next major breakthrough occurred in 1970 when Felix P. Kapron, Donald B. Keck, and Robert D. Maurer of the Corning Glass Works announced the achievement of losses below 20 dB/km in single-mode fibers. The Corning scientists followed this with further improvements until they finally developed fibers that had losses of only about 2–3 dB/km. With this remarkable achievement it became possible to think of an optical communications link with distances of at least a few miles between repeating stations, where the attenuated signals are reamplified and reconstituted.

An intense effort to understand the various loss mechanisms that exist in fibers demonstrated that once the necessary purity of materials had been achieved, the remaining losses are due mainly to scattering from small inhomogeneities in the glass (Rayleigh scattering). These, however, were expected to contribute losses of less than 1 dB/km at optical wavelengths above 1 micron. It was thus expected that optical communications links of the future would use the near-infrared region of 0.8–1.1 microns.

The impurity and scattering losses in fibers as a function of wavelength are lowest at a wavelength of about 1 micron. Because the optimum light sources for driving the fibers emit in the same wavelength region this is a particularly fortunate occurrence.

Communicating through fibers

There are two main modes of communication, analog and coded. In the analog mode the electrical signals originating in the information source are transmitted in their raw form over some kind of channel. One example is the transmission of the voice signal produced by the telephone microphone over a wire line. Another is the transmission of television signals over cables. In coded communication the analog signal is transformed by some coding scheme into a sequence of electrical impulses. At the receiver end these impulses are decoded and transformed back into the desired analog signal.

Coded communication in the 1970s was gaining rapidly over the analog mode because it lent itself much more easily to processing and manipulation by integrated electronic circuits and because it was inherently compatible with the binary (yes-no) logic of computers that invariably became part of any communication link. For these reasons glass fibers that transmit information in the form of optical pulses, a kind of coded communication, were expected to enjoy increasing popularity.

An optical communication link using a pulse-code modulation scheme may be used to transmit a voice message, as illustrated in figure 5. (Pulse-code modulation occurs when a radio wave or signal is modulated so that the intelligence is conveyed by a sequence of

Figure 5. The electrical voice signal from a telephone is converted by pulse-code modulation into a digital optical system that is transmitted through a fiber and reconverted to a voice signal.

50

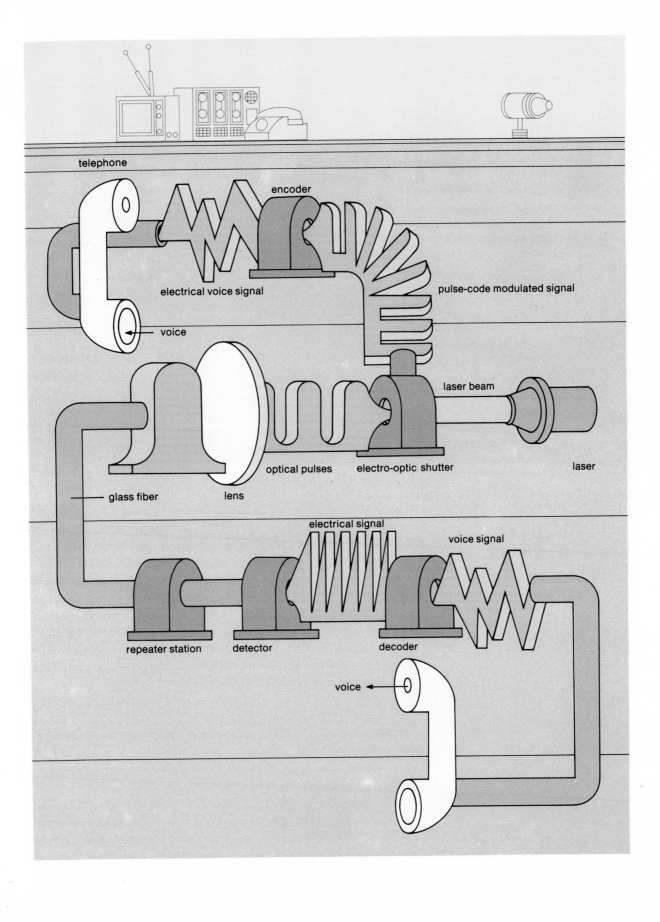

telephone

encoder

electrical voice signal

pulse-code modulated signal

voice

laser beam

optical pulses

electro-optic shutter

laser

glass fiber

lens

electrical signal

voice signal

repeater station

detector

decoder

voice

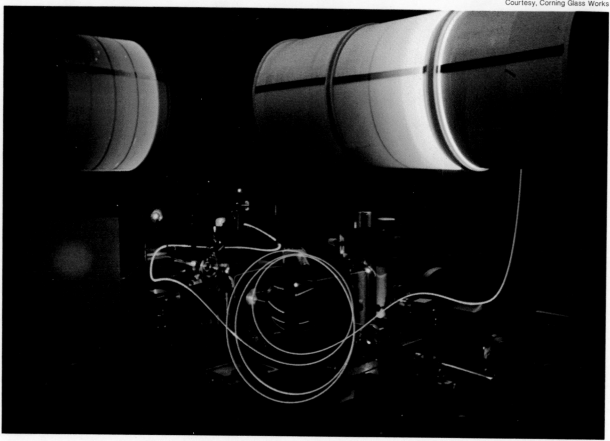

Green light is transmitted in a demonstration test through optical fiber spools 3 km (right) and 1 km (left) in length. Light in the visible wavelengths is used for such tests, but for actual communications systems light in the infrared and near-infrared regions has the most desirable characteristics.

pulses of the wave that are usually of the same size and shape and are transmitted at multiples of a standard time interval.) The amount of information that can be transmitted depends on how many pulses can be compressed into a given time slot. There are a number of fundamental limitations as to how short and how densely packed the optical pulses can be. One of these limitations has to do with the ability to switch the light on and off fast enough. By 1974, however, electro-optic crystal switching could obtain pulses at a repetition rate of 1,-000,000,000 per second. At this information rate, the equivalent of 50 television channels or 100,000 telephone conversations could be transmitted on a single channel. This process is limited, however, by the broadening of the optical pulses as they propagate through the glass fiber.

Pulse broadening in fibers results from two principal mechanisms, the simplest one to understand occurring in multimode fibers. In such cases a given optical pulse excites simultaneously a large number of modes (rays). Referring to figure 2, one can consider two of these rays, such as 1 and 2. It is clear that ray 1, taking a more direct path, will arrive at the end of the fiber before ray 2, which zigzags more steeply. The result of this multiplicity of paths of the pulse is a gradual broadening of the pulse with distance. If the broadening becomes so great

52

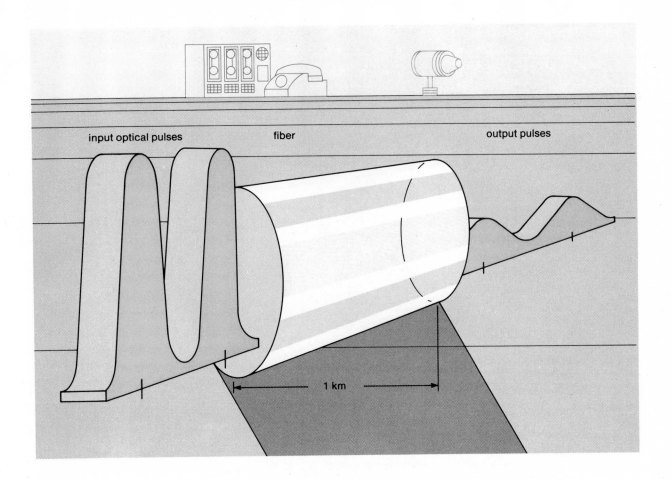

that the distinction between adjacent pulses begins to blur, the information contained in the pulses is lost. This situation is depicted in figure 6. Because of pulse broadening, researchers expected that multimode fibers would be limited to information rates of approximately 50 million bits (pulses) per second.

Pulse broadening also occurs in small-core, single-mode fibers but for a different reason. In this situation there occurs a subtler physical mechanism known as dispersion that is present in all material. The pulse consists of slightly different frequencies that each have slightly different velocities, thus causing the spreading. But this is a much weaker effect than the ray dispersion described above, and so single-mode fibers can be expected to transmit up to 1,000,000,000 bits per second over distances of about one mile. Periodically spaced repeater stations are to be used to reconstitute the pulses, amplify them by electronic means, and launch them anew on their way.

Thus, the following division of tasks can be envisioned: multimode fibers will be excited by diodes emitting incoherent, largely divergent light waves for low-to-medium-capacity (less than 100 million bits per second) communication channels; single-mode fibers will be excited by lasers producing coherent light for the large-capacity (approximately 1,000,000,000 bits per second) channels.

Figure 6. Pulse broadening occurs in multimode fibers because of the excitation of many rays and the consequent multiplicity of paths in the pulse; in single-mode fibers it is caused by the dispersion generated by the slightly different frequencies that make up a pulse. When pulses are broadened, the distinctions between them are blurred and information is lost.

53

Figure 7. A single tiny crystal of gallium arsenide (below) can house optical lasers, detectors, and modulators. Entire optical circuits may some day be prepared on such crystals. In the photograph at right, a solid-state laser (small rectangle atop block) is much smaller than the grain of salt at the far right.

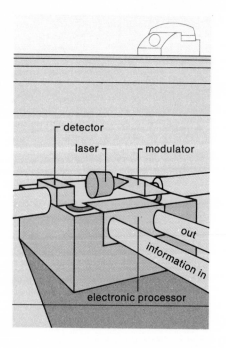

Integrated optics

In order to communicate through fibers many devices must be used. These include lasers and light-emitting diodes to act as sources for the light, modulators to impress the information on the light beam, and detectors that will recover the information from the beam. Moreover, these devices and their interconnections must be compatible in size with the fiber transmission lines.

The need for these new "optical circuits" has given birth to a new field known as "integrated optics." In the course of a few years it has become a substantial area of research and development, straddling the boundaries of optics, quantum electronics (lasers), and semiconductor material technology. Unlike the situation with fibers, the variety of problems that must be solved in this field is very large, and the form and direction of the final solution are not yet clearly defined. One direction, pointed out by a series of related experiments at the California Institute of Technology, is to fabricate many of the devices needed to manipulate the light in single crystalline wafers of gallium arsenide. Lasers, detectors, waveguides, and modulators consisting of this material have been fabricated. It seems reasonable to assume that eventually entire optical circuits, analogous to integrated electronic circuits in silicon, can be produced on single miniature gallium arsenide crystals (figure 7). Other substances under study for use in such wafers include garnet and a compound of aluminum, gallium, and arsenic.

54

F. W. Goro

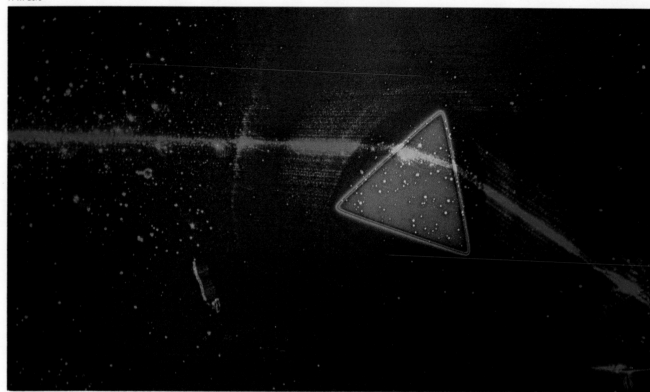

In summary, light pulses propagating in extremely thin glass fibers seem certain to carry an ever increasing share of man's future communications. They will undoubtedly open up new services and affect civilization in ways that can only be surmised at this time.

FOR ADDITIONAL READING:

Cook, J. S., "Communication by Optical Fiber," *Scientific American* (November 1973, pp. 28–35).

Maurer, R. D., "Glass Fibers for Optical Communications," *Proceedings of the IEEE* (Institute of Electrical and Electronics Engineers, vol. 61, April 1973).

Stepke, E. T., "Optical Data Links" and "Integrated Optics," *Electro-Optical Systems Design* (September 1973).

Laser light is refracted by a thin-film prism that was made by depositing a film of zinc sulfide 1 micron (0.001 mm) thick on a substrate of glass and then adding a second layer of zinc sulfide through a triangular mask. The prism, seen above enlarged by 30 diameters, is about 1 mm in height and several microns thick. The bright circular area surrounding the prism is laser light reflected by the microscope lens through which the photograph was taken.

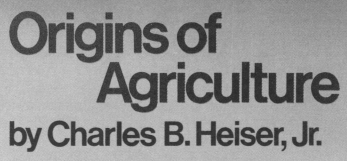

Origins of Agriculture
by Charles B. Heiser, Jr.

The greatest revolution in history occurred when man learned to domesticate plants and animals. Scientists are only now discovering the secrets of where and how it happened.

For most of the two million-plus years of his existence, man has been a food collector or hunter, depending on nature to provide him with enough to eat. It was only about 10,000 years ago that he invented agriculture and thus became a food producer. The keeping of animals and the planting of seeds, although they may seem fairly simple matters in retrospect, represented one of the most drastic changes in the history of mankind. Writing, cities, political organization, monumental art, and architecture, which first appeared a few thousand years after the origin of agriculture, all resulted either directly or indirectly from the new mode of subsistence. Agriculture gave birth to civilization, but at the same time it brought new problems to mankind. When man became a food producer he started making changes in his environment that he has not yet learned to control.

Although the "agricultural revolution," as it has been called, was a fundamental change, it was not a sudden one. Instead, it developed gradually over several thousand years in different parts of the world. Our understanding of the origin of agriculture has increased greatly in the last quarter of a century, making it necessary to alter many earlier ideas on the subject. Most of this knowledge has resulted from the work of the archaeologists who have uncovered the settlements of primitive farmers. Radiocarbon dating techniques have allowed them to assign rather precise dates to many of the events connected with plant and animal domestication.

Where agriculture began

Current studies indicate that agriculture was first practiced in the Near East. Numerous sites in Iran, Iraq, Syria, Jordan, Israel, and Turkey have yielded evidence of early cultivation. At the site of Jarmo in Iraq, where Robert J. Braidwood began investigations in 1948, grains of cultivated wheat and barley and bones of domesticated sheep and goats were found in deposits dated 6750 B.C. Since then, other sites have revealed even earlier dates for these plants and animals.

The idea that agriculture began in the Near East is not a new one, but earlier investigators had supposed that it originated in the Fertile Crescent, the area of fertile land extending from Babylonia up the Tigris and Euphrates to Assyria, westward through Syria, and southward to southern Palestine. Recent investigations indicate that it may have begun in hill country and moved into the river valleys later, after man learned to control water. Irrigation began to be practiced around 5000 B.C., and with it man began one of the great alterations of his environment for which he was to pay the consequences. Irrigation can bring an accumulation of salts to the surface, making plant growth difficult or impossible. That this occurred fairly early is indicated by the archaeological record, for barley, which is more salt-tolerant than wheat, is known to have replaced wheat in some regions. Many of the areas once occupied by agricultural villages are now mostly desert. While natural climatic change may be partly responsible, man probably contributed by his mismanagement.

CHARLES B. HEISER, JR., is professor of botany at Indiana University, Bloomington.

*Cultivation of wheat, one
of the oldest and still one
of the most important cereals,
changed little from its inception
in the ancient Near East until
the development of agricultural
machinery in the 19th century. Old
Testament farmers would recognize
the harvesting methods depicted
on a medieval illuminated calendar
(left) in the Communal Library
at Forlì, Italy,
but not the mechanized harvest
in Washington state (above).*

From the Near East agriculture spread to the Balkans and Egypt. Whether it also spread directly to Southeast Asia is not certain, and it may have originated independently in Southeast Asia at the same time. In 1969 Chester Gorman reported finding plant materials in Thailand — possibly including domesticated species — that dated to about 7000 B.C. In general, the climate of Southeast Asia is not as favorable for the preservation of plant and animal material as that of the Near East. Nevertheless, current research should provide us with a better understanding of the development of agriculture there.

In the Americas evidence of early agriculture comes from Mexico and Peru. In excavations in Tamaulipas, Mex., Richard S. MacNeish found gourds, squashes, beans, and chili peppers—either wild species or incipient domesticates—in levels dated between 7000 and 5500 B.C. Corn, or maize, which was to become the basic cereal in the Americas, was not found in these levels at Tamaulipas, however. In an attempt to find early evidence of maize, MacNeish began investigations in 1961 at Tehuacán, a site in south-central Mexico, and his efforts were rewarded with a remarkable series of plant finds that give us the best record so far known of the transition from food collecting to full-fledged agriculture. Man first appeared at Tehuacán around 10,000 B.C., and for several thousand years he depended on wild plants and animals for his food. About 5000 B.C. the first evidence of domestication appears with remains of maize, squash, chili pepper, avocado, and amaranth. Improved types of these and other plants, including cotton, were found in more recent levels, indicating that people had brought additional plants into cultivation or had acquired them through exchange.

Until recently our knowledge of prehistoric agriculture in Peru was based almost entirely on archaeological finds from coastal areas, the earliest of which was dated around 3000 B.C. Then, in 1970, MacNeish found evidence of agriculture in highland Peru, including maize in deposits dated between 4300 and 2800 B.C. The presence of maize in Peru almost certainly indicates that the people of South America had established contacts with Mexico, where maize appears to have originated. Still earlier evidence of domesticated plants in Peru was reported in 1973, when L. Kaplan, Thomas F. Lynch, and C. E. Smith, Jr., announced that they had found both common beans and lima beans in a highland valley. These beans were apparently being cultivated around 6000 B.C., suggesting that agriculture in Peru is as old as, if not older than, in Mexico. This does not necessarily mean there were contacts at that time, however. Common beans are also represented in the early archaeological record in Mexico, but a similar wild type of bean could have been domesticated independently in both regions.

The earliest reports of cultivated plants in what is now the United States come from the Southwest. At Bat Cave, New Mexico, there is evidence that agriculture was being practiced about 2000 B.C. Maize, beans, and squash became the principal cultivated food plants in this area, and later were to serve as staples for the Indians in the eastern United States. Almost certainly all these plants came from Mexico as domesticates, although the Indians in the east-central United States may have experimented with agriculture before their arrival. Archaeological evidence suggests that the sunflower, the United States' only important contribution to the world's food plants, may have been in cultivation at an earlier date. One other plant known archaeologically, the marsh elder or sumpweed, was probably cultivated but became extinct as a domesticate before historical observations were made. Quite possibly it was displaced by the superior food plants of Mexico.

Diagram on the opposite page shows the areas and approximate dates (B.C.) for the first cultivation of six of man's important domesticated plants.

In addition to these cereals and other plants, early man domesticated a number of starchy crops. Such plants as the Irish potato, the sweet potato, and manioc, the plant from which tapioca is obtained,

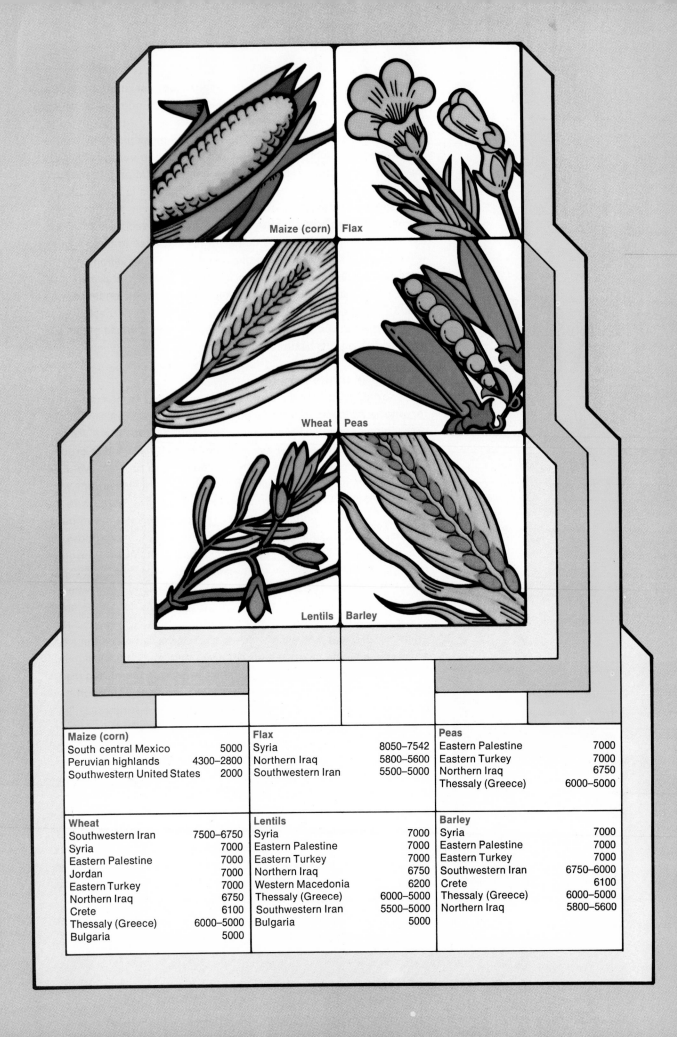

Maize (corn) Flax

Wheat Peas

Lentils Barley

Maize (corn)		
South central Mexico		5000
Peruvian highlands		4300–2800
Southwestern United States		2000

Flax		
Syria		8050–7542
Northern Iraq		5800–5600
Southwestern Iran		5500–5000

Peas		
Eastern Palestine		7000
Eastern Turkey		7000
Northern Iraq		6750
Thessaly (Greece)		6000–5000

Wheat		
Southwestern Iran		7500–6750
Syria		7000
Eastern Palestine		7000
Jordan		7000
Eastern Turkey		7000
Northern Iraq		6750
Crete		6100
Thessaly (Greece)		6000–5000
Bulgaria		5000

Lentils		
Syria		7000
Eastern Palestine		7000
Eastern Turkey		7000
Northern Iraq		6750
Western Macedonia		6200
Thessaly (Greece)		6000–5000
Southwestern Iran		5500–5000
Bulgaria		5000

Barley		
Syria		7000
Eastern Palestine		7000
Eastern Turkey		7000
Southwestern Iran		6750–6000
Crete		6100
Thessaly (Greece)		6000–5000
Northern Iraq		5800–5600

Alpha

were brought into cultivation in the Americas in prehistoric times. Bananas, taro, and breadfruit were domesticated in Southeast Asia, and different kinds of yams were cultivated in various parts of the world. Carl O. Sauer has argued that the cultivation of these plants by vegetative propagation, which often involves no more than putting a piece of the plant in the ground, is far simpler than propagation by seeds and may have been acquired by man even earlier. If this is true, it has not been revealed by the archaeological record. However, the fleshy, edible parts of these plants are less likely to be preserved than are seeds. Moreover, some of these plants may have originated in humid areas, where the climate was not conducive to their preservation. What is clear is that seed agriculture led to man's great cultural advances, since all the early civilizations for which we have good evidence were based on seed crops—wheat and barley in the Near East, rice in Southeast Asia, and maize in the Americas. Three of these plants still serve as the world's basic foodstuffs.

On the trail of the cereals

Our knowledge of how the basic cereals originated is most advanced for wheat (*Triticum*), where the combined efforts of the botanist and archaeologist have given us a fairly complete picture. There are several different species of wheat, and they form three distinct groups. One of the important clues to the understanding of their evolution occurred half a century ago when it was found that these groups were characterized by different chromosome numbers. The group with the most primitive species had two sets of chromosomes or seven pairs. Another group, including the durum wheats, widely grown today for making paste products such as macaroni, had 4 sets of chromosomes or 14 pairs. The third group, the bread wheats, had 6 sets of chromosomes or 21 pairs. The chromosome sets were designated by letters— the primitive wheats as AA, the durum group as AABB, and the bread wheats as AABBDD.

From this, it was evident that the species of the last two groups had originated through hybridization, followed by chromosome doubling. It also became apparent that the primitive wheats had furnished the AA chromosome sets to the other wheats. The intriguing question concerned the source of the BB and DD sets. Through a series of brilliant investigations by researchers in many countries, the plants involved in the origin of the other wheats were identified as species of goat grasses (*Aegilops*), weedy grasses of no use to man by themselves. From the present-day distribution of the wild wheats and goat grasses and from the archaeological record, it is clear that the hybridizations and chromosome doublings occurred in the Near East. Man was not directly involved, except insofar as he may have increased the range of the plants and thus brought them together so that hybridization could take place. Later he improved his cultivated wheats by conscious selection, and in recent years the knowledge of how wheat evolved has enabled plant breeders to make still greater improvements.

For a long time the origin of maize (*Zea mays*) was shrouded in mystery. It had never been found growing without man's aid, and it was generally thought to be extinct as a wild plant. Its closest relatives were known to be two coarse wild grasses, teosinte (*Euchlaena*) and gama grass (*Tripsacum*), both native to the Americas. In 1939 Paul C. Mangelsdorf and R. G. Reeves advanced the hypothesis that maize had originated in South America and was carried to Mexico by man. There hybridization took place between maize and *Tripsacum*, resulting in the creation of teosinte. Maize in turn hybridized with teosinte, giving rise to superior races of maize. The thesis was supported by the fact that teosinte is intermediate between maize and *Tripsacum* in many of its characteristics. The later discovery of pollen dated about 80,000 B.C. below Mexico City and of very primitive ears of maize at Bat Cave—far older than any archaeological maize known from Peru—forced Mangelsdorf to recognize that maize probably originated in Mexico. Still later, MacNeish uncovered very small ears of maize in early levels at Tehuacán. At the time these were regarded as wild maize, but they are now believed to represent a very early cultivated form.

The theory of a hybrid origin for teosinte received considerable support for a number of years, but there were some skeptics. Shortly after Mangelsdorf and Reeves published their original hypothesis. George Beadle questioned the hybrid theory and suggested that teosinte might be the ancestor of maize. One problem with this idea was that teosinte grains are extremely hard, and it was difficult to imagine how man could have used them for food. Beadle, however, showed that they could be popped very much like popcorn. There is also evidence in the early chronicles from Mexico that the grains of teosinte were parched. Later Beadle resumed his studies of maize, and he and a number of others became convinced not only that teosinte is a natural species, rather than a hybrid, but also that it is the "mother of maize." If this is the case, the search for wild maize is at an end. That the Indians were able to domesticate such a seemingly unpromising plant as teosinte and develop it into the staple food grain of the Western Hemisphere is definitely a tribute to primitive man's ability as a plant breeder.

Less is known about the origin and early history of rice (*Oryza sativa*), the world's third major cereal and the one that feeds the most people today. Archaeological evidence of rice dated between 5000 and 3500 B.C. has been found in Thailand. Linguistic evidence from both China and India indicates that rice had become a basic food plant long before the Christian era. Several species of wild rice are known from the tropics (not to be confused with the wild rice [*Zizania aquatica*] of the United States and Canada, an entirely different plant), and according to Japanese botanists one of these, *Oryza perennis*, is the probable progenitor of common rice. Today rice is still considered a sacred plant in parts of the Far East, just as maize is among some American Indians.

Early Americans, like their counterparts in Asia and the Near East, looked on their basic cereal with religious awe, an attitude exemplified by the silver statuette (opposite page, top) of corncobs with gilded leaves. Now in a West Berlin museum, it was found on the northern coast of Peru and is probably of Inca origin. In a more modern vein, shelled corn (bottom) is conveyed into an open silo in Media, Pa.

Domestication of animals

Knowledge of the origin of man's principal domesticated animals has also increased greatly in the last two decades. It has long been thought that the dog was the first domesticated species, and this may be the case, although the dog is not represented among the earliest known remains of domesticated animals. The earliest known bones of the dog were found in Idaho and dated at 8400 B.C. If, as many investigators believe, the dog was first domesticated in the Old World and came to the New World with man when he crossed the Bering land bridge, domestication must have taken place long before that date. It is now well established that the dog is simply a domesticated wolf, however, and since wolves are widespread it may have been domesticated in more than one place. The dog has been used for food, but it probably served other, more important functions in early times just as it does today.

Among the earliest domesticated food animals were several of the ruminants or cud-chewing animals of the bovid family. Goats were domesticated in the Near East before 8000 B.C., and domesticated sheep appeared in the same region and in Greece shortly afterward. Remains of cattle dated at 7000 B.C. have been found in Greece and appear in Turkey before 6000 B.C. There may have been a separate domestication of cattle in India, giving rise to the humped breeds. If so, the appearance of humped cattle in Mesopotamia around 4500 B.C. would indicate that contact with India had been established.

The pig is the earliest among the nonruminant domesticated food animals. It is known from Greece around 7000 B.C. and was also apparently domesticated independently in the Far East. There may have been other, independent, domestications in the Near East and Europe as well. The pig is thought to have been a relatively easy animal to domesticate, and so multiple origins appear likely.

The domestication of other Old World animals occurred later. The horse was probably first tamed in the Ukraine around 4000 B.C. At one time it was an important meat source, and some peoples still use it for food today, but the chief effect of its domestication was to revolutionize warfare. The chicken, still a basic meat source, is thought to have been domesticated in India from the wild jungle fowl. The date of its origin is unknown, but it had reached Persia and Egypt by 2000 B.C.

Very few animals were domesticated in the Americas, and only one of them, the turkey, was to assume any importance in other parts of the world. The turkey was domesticated in Mexico and appears at Tehuacán around the beginning of the Christian era. Whether the turkey had reached South America before the arrival of the Spaniards is not entirely clear. The llama, the alpaca, and the guinea pig were all domesticated in South America. The llama and alpaca, both cameloids, served many purposes, but they were never used to prepare the fields for planting. It was not until the introduction of Old World animals by the Spaniards that animal power began to replace human labor in much of the agricultural work in the Americas.

Domestication of animals paralleled that of plants. Some, like the dog, were primarily companions and helpers while others, such as goats and sheep, were valued as sources of food and other products. Areas and approximate dates (B.C.) for the appearance of six major domesticated animals are shown in the diagram (opposite page). Asterisks indicate dates determined by radiocarbon methods.

64

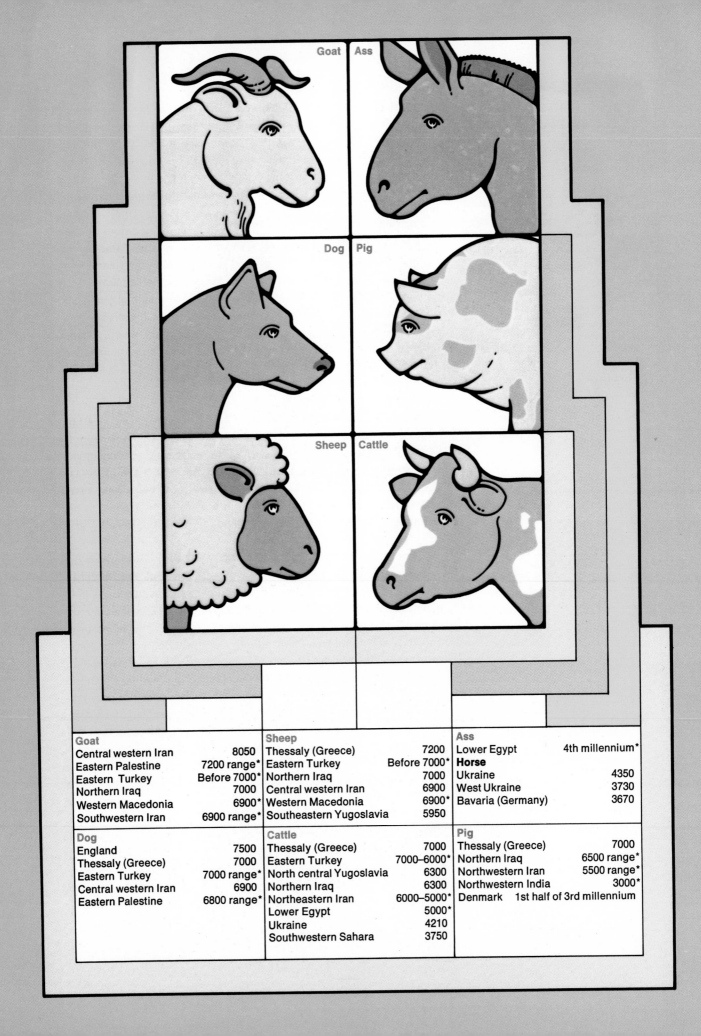

Goat		Sheep		Ass	
Central western Iran	8050	Thessaly (Greece)	7200	Lower Egypt	4th millennium*
Eastern Palestine	7200 range*	Eastern Turkey	Before 7000*	**Horse**	
Eastern Turkey	Before 7000*	Northern Iraq	7000	Ukraine	4350
Northern Iraq	7000	Central western Iran	6900	West Ukraine	3730
Western Macedonia	6900*	Western Macedonia	6900*	Bavaria (Germany)	3670
Southwestern Iran	6900 range*	Southeastern Yugoslavia	5950		
Dog		**Cattle**		**Pig**	
England	7500	Thessaly (Greece)	7000	Thessaly (Greece)	7000
Thessaly (Greece)	7000	Eastern Turkey	7000–6000*	Northern Iraq	6500 range*
Eastern Turkey	7000 range*	North central Yugoslavia	6300	Northwestern Iran	5500 range*
Central western Iran	6900	Northern Iraq	6300	Northwestern India	3000*
Eastern Palestine	6800 range*	Northeastern Iran	6000–5000*	Denmark 1st half of 3rd millennium	
		Lower Egypt	5000*		
		Ukraine	4210		
		Southwestern Sahara	3750		

The unanswered questions

All our major food plants and animals, and most of the minor ones as well, were domesticated in the prehistoric period. The archaeological record has revealed a great deal about where and when this took place. Botanists and zoologists have worked out many of the details concerning the evolution of the plants and animals involved, although much remains to be learned. But there are still several unanswered questions.

It had been thought that plants were domesticated before animals, except, perhaps, for the dog. Yet the remains of domesticated animals dating from before 8000 B.C. that have been found in the Near East far precede any presently known evidence for domesticated plants. It may be that some people in the Near East specialized in animal domestication while others concentrated on plants, but further investigations are needed to give us a clearer picture for this part of the world. It seems fairly evident that plants were domesticated much earlier than animals in the New World.

One beginning or many?

Did agriculture have a single origin or more than one? This is part of a larger question concerning all cultural development. Almost from the time the Americas were discovered, there were some who maintained that the New World civilizations were derived from those of the Old. There are, of course, many similarities between the two areas — metallurgy, writing, monument building, and political organization, as well as agriculture. Did these develop independently in the New World, or were they the result of diffusion?

It is generally accepted that man came to the New World over a land bridge across the Bering Strait, but these migrations are thought to have taken place long before agriculture had developed. So if the germ of civilization came to the New World from the Old, it would have to have been by an ocean voyage after man was established in the Americas and the Bering land bridge had disappeared. Thor Heyerdahl's voyages in primitive craft from South America to Polynesia and from Morocco to the Caribbean were designed to prove that such trips were possible. He succeeded, but this, of course, does not prove that voyages of this kind actually took place in pre-Columbian times. Since agriculture in the New World developed at least 1,000 years later than in the Old World, there would have been ample time for such voyages, perhaps accidental ones. But if man brought the idea of agriculture to the New World from the Old he did not bring the plants with him, for the first cultivated plants in the two hemispheres were completely different.

The prevailing view today is that agriculture probably developed separately in the Old and New Worlds. Possibly there was a single point of origin in each hemisphere — in the Near East and Mexico, for example — from which the practice diffused to other areas. It seems more likely, however, that agriculture originated independently in

66

many places throughout the world, with different species being brought into domestication in different areas.

This does not rule out the possibility of transoceanic voyages in prehistoric times, but it seems likely that such contacts, if they occurred, took place after agriculture was well developed. Supporters of the theory of transoceanic contacts often cite, as evidence, the fact that certain species of cultivated plants were present on both sides of the Pacific when the Spanish arrived in the New World. The sweet potato, native to the Americas, was cultivated not only there but also in some of the Pacific islands at the time of Columbus. The bottle gourd, apparently native to Africa, was present in both Thailand and the Americas by 7000 B.C. The coconut, indigenous to the Indo-Pacific region, was established in Panama before the 16th century. The fruits of the gourd and coconut will both float in salt water for long periods without losing seed viability, so it is not necessary to invoke the aid of man to explain their wide distribution. On the other hand, the sweet potato could not have been transported by ocean currents. It is fairly widely accepted that man must have carried it to the Pacific region, although the suggestion has been made that birds may have transported the seeds. In any case, the botanical evidence indicates that one plant, at most, may have been introduced across the Pacific by man prior to the Spanish voyages.

The role of the environment

How did the environment figure in the development of agriculture? It had been thought that the climate in the Near East remained remarkably stable for a long period before the beginnings of agriculture, but in 1968 H. E. Wright, Jr., demonstrated through studies of fossil pollen that the climate in Iran shifted about 11,000 years ago. Before that date the plant life had been mostly herbaceous or grassy. After that time

Butcher plies his trade (opposite page) in a 5th dynasty (2494–2345 B.C.) sculpture from Saqqarah, Egypt. Livestock raising in 20th-century America bears the hallmarks of a mass-production industry, with cattle fattened for market in huge feedlots, such as this one (above) in Porterville, Calif.

67

Rice plant (above) flowers in Bali, Indonesia. The staple food of more people than any other cereal, rice is still cultivated throughout the Far East by intensive labor, much in the manner shown in the mid-16th-century scene (top right) in the Tokyo National Museum. In stark contrast is the almost surrealistic aerial view (right) of rice fields in the Colusa-Grimes area of California.

trees appeared, and Wright interpreted this as indicating a change from a cool steppe to a warmer, and perhaps moister, savanna. Exactly how such a climatic change may have figured in the origin of agriculture is, however, unclear.

Certain similarities may be noted between the Near East and Mexico, the areas of early agriculture for which we have the most knowledge. In both areas, the manipulation of plants appears to have begun in hilly or mountainous regions. Such a terrain would have provided a number of microclimates characterized by different kinds of plants and animals, so that a man living there would have had a variety of foods within easy reach. Thus, he could have been more or less sedentary, at least during part of the year—a probable prerequisite for plant cultivation. Moreover, both of these areas are somewhat arid. At first glance, this might seem detrimental to the development of agriculture, but it probably proved advantageous as long as there was sufficient rainfall at certain times of the year. There would have been less vegetational cover to remove in preparation for planting, and there would have been fewer weeds, insects, and fungal diseases to plague man's new crops than would have been the case in more moist areas.

Why did it happen?

A very basic problem in human culture, as Kent Flannery has pointed out, involves the question of why man changed his mode of subsistence at all. So a final question to be asked is: "Why was agriculture invented?" Man as a food collector was certainly well acquainted with the plants and animals in his environment. He probably knew the habits of the animals and that seeds gave rise to new plants like their parents, but he had to take this knowledge and apply it. Braidwood has stated that agriculture began "as the culmination of the ever increasing cultural differentiation of human communities." According to this line of reasoning it was "human nature" to invent agriculture when man reached a certain level. Agriculture arose late in the history of man because he was not ready for it until about 10,000 years ago.

In 1968 Lewis R. Binford rejected Braidwood's arguments and postulated, instead, that an increase in population was responsible. Demographic stress would have provided a strong selective pressure for bringing about more efficient means of securing food. More recently, Flannery suggested that people at the edge of the optimum zone for wild cereals in the Near East attempted cultivation in order to duplicate the abundant stands of cereals found within that zone. Similarly, in Mexico, a genetic improvement in wild maize could have led to a conscious effort to introduce it into new habitats.

Some of the older ideas concerning the origins of domestication, once largely ignored or dismissed, have been the subject of renewed interest. Eduard Hahn, a 19th-century German geographer, proposed that cattle domestication originated through a need to have animals available for sacrifice in connection with fertility ceremonies. Some substance was given to Hahn's thesis by James Mellaart's discovery of

Though not one of the most important cereal crops today, barley (below) was one of the earliest domesticates. Together with wheat, it was the "corn" of the King James Bible. Reminiscent of the story of Joseph and his brothers is the Middle Kingdom (c. 2040–1786 B.C.) sculpture (right) of an Egyptian scribe and his attendants in a granary. The modern equivalent is the giant Farmers Cooperative Co. elevator (bottom right) in Kansas, capable of storing approximately 17 million bu of grain from more than 50,000 farms in central Kansas, southeastern Colorado, and northern Oklahoma.

what appears to have been an advanced cattle cult at the archaeological site of Çatal Huyuk in Turkey around 6000 B.C., although the presence of cattle worship at this date does not necessarily mean that animals were domesticated in connection with it.

Grant Allen, a Canadian philosopher and science writer, postulated in 1897 that seed agriculture arose as an adjunct of primitive burial practices. Man may have observed that plants grew well around newly made graves, and this could have led him to dig up increasingly large areas. Since man would associate the improved plant growth with the newly interred body, Allen reasoned that human sacrifice thus became connected with seed planting; certainly there is no doubt that human sacrifice was often associated with primitive agriculture, although this may have come about for other reasons. Many objections can be made to Allen's hypothesis, but the possibility that agriculture had an accidental origin does merit consideration. Primitive man recognized spirits in plants and often attached special significance to the firstfruits or the last sheaf of the harvest. Perhaps primitive seed collectors returned these seeds to the ground, and the first intentional plantings were a propitiation to the gods rather than a direct attempt to produce more food.

In the final analysis, it must be admitted that we do not know exactly why agriculture came into existence. The archaeological record has revealed a great deal to us, but it cannot be expected to provide all the answers. What is clear is that about 10,000 years ago seed planting and animal domestication began. Man became a food producer for the first time, although—for all his advances since then—he has yet to eradicate hunger from the world.

FOR ADDITIONAL READING:

Baker, H. G., *Plants and Civilization,* 2nd ed. (Wadsworth, 1970).

Braidwood, Robert J., *Prehistoric Men,* 7th ed. (Scott, Foresman, 1967).

Harris, David R., "The Origins of Agriculture in the Tropics," *American Scientist* (March-April 1972, pp. 180–193).

Heiser, Charles B., Jr., "Some Considerations of Early Plant Domestication," *BioScience* (March 1969, pp. 228–231); *Seed to Civilization: The Story of Man's Food* (Freeman, 1973).

Isaac, Erich, *Geography of Domestication* (Prentice-Hall, 1970).

Sauer, Carl O., *Agricultural Origins and Dispersals* (American Geographical Society, 1952).

Schery, Robert W., *Plants for Man,* 2nd ed. (Prentice-Hall, 1972).

Ucko, Peter J., and Dimbleby, G. W. (eds.), *The Domestication and Exploitation of Plants and Animals* (Duckworth, 1969; Aldine·Atherton, 1971).

Sun, Wind, and Earth: Energy Sources for Tomorrow?

The energy crisis of 1973-74, caused basically by the accelerating consumption of finite petroleum resources and intensified by cutbacks in oil production in the Middle East, has generated renewed interest in alternative sources of power. Among these, coal and nuclear fission both raise problems of environmental pollution, and man's technology has not yet succeeded in harnessing nuclear fusion. Three sources that are both clean and readily accessible are the sun, the wind, and the heat from the earth's interior. On a small scale all have been used in the past. The three articles that follow examine each one in terms of its potential for producing significantly large contributions to the world's power supply in

Power from the Sun

by John F. Henahan

The sun can provide an almost inexhaustible supply
of clean, inexpensive energy if man can learn
to tap it efficiently.

Man has good reason to look to the sun—a clean, powerful, and essentially limitless energy reservoir—as an alternative source of power to conventional fuels. George Löf, a scientist from Colorado State University, maintains that if coal, oil, and natural gas had not been cheap and readily available, the sun would probably already be used to heat and cool homes, to cook meals, to distill water, and perhaps even to generate electricity. In fact, the technology for doing so already exists. Löf has been living in his own sun-heated house for the last 15 years, reducing his space-heating bills by 30% in the process. In addition, thousands of people in Japan, Australia, Israel, and Florida have been using solar energy to provide warm water. In support of Löf's argument, it has only been since the late 1950s, when the Japanese gained access to "more convenient fuels" such as petroleum, that the use of solar water heaters began to decline in that country.

It is the cost of solar power that has prevented its widespread utilization. Even though man has proved many times that he knows how to tap the energy of the sun for heat and electricity, the cost of making it competitive with coal, oil, and natural gas has been beyond his reach. By 1974, however, in the midst of an energy crisis in which traditional fuels were becoming more expensive and scarcer, more manpower and money than ever before were being marshalled to break down the economic obstacles to the large-scale use of solar power.

JOHN F. HENAHAN, former editor of Chemical and Engineering News, *is a science writer.*

76

Costs aside, solar power is an attractive energy option. The solar energy that falls on the earth's surface in one year is approximately 30,000 times the amount required annually to power all the devices that man uses to make his life more comfortable. Put another way, the roof of an average house absorbs about 500,000 calories of heat energy from the sun during a clear eight-hour day. This is equivalent to the power generated from the burning of about 14 gal of gasoline. Translated into electrical energy, it could produce about 56 kw-hr of electrical power, approximately three-quarters of the house's power needs.

Among the difficulties in making solar energy more economically acceptable are the facts that the sun does not shine all the time and that even when it does its energy is spread so thinly across the earth's surface that expensive devices of varying complexity must be developed to collect, concentrate, and store it. In spite of this, solar power for some applications is not expensive. For example, some solar water heaters used in Japan are as simple as a large plastic water tub covered with a glass or plastic sheet and painted black on the bottom to absorb the sun's radiation. In such cases costs can be as low as $20 for a unit that produces bathwater as hot as 100° F. But in order to generate electricity on a large scale, either by using solar cells or other techniques, the costs are well beyond economic feasibility at this time.

An object of man's attention since his earliest days, the sun has been worshipped by many peoples, including the Incas of Peru (opposite page). In the 20th century the effort to utilize it as a source of power has led to the construction of such facilities as the solar furnace at Odeillo, France. There, an array of parabola-shaped mirrors focuses the sun's rays onto a collector, where the solar energy is transformed into electrical power.

77

Solar heating and cooling

In a major attempt to define the best, fastest, and in the long run most economical way to harvest the sun's energy, the U.S. National Science Foundation (NSF) and the National Aeronautics and Space Administration (NASA) organized a panel of solar energy experts. In their 1972 report, *Solar Energy as a National Energy Resource,* they judged that with the right kind of research effort and an adequate expenditure of money, the solar heating and cooling of buildings could be a reality in the U.S. within five to ten years.

Their optimism was encouraged by the fact that in the U.S. in 1974 more than 20 individuals were heating their homes and/or their water supplies with solar energy. In several such houses the sun's heat is taken up by flat rooftop collectors, which usually consist of a black metal sheet to which a number of hollow tubes are bonded. The collector is covered by two transparent glass or plastic plates. When exposed to sunlight, the black sheet absorbs the heat and transfers it to the fluid within the pipe. The warm air or water is then circulated throughout the house. Heat that is not used immediately can be stored in insulated tanks containing gravel or some other good thermal absorber. The glass plates on the solar collector allow the radiation of the sun to pass through and heat the metal sheet, but at the same time the plates trap the infrared radiation that is given off by the metal.

Other solar heating units vary in complexity, depending on how much money the designer wants to spend, where he lives, and his

Heating and cooling of the "Solarchitecture House" in California are accomplished by placing large plastic bags full of water on the roof under a sliding insulated panel. In the summer (below) the panel covers the water bags by day to block the sun's rays and then is slid back at night so that the water can be cooled by the night air; this water then acts to cool the house the next day. In the winter (opposite page) the panel is slid back during the day so that the water can absorb the heat from the sun and is used as a cover during the night to keep the heat inside. The arrows indicate the direction of the heat.

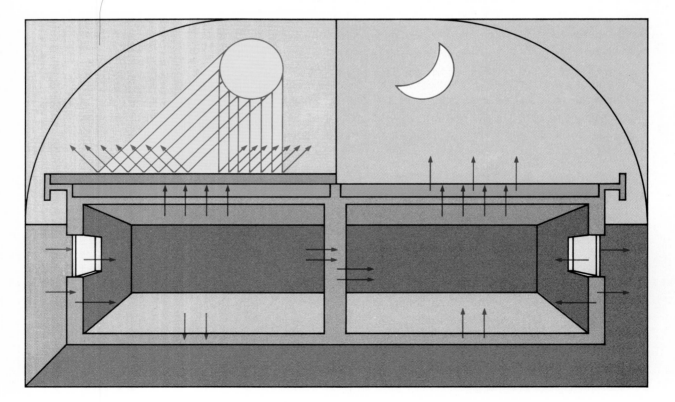

ingenuity. For example, in a solar house designed and inhabited by Steve Baer of Corrales, N.M., the entire south wall consists of a stack of water-filled 55-gal drums arranged on the inside of a glass window. The sun heats the water in the drums, and the amount of heat supplied to the house during the day can be controlled by drawing or opening a curtain between the room and the drums. To warm the house in the evening, an insulated panel is drawn over the outside of the window. The drum wall can also double as a summer cooling system if the window is opened and the night air allowed to cool the warm drums. According to Baer, the drums can be installed for less than $5 a square foot. He suggests that solar houses of his design are most effective at latitudes between 30° and 45°, that is, in all but the northern reaches of the U.S.

In the "Solarchitecture House" built by Harold Hay in Atascadero, Calif., the roof is covered with large plastic bags full of water. They absorb solar energy on sunny winter days and heat the interior of the house. On winter evenings the bags are covered with a sliding insulated panel to keep the heat inside, while in the summer the panels are slid back so that the temperature of the water is lowered by the cool night air. During summer days, the house is kept cool by closing the panel to block the sun's rays. Hay estimates that his system would meet all of the house's cooling needs and 90% of its heating requirements, maintaining an inside year-round temperature of 70° F. In Atascadero, temperatures during an average year range from 10° F to more than 100° F.

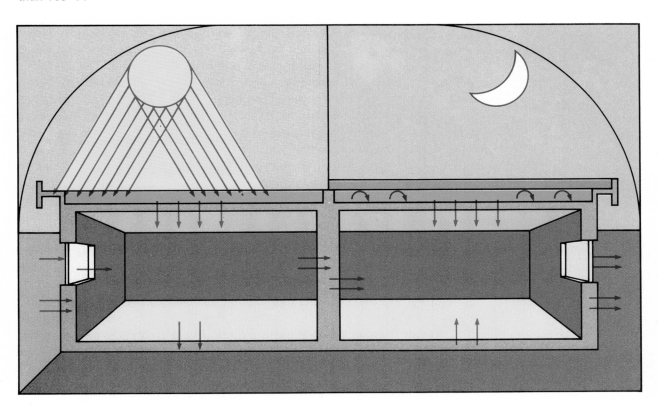

Solar furnace in Massachusetts comprises a plane mirror (foreground) and, reflected in it, a concave mirror and target hut.

Another experimental solar house, under study at the University of Delaware's Institute of Energy Conversion, goes beyond other installations in that it employs an array of solar cells on the roof to make its own electrical supply. Solar cells, used for many years in the space program, are wafers of highly purified silicon or some other material. When photons (particles of light) encounter the crystal lattice of the silicon, they release electrons, which in turn produce electric current. The power generated can be used directly or stored for use on cloudy days in the same kind of lead acid battery found in automobiles.

As of 1974 silicon cells were about 100 times too expensive for anything other than specialized small-scale uses, as in the space program, but Karl Böer, the director of the Delaware project, believes that mass production techniques for a different type of solar cell—containing cadmium sulfide and copper sulfide—could reduce the cost of electricity to a competitive two cents to five cents per kw-hr. The Böer technique consists of preparing a solar cell "sandwich" by first vaporizing the cadmium sulfide and depositing it as a thin film on a metal base. On top of it a layer of copper sulfide is electroplated, and the sandwich is then covered with a metal grid that allows sunlight to pass through and activate a flow of current. To protect the solar cell from dust, rain, snow, and other disturbing factors, the entire assembly is coated with transparent plastic.

According to Böer, his solar house could act in tandem with a commercial electric utility grid. That is, during peak hours of sunlight surplus power produced by the rooftop solar cells could be fed back to the power grid. The utilities could then use the excess power to meet peak demands on hot summer days, when air conditioners cause a heavy drain on the available electrical power.

Going beyond the individual solar house concept, scientists from the California Institute of Technology, the University of California, and the Southern California Gas Co. are exploring the commercial feasibility of installing solar heating units in apartment buildings in the Los Angeles area. They calculate that a solar collector with an area of about 60–70 sq ft could supply about 75% of the building's hot water needs, while a collector about five times larger could also supply up to three-quarters of the space-heating needs.

In addition, NASA planned to equip its 53,000-sq-ft Systems and Engineering Building at Langley Research Center in Hampton, Va., with a solar heating and cooling system. According to the construction schedule, the system was expected to be in action by about mid-1975. The solar complex was to be equipped with tanks to store hot or cold water for use during days when the sun does not shine. On other days when the sunshine is inadequate, the system was to be supplemented by Langley's central steam system.

In regard to costs, Löf made a survey of eight U.S. cities in 1970 with widely varying climates and calculated that solar heating would be cheaper than electrical heating in all but one (Seattle). For example, in Albuquerque, N.M., where electrical heating costs about $1.58 per 100 kw-hr, solar heating, using available technology, would cost about 59 cents per 100 kw-hr. In sunny areas, such as Santa Maria, Calif., solar heating might even cost less than natural gas, the cheapest and cleanest fuel available.

The solar farm

As important as the solar heating and cooling of buildings might be in reducing demands on dwindling fuel supplies, the large-scale production of electricity from the sun would have an even greater impact. There are two major ways to use the sun to make electricity, and both have already been demonstrated on a small scale. In the thermal approach the sun's heat is concentrated by parabola-shaped mirrors onto the surface of a pipe containing air and water. When the pipe becomes hot enough, the heated air vaporizes the water for use in a turbine, which then runs a generator. The disadvantage of this method is that the parabolic mirror must be continually rotated to follow the sun across the sky. An alternative approach is to use flat plate collectors similar to those used in several experimental solar heating systems. They do not have to be rotated and can be used on hazy or cloudy days.

Among the strongest partisans for the thermal approach are Aden Meinel and his wife Marjorie, who proposed the creation of huge "solar farms" in 1971. According to the Meinel plan, solar heat would be concentrated by large arrays of collectors covering as much as 15,000 sq mi in the southwestern U.S. or other sunny areas. Acting as an energy source for electrical power plants, solar farms of this type could theoretically meet all of the electricity needs of the U.S. by about 2000, the Meinels' estimate.

At present, the Meinel idea has technological bottlenecks, including the fact that the heat pipes of experimental models of the collectors do not get hot enough (about 1,000° F) to build up the steam pressures required for the economical operation of commercial power turbines. To solve that problem, Meinel and his colleagues are attempting to develop selective coatings for the pipes that will absorb the sun's heat without losing much of it as infrared radiation.

In addition to the technological shortcomings of the solar farm, authorities such as George Löf indicate that it is too expensive to consider for any large-scale use in the foreseeable future. Meinel agrees that even under the most ideal conditions electricity produced on a solar farm would cost about 30% more than electricity from a conventional power plant. However, he believes that the rising costs of coal, oil, and natural gas should diminish this cost differential in a few years.

The photovoltaic approach

The "photovoltaic approach" to generating electricity utilizes the solar cell, which as mentioned earlier, was as of 1974 too expensive to be used for large-scale power production. The reason for its costliness is that each cell is hand made at a cost of about $1 per sq cm of cell area. Before solar cells can be taken seriously for use in large electric power plants, their costs must be reduced to about 0.7 cent per sq cm, including the cost of installation.

Eugene Ralph, a solar cell expert from Heliotek in Sylmar, Calif., believes that solar cell costs can be reduced significantly by increasing the efficiency of the cells and by using mirror devices to concentrate more sunlight onto the solar cell surface. Finally, Ralph predicts that a mass production system being used to make silicon in the form of long, thin ribbons can reduce production costs from 30 cents per sq cm and installation costs of $30,000 per kw to a more competitive 0.25 cent per sq cm at an installation cost of $250 per kw.

Orbiting power plants

Even with significant cost reductions, power plants that draw electricity from solar cells could not be operated at night or on cloudy days. Therefore, electrical power must be stored or drawn from some other source during the "down" period. As an alternative, Peter Glaser of Arthur D. Little, Inc., in Cambridge, Mass., visualizes huge satellite solar power stations (SSPS), which would orbit the earth at an altitude of 22,300 mi and be exposed to the sun nearly 24 hours a day.

In Glaser's plan, electrical power in an SSPS would be produced by a battery of solar cells ten square miles in area and surrounded by reflecting mirrors. The electric current would then be converted to a beam of microwaves, transmitted back to a receiving antenna on earth, and reconverted to electricity. Glaser believes that an SSPS, which would weigh five million pounds, would generate 10,000 Mw of electrical power each year, equivalent to the annual energy requirements of ten million people.

Russ Kinne from Photo Researchers

But Glaser's high-flying idea generates new problems. To lift the many components of the SSPS into orbit would require about 500 two-stage trips of the space shuttle, scheduled by NASA to go into operation during the 1980s. This, of course, would cost billions of dollars at a time when public enthusiasm about the space program has begun to wane.

Solar sea power

One of the most fanciful applications of solar power, now under study at Carnegie-Mellon University in Pittsburgh, Pa., involves using the varying temperatures of sea water to create a huge "heat pump" for manufacturing electrical energy. This concept is based on the principle that a fluid can be made to expand by absorbing heat that is at a high temperature and to contract by liberating heat that is at a lower temperature. The energy produced when the fluid expands can be used to move a piston or rotate a power turbine. Fortuitously, the energy for operating such a heat pump is available in areas where the surface waters of tropical seas, heated by the sun, are from 25°–40° F warmer than the water a few thousand feet below. When the cold water is sucked up to the warm-water level, the energy produced by the expansion of the cold water is great enough to vaporize water or some other fluid. This gas can then be used to operate a low-pressure turbine.

Full-sized model of the solar cell in the U.S. Explorer 7 spacecraft (opposite page). The device supplied power to the spacecraft by converting radiant energy from the sun into electricity. The rooftop apparatus in Israel (above) uses solar energy to provide a constant supply of hot water to the building below.

83

The potential of a solar sea power plant cannot be fully evaluated until day-to-day operating costs and various technical problems are confronted and solved, but Clarence Zener of Carnegie-Mellon estimates that such a facility could be built off the Gulf Coast of the U.S. more cheaply than a nuclear power plant or any other type of solar power plant now under consideration. For example, while it cost about $400 per kw to build the cheapest nuclear power plant available in 1974, a solar sea power plant could conceivably be built at a cost of $300 per kw. The electricity from the sea power plant, built offshore, might be used to break water down into hydrogen and oxygen. Piped back to the mainland, the hydrogen could be used in a fuel cell to produce electricity. As an example of the power available to a solar sea power plant, the NSF-NASA solar panel estimated that 26 trillion kw-hr per year of energy could be recovered from the Gulf Stream off the southeast coast of the U.S., enough to meet the nation's projected energy needs in the year 2000.

Photosynthetic solar energy

A growing number of solar energy experts believe that the best way to put the sun to work would be to let it do what it does best, carry out the natural process of photosynthesis. This is the process by which sunlight, water, and carbon dioxide react to produce carbohydrates and free oxygen. Because photosynthesis is the ultimate source of man's present supply of coal, oil, and natural gas, some scientists would like to exploit this age-old process for today's immediate energy needs.

Among the apparently most simple suggestions for exploiting photosynthesis are the "energy plantations" proposed by George Szego and his colleagues at InterTechnology Corp. in Warrenton, Va. They would comprise large areas of land set aside for the cultivation of trees that would eventually be chopped up and used instead of coal or oil to fire the steam boilers of large power plants. Szego notes that wood fuel is a renewable resource which burns cleanly, and that cutting down trees is much less environmentally devastating than drilling oil wells or strip mining coal. To these advantages, he adds the fact that the forest areas of an energy plantation could also be used for nonconflicting applications, such as recreation, water retention, erosion reduction, and wildlife conservation. In its report, the solar energy panel estimates that energy plantations covering less than 3% of the U.S. would produce stored energy equivalent to the total anticipated annual needs of the U.S. in 1984.

In another approach, Martin Kamen of the University of California in San Diego proposes that if the photosynthetic "machinery" of algae or other plants could be isolated in large quantities it could be used to break down water into hydrogen and oxygen, both of which can produce electricity when combined in a fuel cell. In laboratory experiments Kamen and his colleagues have been able to separate this machinery, chloroplasts, from common spinach plants and mix it with an enzyme called hydrogenase and another substance called ferredoxin. When water is added to this mixture, hydrogen is produced.

Other researchers have shown that many organic products of photosynthesis can be converted to a mixture of burnable gases, liquids, and solids by heating them at temperatures ranging from 900° to 1,600° F. Such processes could eliminate growing piles of organic wastes and at the same time produce useful fuel.

Future prospects

The NSF-NASA solar panel estimates that 20% of the total energy needs of the U.S. in the year 2020 could come from solar power if about $3.5 billion is spent on developing it over the next 15 years. This would require federal spending well beyond the present level ($13.8 million in fiscal 1974), but both Congress and the administration seem to be taking steps in that direction. For example, in his 1975 budget message, Pres. Richard Nixon asked for an expenditure of $50 million for research and development on solar energy. In spite of that more than fourfold increase, frustrated solar energy experts argue that if half of the money spent to bring nuclear energy to its presently limited level of use had been spent on solar energy since World War II, many of the present energy problems would already have been solved. Yet, in the same budget, President Nixon increased his requests for nuclear energy funds from approximately $630 million in fiscal 1974 to almost $900 million in 1975.

A 1972 report of the Rand Corp. (*California's Electric Quandary*), suggested that solar energy development could be accelerated by providing tax relief for companies that make, sell, or install solar water heating and space-heating equipment. As a last resort, the report recommends that electrical power devices of many types be taxed highly and, if necessary, banned for the duration of an energy crisis.

In addition, many have suggested that architects and construction firms should be taking steps to ensure that new housing units and apartment buildings are designed to accommodate solar heating and cooling devices. Similarly, the Federal Housing Administration and Veterans' Administration could provide encouragement by extending loans for the installation of domestic solar power systems.

Power from the Wind
by Samuel Moffat

**One of man's oldest sources of energy is being
reexamined for applications in the 20th century.**

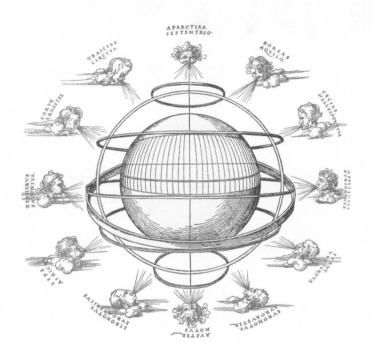

Throughout North America, in government and industrial laboratories and more than a dozen universities, scientists and engineers in 1974 were urgently attempting to update an ancient technology for extensive application in the last quarter of the 20th century. Their purpose was to capture some of the elusive strength of the winds that swirl over the earth, in order to provide more power for an energy-deficient America. Their efforts were being paralleled in Israel and in several European nations, particularly Sweden.

No one knows who invented the first windmill. There is no firm evidence that one existed until after the time of Christ, since earlier historical references are unsubstantiated. About A.D. 400 the Chinese are known to have had a primitive wind device in which air scoops turned a vertical axle. The Persians, however, were apparently the first to make what today would be called windmills. The first reference to them is in a 10th-century document. Later descriptions indicate that the Persian device consisted of vertical sails made of cloth with arms that moved horizontally in a circle about a vertical axis.

The windmills with which the Western world is familiar are quite different. Their sails rotate in a vertical plane around a horizontal axis, and they are more efficient than the Persian type. It is quite possible that they were invented separately rather than adapted from the earlier design. They underwent a long period of development, particularly in The Netherlands and England. There were thousands of windmills in Europe in the 18th and 19th centuries. They were used to provide power to pump water, grind grain, saw wood, and perform similar work. Several hundred thousand were said to be operating on farms in the United States late in the last century.

SAMUEL MOFFAT has written extensively in the area of applied science and technology.

Windmills, however, like the great sailing vessels that once plied the seas, began to lose their dominance as first the steam engine and then gasoline and diesel engines and electric motors took over supplying the power for most of man's work. Although modern windmills may be built with carefully designed turbine blades and can generate electricity and operate more efficiently than ever before, they have not been able to compete economically with these other engines, which provide power whenever it is needed (the wind not being dependable) and use cheap fossil fuels or hydroelectric energy. By about 1950 interest in wind power had begun to decline drastically, and engineers seeking financial backing for windmill projects found little money available. In the non-Communist countries during 1973 there were only six firms making windmills designed to generate electricity.

Reviving interest

And yet there persists a strong interest in utilizing wind power, prompted by the need for additional sources of energy. In 1971 Oregon State University began a three-year, $132,000 research project to assess the feasibility of supplementing electric power in the Pacific Northwest with wind generators. The study was funded by four Oregon utility districts. In 1972 a panel of specialists organized by the U.S. National Science Foundation (NSF) and the U.S. National Aeronautics and Space Administration (NASA) to determine the potential value of solar energy and wind energy recommended that the federal government fund programs to make wind power economically feasible. The panel concluded that wind-driven generators could provide nearly one-fifth of the electricity the U.S. would need by the year 2000.

Twelve wind gods surround an armillary sphere (opposite page) in a woodcut by German artist Albrecht Dürer (1471–1528). The multivane fan windmill (above), invented in the United States in the mid-19th century, could lift water to greater heights than the Dutch-type windmill and could operate in low winds. The tail vane keeps the wheel facing the wind.

89

In 1973 the NSF and NASA sponsored a major workshop on wind-energy conversion systems. During the same fiscal year the NSF began financing two substantial wind research projects, at the University of Montana and at Oklahoma State University. The total NSF expenditure for wind research in the 1972–73 fiscal year was $200,000. In 1973–74 it was about $1 million.

Why this sudden new concern with an ancient, apparently outmoded technology? There are at least three reasons: shortages and rapidly rising costs of fossil fuels, the expenses of making power plants meet more stringent environmental standards, and the recognition that the U.S. should become more independent of outside energy sources. These factors combined to stimulate interest in developing alternate methods of power production, including wind power. As the report of the 1973 wind-energy workshop noted, "future energy needs will have to be supplied from a variety of sources such as coal, oil, nuclear, geothermal, direct solar, and, of interest here, the wind." Wind and direct solar are particularly important because they neither pollute the environment nor consume any nonrenewable resources.

*Windmills of various designs
have been used in many parts
of the world to supply power.
On the island of Majorca (opposite
page, top) tail vanes turn the blades
of a series of mills into the wind,
while on the Greek island
of Míkonos (opposite page, bottom)
a stone tower mill is equipped
with 12 jib sails. At Enfe
on the coast of Lebanon (left)
windmills pump water out of salt pans.
The traditional Dutch-type
windmill (below) has been used
in The Netherlands for centuries
to drain land for agriculture.*

Technological know-how

In light of wind power's virtual neglect for two decades, it is rather surprising to hear most experts say that there are no technical problems standing in the way of using it to generate electricity on a reasonably large scale. Half a dozen speakers at the 1973 NSF-NASA workshop made just that point. Karl H. Bergey of the University of Oklahoma summarized the situation: "The subject is old. It has been reviewed and investigated by innumerable competent people over the years and . . . the technology itself has reached a high level of development in Germany and in other parts of the world." This means that making a windmill today is "like making an electric motor or an airplane," according to Peter B. S. Lissaman of the Pasadena, Calif., firm AeroVironment, Inc. "We know how to do it."

The technical feasibility of windmills in certain situations is well established. For example, they can be installed in out-of-the-way or even completely isolated locations where quite small amounts of electricity are needed. In such cases they may power meteorological recording devices, radio relay stations, or small navigational aids. A windmill

for such an application might require a rotor blade about one meter in diameter. It could have a rated output of 50 w and produce a modest 5 kw-hr per month.

In a second situation, a larger capacity would be required to provide electricity for a home or a small factory. For example, Henry Clews, his wife, and their two children live in a three-room house in the country-side near North Orland, Me. All their electricity is generated by wind power. At first they had only one windmill, with a capacity of 2 kw. But even that produced enough electricity for lighting; household appliances, including a vacuum cleaner; sewing machine; stereo phonograph; television; and power tools. Driven by winds averaging 8.6 mph, it provided 120 kw-hr per month.

The complete installation for the Clews house cost less than the price quoted by the local electric company for running a power line into the property (they live five miles from the closest paved road), and there is no monthly utility bill. Clews estimated that over the next 20 years his electricity would cost only about half what it would if he had installed a gasoline or diesel generator. Furthermore, the only maintenance his system requires is a quart of oil once every five years and a new set of batteries in about ten years (the batteries store up electricity as a precaution against calm periods). The Clewses liked the first windmill so much that they later added a second one, which helps service a guest house on the property. Both houses have "more than a minimum of modern conveniences," Clews reports.

According to Clews his experience demonstrates that where electricity is not easily available "wind-generated power can actually represent the cheapest available means of generating power." Wind generators of a few kilowatts have also been installed in such varied locations as the Arctic Circle, Antarctica, the Alps, and in some deserts. They have provided power for weather stations and lighthouses. Solid-state electronics has made it possible to reduce the power demand for many applications to levels where a windmill can provide all that is required. A large lighthouse, for instance, may only require 1 kw to power its bright lamp.

Large systems

Thus, it is clear that small- and medium-sized windmills may be not only technically but also economically justified under the proper circumstances. What has not been demonstrated is that large wind-powered installations, producing hundreds of thousands or even millions of watts of power, can be operated profitably as part of a major utility network.

The most dramatic attempt to prove the financial value of the large-scale use of wind power occurred in the United States during World War II. A giant wind turbine with two blades forming a propeller measuring 175 ft across (more than half the length of a football field) and weighing 16 tons was built on 2,000-ft Grandpa's Knob, west of Rutland in the Green Mountains of Vermont. The generator became operational in October 1941 and produced power for the Central Vermont Public Service Corp. for 16 months until a main bearing failed. Wartime material shortages prevented the replacement of the bearing for two years. A month after the machine was repaired one of its blades broke off because of a weld defect, and it was never operated again.

The generator at Grandpa's Knob was built for the S. Morgan Smith Co. of York, Pa., and was designed by Palmer C. Putnam, and therefore is known as the Smith-Putnam Wind Turbine Project. Technically it was a remarkable achievement. The best engineering talent of the time, including the world-famous California Institute of Technology aerodynamic expert, Theodore von Kármán, contributed to the design. After World War II, the company undertook an economic analysis of the project and determined that units like that at Grandpa's Knob would cost about 65% more per kilowatt to build than the Vermont utility could afford to pay. Putnam suggested several ways of reducing costs by improving the design, but the Smith Co. decided not to invest further capital in the project, which had already cost $1,250,000.

Since that time there has been no major effort to develop individual wind generators that would be able to provide large amounts of power at prices competitive with more conventional sources, such as water or steam produced at plants heated by coal or oil. Another economic difficulty with windmills is that, even when they are big, they still do not generate vast amounts of power. The Smith-Putnam turbine, the largest such machine ever built, only produced sufficient electricity "to light a town," in Putnam's words. The rated output of 1.25 Mw represented less power than that under the hoods of five 1974 Cadillac engines run at maximum output. Thus, many large wind generators would have to be joined together to make a major contribution to a utility network.

Five-year plan

The National Science Foundation recognized that economics was the principal issue in developing wind power and formulated its program accordingly. "Our five-year objective," said Louis Divone, manager of NSF's Wind Energy Conversion Program, "is to have experimentally verified the technical and economic feasibility of large wind-energy systems. We will also do research on the legal issues and the aesthetic, environmental, and social impact—all the major considerations."

NASA, particularly its Lewis Research Center in Cleveland, was collaborating with NSF. Early in 1974 a 100-kw experimental system was being designed by NASA for NSF. Joseph M. Savino, technical director of the NASA effort, said, "We're going after the wind generator systems first. We can build wind generators right now, and we can probably build big ones. The question is whether we can build big plants which reliably supply power to the customer on demand and at a cost about the same as from other sources such as fossil fuel or nuclear plants. That we don't know." It is expected that NASA, working with industrial contractors, will design, build, and test a number of experi-

Smith-Putnam wind turbine, built on Grandpa's Knob in the Green Mountains of Vermont, generated electric power for 16 months in the early 1940s. The two blades measured 175 ft in length; after one broke off, the machine was never operated again.

93

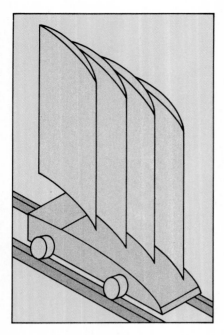

Vehicles bearing airfoils that resemble airplane wings set on end would be moved by the wind around the elevated vertical track structure shown on the opposite page. The axles of the vehicles would turn electric generators. A vertical track allows the airfoils to be positioned both above and below the vehicles, thereby reducing the bending moment on the vehicle carriages. Above, the vehicle with airfoils extending in only one direction is designed for use on a horizontal track.

mental systems in an effort to find the configuration that will ultimately provide the lowest cost per kilowatt-hour.

NSF planned to support three other research activities that relate to large systems, as well as smaller applications. First, there was to be a research and development effort to improve the performance characteristics and reduce the costs of components for wind systems. Although others had demonstrated that large wind machines worked, the essential engineering data with which to analyze or reproduce them often were not available. Research was planned not only on advanced rotor-type systems but also on innovative or unconventional designs. Second, NSF would investigate requirements for different applications, including methods of storing electricity generated by windmills in order to provide a consistent source of power.

NSF also planned to investigate the problems of measuring wind energy, not only for the purpose of picking the best locations for future wind generators but also to predict the power output of a site. Most of the existing wind data were obtained from airports, weather stations, or other reasonably accessible locations; but these are frequently not the best sites for wind machines. It is often possible to find locations that have annual average wind speeds 50% higher than those in the same general area. Such sites can produce considerably more electricity than less windy ones, for the power in the wind increases with the cube of the wind velocity.

Even before the federal effort in the U.S. was under way a number of universities and corporations had become interested in exploring the potentialities of wind power. Following is an abbreviated summary of some of the major recent developments.

Windmills

Brace Research Institute at McGill University in Canada began investigating wind-power technology in the early 1960s, particularly in regard to small windmills that could be used in underdeveloped rural areas. They offered plans for "a cheap wind machine for pumping water" whose vertical shaft rotor (a Savonius rotor) is made of two oil drums cut in half and then welded together so that the drum segments form wind scoops. Brace also perfected a 10-hp wind turbine that utilizes the latest aerodynamic theory in its design and can be built from readily available materials.

Engineering students and faculty at the University of Oklahoma built a wind-driven generator designed to charge completely the batteries of a two-seat electric "Urban Car" in eight hours under average wind conditions. The car has a range of 25–50 mi before needing another charge. A "sailwing" windmill utilizing an aerodynamic surface of sailcloth was developed at Princeton University. It proved to be efficient and is also lightweight. And a Wisconsin commune, Windworks, designed inexpensive propeller-type blades made of easily shaped paper honeycomb which is then covered with fiberglass cloth and resin.

Among the more innovative approaches being explored was that

94

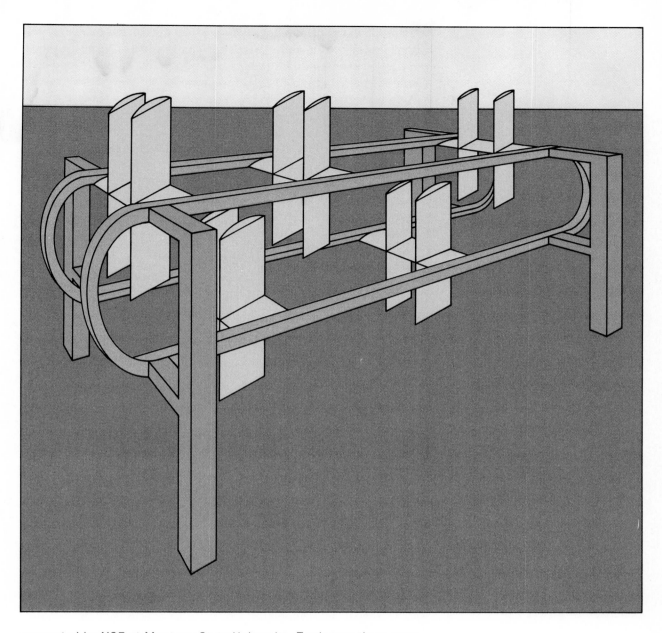

supported by NSF at Montana State University. Engineers there were attempting to design airfoils, resembling airplane wings set on end, that would be attached to railroad cars. The wind would then move the cars around a 5- or 10-mi-long closed track, and the axles of the cars would turn electric generators. Target output of the design is 10–20 Mw.

Energy storage

The conventional way of storing energy produced by a wind generator is in batteries, as at Henry Clews's house in Maine. New combinations of electrodes and electrolytes may improve battery technology, but their future is uncertain.

Investigations of a "superflywheel" that would store energy in a rotating wheel were carried out at Johns Hopkins University. Several flywheel materials (glass, fiberglass, wood, and certain proprietary materials) were expected to provide cheaper storage possibilities, in watt-hours per dollar, than the lead-acid battery, which is the best present device for bulk storage uses.

Water in a reservoir represents stored energy. In the same way, air can be compressed for later release. Air could drive a turbine directly, or it might be combined with a small amount of fuel to increase efficiency. At the Hydro-Quebec Institute for Research, Quebec, Can., researchers were designing a prototype system, driven by a wind machine, to compress air and store it in underground salt caverns. The investigators believed that the system could be enlarged to have a capacity of several megawatts.

Early in this century Danish wind experts first demonstrated another solution to the storage problem: Do not try to accumulate the energy at all, but feed electricity into a power network as it is generated. This would require a backup power source for the times when wind generation could not meet the need, or else a large enough wind-generating system so that there would always be a sufficient amount of electricity produced somewhere in the complex to satisfy customers' requirements.

Wind strength and energy potential

Although great masses of wind data have been collected at many locations throughout the world, as of 1974 experts still had only limited ability to predict accurately the energy output of potential windmill sites or even to estimate the total amount of energy that can be extracted from the wind. There have been numerous estimates of the latter but no systematic study of the situation. NSF expected to spend at least two years getting a much better estimate.

Experts do not all agree about the minimum wind velocity required for power generation. It can be quite low, as Clews's experience in Maine shows. According to one evaluation of commercially available small windmills, an "average wind speed of at least 12 mph is necessary for most applications" and there is "no output at all below about 6 mph." In general, larger-capacity wind generators require greater velocities.

Some of the areas in the U.S. where strong, steady winds might be utilized most effectively include the Great Plains states. An analysis of wind records at Amarillo, Tex., over ten years revealed that there were only about 100 hours a year when there was no wind at all. The 31-year mean at Amarillo was 13.7 mph. The Oregon State University study was pointing toward locations either in the Columbia River Valley or two to three miles off the Oregon coast in the Pacific Ocean. A study undertaken by the University of Alaska revealed that winds capable of generating power were available more than three-fourths of the time at sites in the Aleutian Islands.

Among the most imaginative suggestions for capturing the energy of the wind are those of William E. Heronemus of the University of Massachusetts. Heronemus has proposed massive complexes of windmills for such locations as the continental shelf off New England, in Lake Michigan and Lake Superior, and over northern Wisconsin and upper Michigan. He has envisioned platforms carrying three 2 Mw generators (each more powerful than the Smith-Putnam machine) or 200 generators of 20 kw each. The platforms could be anchored offshore or be planted in the ocean or lake bottom. Heronemus estimated that the 200 generator units might cost between $1,268,000 and $1,-413,000 each. He has also described towers rising 800 ft above the ground, half a mile apart, with rows of wind machines hanging from connecting cables like those in a great suspension bridge. A unit a mile long, producing 19.2 Mw, would cost $2,880,000. Allowing for inflation (and for the federal government to defray development expenses) Heronemus said that these generators ought to cost only about half the amount per kilowatt of the Smith-Putnam machine. Systems combining large numbers of these units could generate tens or hundreds of billions of kilowatt hours of power, according to his calculations, without modifying the weather. Complexes of this sort could obviously make significant contributions to a nation's energy needs.

FOR ADDITIONAL READING:

Golding, E. W., *The Generation of Electricity by Wind Power* (Philosophical Library, 1956).

Heronemus, William E., "The U.S. Energy Crisis: Some Proposed Gentle Solutions," *Congressional Record* (Feb. 9, 1972, pp. E1043–1049).

McCaull, Julian, "Windmills," *Environment* (January/February, 1973, pp. 6–17).

Meyer, Hans, "Wind Generators: Here's an advanced design you can build," *Popular Science* (November, 1972, pp. 103–105, 142).

National Science Foundation and National Aeronautics and Space Administration, *Wind Energy Conversion Systems: Workshop Proceedings June 11–13, 1973* (National Technical Information Service Publication No. PB 231-341 650 AV NTIS, U.S. Government Printing Office, 1974).

Putnam, Palmer C., *Power from the Wind* (Van Nostrand, 1948).

Singer, Charles, *et al.* (eds.), *A History of Technology,* vol. II, pp. 614–628 (Oxford, 1957).

Power from the Earth
by Robert W. Rex

The heat generated beneath the earth's surface
constitutes a source of energy that could
contribute significantly to the world's supply
of power.

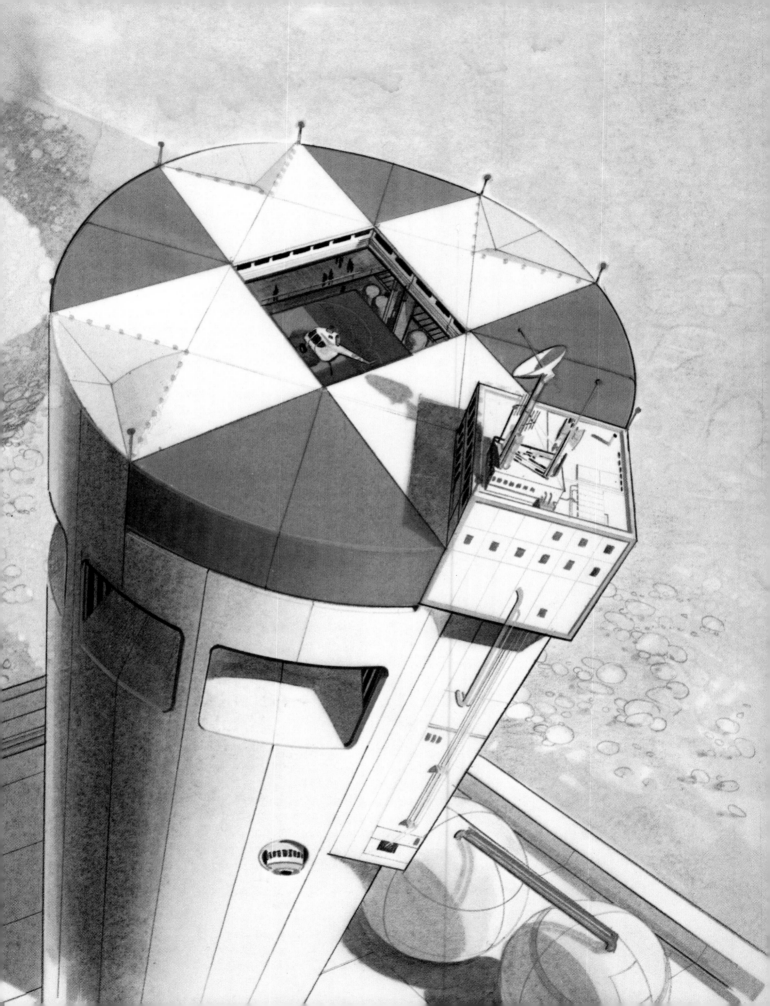

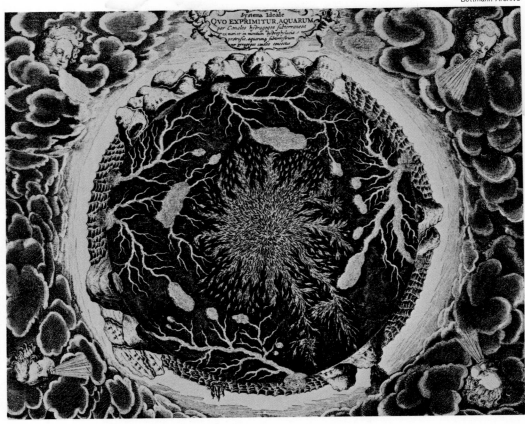

If a person could look with a telescope directly into the interior of the earth, he would see that it is white hot. This intense heat has been built up over billions of years by the accumulation of the thermal energy remaining from the decay of natural radioactive elements and from the frictional dissipation of energy released by the gradual segregation by density of materials within the earth's interior. This white-hot interior is protected by the excellent insulating qualities of the rocks of the earth's mantle and crust. The thermal conductivity of most rocks is so low that they have kept the molten core and the mantle of the earth from cooling, even over geological time. The long history of production of the earth's internal energy and its escape to space has resulted in the establishment of natural heat flow throughout the globe. The heat flow unit (HFU) is one microcalorie per sq cm per second. The average heat flow over the continents and the oceans ranges from 1.7 to 1.9 HFUs. This average is remarkably constant over nearly half to two-thirds of the earth's surface.

There are, however, areas where natural heat flow is substantially higher, often by more than a factor of 10, than the global average. Examples include areas of volcanoes and rift valleys where molten rock rises from deep within the earth's interior to the surface. In these regions large bodies of molten rock push their way upward, rising

ROBERT W. REX *is president of Republic Geothermal, Inc., Whittier, Calif.*

100

buoyantly and transferring very large quantities of mass and thermal energy from deep within the earth's mantle to the surface or near to the surface.

The natural thermal energy of the earth's interior is called geothermal energy. It is often transferred to water in the subsurface, and the energy content of this water can be recovered. If the natural pressures are low, the water may boil underground and steam permeate the natural porosity of the rocks. A well drilled into one of these steam zones produces large quantities of geothermal steam. This steam can be fed into turbines and used directly to generate electricity. If pressures in the water are sufficiently high, boiling is prevented and large quantities of hot water accumulate in the subsurface. This hot water, on flowing into a well, will quickly boil, producing much the same effect as a bottle of soda water being uncorked, and the mixture of steam and boiling water will flow to the surface with great force. The steam can be separated and utilized in turbines in the same manner as dry steam. Under certain circumstances the hot water can be used directly without boiling, and this promises to become a new and major source of geothermal energy.

Recognized for thousands of years, geothermal energy has been used since man first discovered boiling springs and used them for hot

Seventeenth-century concept of the earth's interior (opposite page) is depicted in the engraving "Systema Ideale Pyrophylaciorum" ("Concerning Volcanic Activity"), which appeared in the book Mundus Subterraneus. *Above, steam, which can be fed into turbines to generate electricity, rises from hot springs in Iceland.*

101

baths. More recently, at the start of the 20th century, natural steam began to be harnessed for the generation of electricity; by the 1970s many nations throughout the world were producing electricity from this source. The rapid rise in the cost of fossil fuels, and the recognition that they pollute the atmosphere, caused an increased awareness among scientists of the potential of geothermal energy. This resulted in large-scale development of new geothermal fields throughout the world. Though geothermal energy is primarily considered in terms of the heat content of subsurface water, an even larger amount exists in hot rock where little water is present. Technology is not yet sufficiently advanced to develop commercially the energy in hot, dry rock, but in 1974 a significant research effort in this area was under way by the U.S. Atomic Energy Commission.

Formation of geothermal regions

One process for rapidly transferring heat to the surface occurs in areas of rifting. Rifting takes place where there is dilation or separation of two of the huge tectonic plates that form the earth's crust, as, for example, along the mid-Atlantic Ridge. One of the important areas of geothermal activity is Iceland, which lies along the crest of that ridge. Along the Great Rift in East Africa there is substantial geothermal potential in such countries as Kenya and Ethiopia. Another area of rifting is the Gulf of California and the Mexicali and Imperial valleys at the head of the Gulf; there the East Pacific Rise has caused a rift valley to form, splitting that part of California west of the rift and its associated San Andreas Fault away from the rest of North America.

In rifts there appear to be deep zones of tension that penetrate hundreds of miles into the mantle, fracturing it and releasing pressure. Molten basalt from partially melted rocks of the earth's mantle flows into these fractures and then squirts to the surface with flow rates approaching nearly the speed of sound, according to some investigators. In some cases this basaltic lava will flow out at the surface, while in others it may be frozen by contact with subsurface water and never appear through the overlying cover of wet sediments. However, the thermal energy content, or enthalpy, of the molten basalt, which rises from the mantle at temperatures between 1,000° and 1,100° C, is transferred to the overlying waters. In this way massive amounts of thermal energy are rapidly pumped from several hundred miles within the mantle to the surface of the earth.

Spreading of the earth's crust can also take place without any actual large-scale break in the surface. This results in a process of crustal thinning, or necking, and is accompanied by an upward rise of the rocks of the mantle. The hot rocks of the mantle, therefore, come much closer to the earth's surface in an area of crustal thinning, and there is an increase in heat flow in such a region. This increased flow can be trapped by the insulating qualities of shale in sedimentary basins. Consequently, the water in large sedimentary deposits formed over an area of crustal thinning can have very high temperatures. This hot, high-pressure water is called geopressured geothermal water. It is abundant in sedimentary basins such as the Gulf Coast of the United States, where it occurs in large quantities offshore and onshore in Texas, Louisiana, Mississippi, Alabama, and northern Florida. It is also found in California and in many places in Africa and the North Sea.

The potential for a geothermal energy source is also high in an area where two crustal plates impinge. In this type of collision there may be plate annihilation, in which the oceanic plate is dragged down under the plate that is more continental. Active down drag along the contact between the buoyant continental plate and the sinking oceanic plate creates the deepest holes on earth, the deep ocean trenches. The downward plunging crust in a subduction zone carries large quantities of sediment and water with it. This rock-water mixture has a substantially lower melting temperature than is normal for mantle rocks and, consequently, the subduction process transfers large quantities of rock material into the mantle in an environment where it should be liquid instead of solid. The enormous size of the subduction plates gives them thermal inertia and so it takes some time for subductive material to undergo melting. The molten material thus appears to accumulate in gigantic "droplets" measuring miles to tens of miles in diameter, which then rise buoyantly, melting their way through the overlying mantle and crustal rocks. These gigantic droplets, when crystallized, constitute coarsely crystalline bodies known as batholiths and stocks. If one of the bubbles breaks through to the surface, it will feed a volcano. A long linear subduction zone can in this way give rise to a chain of volcanoes a number of miles toward a continent from an ocean trench. For example, the belt of subduction around the edges of the Pacific Basin is matched by chains of volcanoes located behind the trenches.

The thermal energy carried from within the mantle to the surface by the above-described large masses of molten rock provides an enormous reserve of geothermal power. By the early 1970s scientists understood how to tap that portion of the energy that is present as steam and could utilize in part that portion of the energy present in hot water, but they had not yet developed the technology to extract thermal energy from the molten rock itself.

Geothermal supplies in the U.S.

The resource base for the geothermal energy potential in the United States consists of all of the heat in the ground that theoretically can be recovered. This base is estimated to be exceedingly large, far larger than the conceivable energy requirements of the U.S. for many thousands or even tens of thousands of years. However, the size of the total resource base has relatively little meaning in itself. What is meaningful is the amount of energy that can be recovered through technology that is either presently available or may reasonably be expected to be available in the near future, and also the cost of recovering this energy.

Georg Gerster from Rapho Guillumette

Salt brine brought up with natural steam from beneath the earth's surface is condensed in a tank at the Geysers field in northern California.

103

Amount of geothermal energy in the U.S. in units of Mw-centuries of electricity producible from geothermal energy			
energy price (in mills per kw-hr*)	known reserves	probable reserves	undiscovered reserves
2.90–3.00	2,000	6,000	10,000
3.00–4.00	30,000	400,000	2,000,000
4.00–5.00	—	600,000	12,000,000
5.00–8.00	—	—	20,000,000†‡
8.00–12.00	—	—	40,000,000†§

*Energy price in present dollars.
†Hot, dry rock systems development based on hydraulic fracturing or cost-equivalent present drilling technology.
‡Hot, dry rock at less than 20,000 - ft depth.
§Hot, dry rock at less than 35,000 - ft depth.

The reserves that are known are relatively small, probably equivalent to approximately 32,000 Mw-centuries of electricity. (A Mw-century is a megawatt of electricity produced continuously for a century.) By comparison, the probable geothermal reserves are at least 600,000 Mw-centuries; while the undiscovered reserves are believed to be larger than 40 million Mw-centuries. (Undiscovered reserves are those reserves estimated for known prospective areas by means of geophysical techniques and subject to a discount for the risk factors involved.) This broad range of geothermal potential has to be considered both from the perspective of the degree of uncertainty as to the existence of these reserves and as a function of the cost of energy production in mills per kw-hr (see Table). As the market price for energy increases, a larger and larger proportion of the resource base becomes economically feasible to be used.

Because of the interaction of exploration risk and market price there is no simple answer to the question of the development potential of U.S. geothermal reserves. The probable reserves, if they were developed quickly and at a cost that would be competitive with that of nuclear- and fossil-fuel energy sources, would be sufficient to meet all of the electrical generating needs of the U.S. within the next 30 years, provided that they were found within 500 to 800 mi of their market area. Unfortunately, this is not presently the case for the northeastern and north-central portions of the U.S., and for this reason present geothermal energy technology does not appear to be capable of meeting all of the U.S. electrical energy requirements by the year 2000. However, if the technology to exploit the hot, dry rock geothermal potential were to be successfully developed, the price of electricity at the generating site in the northeastern U.S. would range from 12 to 16 mills per kw-hr compared with electricity from oil and nuclear sources ranging from 12 to 20 mills per kw-hr presently. Coal prices for environmentally acceptable grades are rising rapidly and appear to range from 12 to 18 mills. There appear to be sufficient undiscovered hot, dry rock geothermal reserves to meet the total need for electricity for the entire nation for many thousands of years. But because the U.S. is so far from actually developing the hot, dry rock technology it is impossible to base the nation's energy strategy on the potential of geothermal energy, no matter how attractive it may be economically.

It is evident, however, that for the Gulf Coast and the western U.S. geothermal energy is a leading contender for electricity generation, probably for at least the next 30-year period. In order to calculate the cost of electricity from the table, between 3 and 4 mills per kw-hr should be added to the energy price in order to pay for the cost of the generating and turbine facilities and for operating expenses.

Harnessing geothermal energy

The earliest development of geothermal energy involved creating ponds from natural hot springs and piping the waters to baths. This was followed by drilling shallow wells in areas of natural hot springs in order to increase or stimulate the natural water flow. These wells were often successful, and many former hot springs were replaced by hot flowing wells. Deeper drilling resulted in the production of fluids at higher and higher temperatures.

As more research was done, it became evident that there were different kinds of geothermal resources and that each one involved a different technology for energy extraction. For example, dry steam geothermal energy at the Geysers in northern California is produced by drilling with an oil-well type of rig into fractured metamorphic, sedimentary, and igneous rocks of the Franciscan formation of Jurassic-Cretaceous age. Drilling in these rocks is initially carried out with a fluid containing mud, and special additives are used to keep the mud from gelling at the temperatures encountered. When substantial quantities of drilling mud drop from the well into fractures in the rock, the drilling operation is stopped. Well casing is cemented in place, and the mud is pumped out. Compressed air, with chemical additives, is then used to drill into the steam zone. Drilling proceeds as steam is encountered, and the combination of steam plus air serves to blow the drill cuttings out of the hole. When steam pressure becomes so great that further drilling is not practical, the well is completed.

At the Geysers steam is produced from natural fractures in the rock. The wells have two valves set on the casing head, and then a pipe connects the wellhead with a separator that removes bits of rock, grit, and debris from the flowing steam. The steam then passes through insulated gathering lines to the main generating station, where electricity is produced. The steam is delivered to the generating facility at a pressure slightly in excess of 100 psi, and an attempt is made to keep the temperature at or close to 237° C. The steam turbines are rated at 55 Mw each, and two turbines are usually housed together in one module in the generating plant. The newest approach is to combine two turbines with a single generator, with projections calling for two 70-Mw turbines driving a single 140-Mw generator.

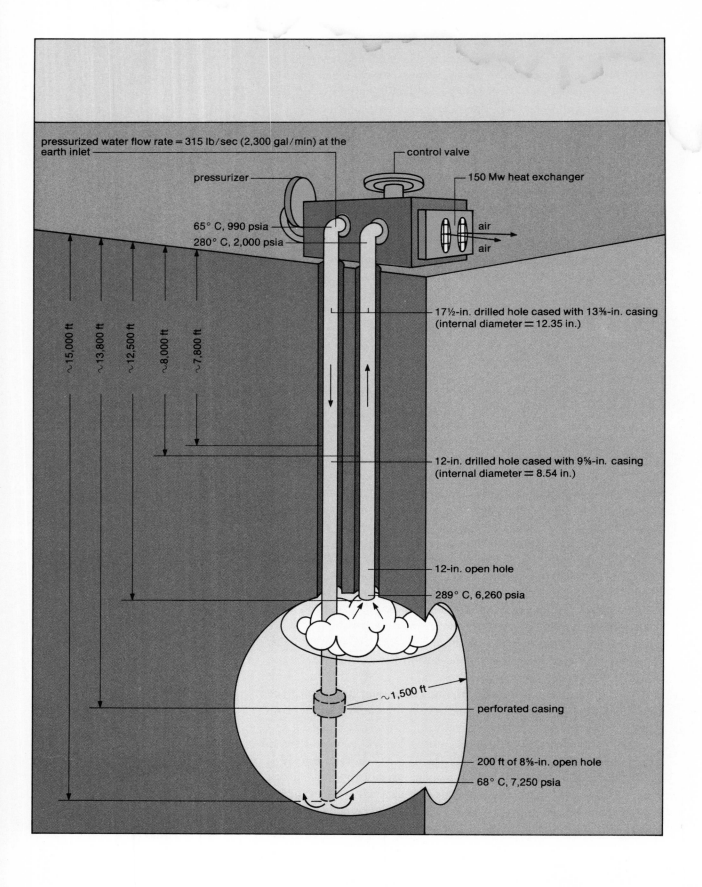

pressurized water flow rate = 315 lb/sec (2,300 gal/min) at the earth inlet

control valve

pressurizer

150 Mw heat exchanger

65° C, 990 psia

280° C, 2,000 psia

air

air

~15,000 ft

~13,800 ft

~12,500 ft

~8,000 ft

~7,800 ft

17½-in. drilled hole cased with 13⅜-in. casing (internal diameter = 12.35 in.)

12-in. drilled hole cased with 9⅝-in. casing (internal diameter = 8.54 in.)

12-in. open hole

289° C, 6,260 psia

~1,500 ft

perforated casing

200 ft of 8⅝-in. open hole

68° C, 7,250 psia

After driving the turbine the steam is condensed in barometric condensers, and the water circulated through forced-draft cooling towers. Approximately 75% of the water is evaporated to the atmosphere, and this serves to reject thermal energy from the system. The remaining 25% carries with it the boron, much of the sulfur, and the ammonia produced with the geothermal steam and is injected back into the top of the steam zone. In this way the geothermal plant operates almost entirely on a closed cycle. Small amounts of hydrogen sulfide and large amounts of water vapor escape, and experimental work is under way to trap the hydrogen sulfide. The amount of water rejected by evaporation is comparable to the normal evapotranspiration load of a forest area such as that located in the vicinity of the Geysers.

Another type of geothermal resource is the hot-water system. The largest commercial development site of hot-water geothermal fluids is the Wairakei field in New Zealand. There, large amounts of hot water are produced from wells that range in depth from about 900 to 3,000 ft. Much of the water boils in the well casing, and a mixture of steam and water comes from the wells. Centrifugal separators separate the steam from the hot water at the surface. The steam is collected by gathering lines similar to those used for dry steam fields, and the hot water goes to silencers where it is allowed to boil off excess heat at atmospheric pressure. A substantial portion of the thermal energy is rejected to the atmosphere in this way, and the residual hot water is transmitted to the local river drainage system and flows to the ocean. Thus, there is a substantial loss of fluid from the source, and only approximately 60% of the thermal energy in the geothermal fluid is transmitted to the gathering system. The proportion of steam to water is 1 to 4, indicating that about 80% of the mass of the produced fluid is being wasted. Many geothermal specialists believe that most future hot-water-system developments will require reinjection of the unused water in order to extend the life of the fields and maximize the total recoverable heat reserves.

Abundant geothermal resources occur in many places in the world where there is insufficient thermal energy content in the water to produce a large quantity of steam by boiling or where the waters may contain chemical constituents that are precipitated on vaporizing. For this reason extensive research was under way in 1974 to develop technology that would permit the transfer of thermal energy through a heat exchanger to another working fluid, such as freon, or to a hydrocarbon, such as isobutane. This secondary working fluid would then run through a special turbine, which, in turn, would drive an electrical generator. By appropriate selection of materials and components, the binary fluid system can operate with much smaller turbines than can a steam system, thereby allowing the turbine system economies to compensate for the cost of the heat exchangers.

A successful binary fluid system has been developed and tested at Paratunka in Kamchatka, Soviet Union, utilizing water of a temperature of 80° C and freon as the working fluid. The principal problem

Proposed method of generating power from the hot, dry rock beneath the earth's surface is shown in the schematic diagram on the opposite page. Wells are drilled into the rock, which is fractured hydraulically so that two wells are connected. Cold water is then allowed to flow down one well into the rock, where it is heated; this causes it to rise up the second well. In a heat exchanger the heat from the water is transferred to isobutane, which then vaporizes and is used to drive a turbine and thereby generate electric power.

107

proved to be leakage of freon from the system. Because of the very high cost of freon this was an important economic constraint, but once this problem was solved, the plant operated successfully. The value of this pilot plant demonstration was substantial because hot waters in the 80° C temperature range are widely distributed throughout the world.

Geothermal water of the geopressured oil-field type is found in many sedimentary basins. On the Gulf Coast of the U.S. there are very large quantities of salty water ranging in pressure from 5,000 to 15,000 psi and occurring at depths of about 6,000 to 30,000 ft. The temperatures of these hot waters range from 150° to 260° C, and they contain substantial quantities of methane in solution. The technology existed in 1974 for accomplishing separately the recovery of mechanical energy from water at high pressures, the recovery of thermal energy from water at the above temperature, and the separation of methane from water. However, nobody had yet demonstrated that this combination of variables could be handled successfully and economically for the recovery from such waters of electrical, mechanical, and chemical energy.

Perhaps the most important geothermal resource of the world from the long-range point of view is the energy in hot, dry rock. Though no demonstrated technology existed in 1974 for its extraction, some reasonable concepts were formulated by various laboratories. The University of California's Los Alamos Scientific Laboratory proposed that wells be drilled into hot, dry rock and that hydraulic fracturing then be used to generate large-scale fracture systems that would connect two wells. Water would then be allowed to flow down from a cold source

108

Wairakei field (left) in New Zealand
is the site of the world's largest
commercial development
of hot-water geothermal power.
The hot water, drawn from deep wells,
becomes mixed with steam
before reaching the surface.
Centrifugal separators (above)
separate the steam from the hot water
at the surface, and the steam
is then transported through pipes
(opposite page) to a power plant
where it will be used to generate
electricity.

Steam pours from a fissure in the earth's surface in Iceland, one of the world's major sites of geothermal activity. The island lies along the crest of the mid-Atlantic Ridge, where rifting causes heat to be transferred rapidly to the earth's surface.

well, and as it passed through the fracture it would be heated by the rock and then withdrawn from the second well. The hot water would be more buoyant than the cold water, and so would rise naturally and in turn transfer heat to isobutane in a heat exchanger. This would cool the water and increase its density. The cool water would then sink back through the cold-water supply well and be reheated. This process would set up a cyclical convective system that would serve to extract the heat from the hot, dry rock over long periods of time. The hot isobutane in the heat exchanger would vaporize, and the resulting vapor would drive a turbine, which, in turn, would drive an electrical generator. The isobutane would be cooled by air-cooling towers.

The University of California scientists believed that there was a reasonable possibility that the artificial fractures produced hydraulically would continue to expand as the rock cooled by the process of heat extraction. When hot rock is cooled along fractures, thermal contraction strains the chilled rock and thereby causes the outward propagation of fractures from the existing system. These new fractures would provide additional surfaces for heat exchange, causing this type of rock fracture system to grow in total size and in its yield of thermal energy. A quantitative model of this system suggested that a single pair of wells would have a useful life of approximately a century. After a century a new pair would be established by drilling two wells approximately 1.25 mi deeper than the existing wells. Therefore, over a period of many centuries this type of geothermal operation could continue in any given area.

By 1974 the most important problem in regard to hot, dry rock sources was the development of technology to generate hydraulic fractures in the rock at the depths necessary for economically significant energy recovery. The technology that had been accomplished was focused primarily on the fracturing of sedimentary and metamorphic rocks under conditions of intermediate and low overburden pressures. Little or no experience had been gained with fracturing granitic and similar crystalline rocks under conditions of burial to depths in excess of about 6,000–18,000 ft.

Environmental impact

The environmental impact of geothermal energy is almost entirely at the site of the field. The thermal energy dissipated in the course of production of electricity is released to the atmosphere by evaporation of cooling water. This water is derived from condensation of the geothermal steam itself. Consequently, a geothermal power plant requires no external source of water, and any excess water produced in the operation may be injected back into the steam field in order to minimize ground subsidence and to conserve water. Excess chemical components, such as boron and hydrogen sulfide, are dissolved in the condensate water and reinjected into the geothermal zone, thereby eliminating any surface disposal of chemical-containing waters. Thus, there is no environmental hazard from brines produced from geothermal systems. However, a small amount of hydrogen sulfide escapes from the evaporative cooling towers in a geothermal plant. In the U.S., efforts were being made by the electrical utilities to devise techniques for inhibiting the escape of this noxious gas, and researchers expected that in a few years the situation would be under control.

An early problem in geothermal development was the noise produced during the testing of steam wells. In recent months extensive use of mufflers began, and they were reducing noise problems substantially. During actual commercial production the steam wells generate such low levels of noise that they are not audible beyond a distance of about 600–3,000 ft.

Little information is available concerning the seismicity associated with geothermal steam production operations. Theoretical and experimental work on the mechanisms of generation of earthquakes indicate that the presence of hot water along a rock fracture system tends to cause rock to move by creep and not by the process known as stick-slip, which results in seismic shocks. This suggests that an area capable of producing geothermal steam is not likely to be one that generates earthquakes, and also that the process of bringing heat from the earth's interior up to the surface actually tends to diminish shallow seismic hazards in that particular locale.

The main environmental impact of a geothermal plant is simply its physical existence, which involves not only a cluster of wells but also associated service roads and facilities. This type of impact is remarkably modest considering the massive effect of almost all alternative energy sources. Research and development in the U.S., Mexico, and other nations is demonstrating that more and more of a geothermal plant can be either placed entirely below ground or at least partially buried in order to minimize the visual impact of the installation.

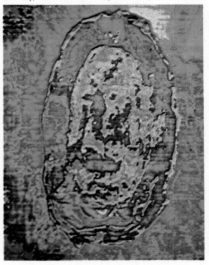

Infrared photography reveals a geothermal source associated with a volcanic crater in the Rift Valley in Ethiopia. The hottest areas are orange, and the coolest are blue. This photographic technique provided the first measurement of the extent of the source and the range of temperatures in the area.

111

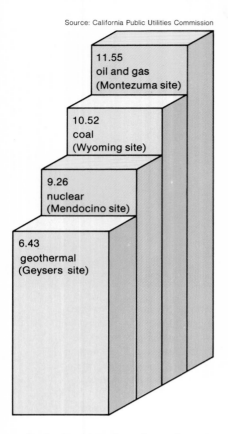

Source: California Public Utilities Commission

11.55
oil and gas
(Montezuma site)

10.52
coal
(Wyoming site)

9.26
nuclear
(Mendocino site)

6.43
geothermal
(Geysers site)

Cost advantage of geothermal power over other energy sources is demonstrated by estimates for 1974 prepared by the Pacific Gas & Electric Co. of California.

Commercial production

The U.S. as of 1974 was the second largest producer of electricity from geothermal energy and was the nation that was expanding its production most rapidly. All of the commercial production was at the Geysers field in Sonoma and Lake counties north of San Francisco in California, where 400 Mw of power were in production, sufficient to meet the electricity needs of a city of 400,000 persons. The second largest geothermal area under development in the U.S. was in the Imperial Valley of California, where several geothermal fields had been discovered and were undergoing test development by both the federal government and private industry. Additional fields were discovered in the Long Valley and Surprise Valley areas of California, and exploration was occurring throughout the states of California, Oregon, Washington, Idaho, Utah, Nevada, Arizona, New Mexico, Colorado, Montana, and Wyoming. A major geothermal field was discovered by the U.S. Geological Survey in Yellowstone National Park, where both dry steam and hot-water geothermal resources are abundant. However, the national park status of the Yellowstone area protected this field from commercial development.

The world pioneer in geothermal development was Italy. Substantial large-scale geothermal development began early in the 20th century in the area of Larderello; this was later expanded into nearby areas so that by the early 1970s geothermal capacity in Italy totaled 384 Mw. Italian production focused on the development of dry steam, and a major exploration program was under way throughout the country to increase the number of commercially developed geothermal producing areas.

For many years the second most active nation in the world in the development of geothermal energy, New Zealand recently slipped to third behind Italy and the U.S. The New Zealanders discovered at Wairakei that they had a major hot-water geothermal field for which no producing technology existed. They then proceeded in the 1960s to pioneer in the development of such technology at other fields in New Zealand, at Kawerau, Rotorua, and Broadlands. As of 1974 the Wairakei field was producing approximately 160 Mw.

The widespread use of hot springs for baths caused the Japanese to become among the world's most vigorous utilizers of geothermal energy, even if on a small scale. However, the presence of about 5,000 hot spring spas developed for recreation in Japan has actually served to significantly inhibit development of many potential hot spring areas for their power potential. Commercial production of dry steam at the Matsukawa field totaled 20 Mw in 1974, and another 30 Mw were developed in the hot-water geothermal area at Otake.

Iceland has used geothermal energy on a large scale, but as of 1974 only a small fraction was being utilized for electricity production. Instead, the geothermal hot water was used directly for heating. The capital city of Reykjavik solved its smoke and smog problem when it shifted from burning peat to using geothermal hot water for heating.

In addition, geothermal hot waters are widely used for greenhouse farming and for certain industrial processes.

In the U.S.S.R. there is an abundance of natural gas and other fossil fuels, and consequently geothermal energy development has been of relatively low priority. However, there is substantial use of hot water for agriculture in the Georgian Republic in the Caucasus. A moderate-sized geothermal power plant was operating in 1974 at Pauzhetka in Kamchatka, and the binary fluid freon-based pilot plant was being successfully run at the Paratunka area of Kamchatka.

A Mexican geothermal power plant at Cerro Prieto in Baja California was producing 75 Mw at the end of 1973. Commercial plants were planned for construction in the mid-1970s in El Salvador and possibly in Kenya, Turkey, Greece, the Philippines, Indonesia, and Taiwan.

Cost comparisons

Electricity produced from steam at the Geysers field proved to be the least expensive new source of energy within the Pacific Gas & Electric Co. utility service area (see graph). Geothermal energy at 90% load factor (ratio of average to maximum load) produced electricity at a cost of 6.43 mills per kw-hr. Oil and gas as an energy source yielded electricity at a price of 11.55 mills per kw-hr. The cost of electricity derived from Wyoming coal that is transmitted to California in 1974 was 10.52 mills per kw-hr, delivered within the service area. The projected costs for a nuclear power plant at Mendocino were 9.26 mills per kw-hr. After the presentation of these figures to the California Public Utilities Commission, Pacific Gas & Electric Co. withdrew its application to construct the nuclear reactor at Mendocino.

It is evident from these data that geothermal energy represents the least expensive source of new power available within areas where dry steam geothermal resources are known to be present. The problem as of 1974 was the need to expand the number of known dry steam geothermal fields by making exploration and production more attractive. The major stumbling block was that the majority of new dry steam prospects were on federal lands that were not yet available for leasing, although a federal geothermal act of 1970 was designed to eventually open some federal lands for geothermal development.

The relatively low cost of producing geothermal energy compared with the cost of producing energy from competitive sources, and also its remarkably small environmental impact, combine to make geothermal energy one of the most attractive new energy sources available. If the technology for exploiting the widespread fields of hot, dry rock can be developed, geothermal power can make a major contribution to the supply of power throughout the world.

Growth Regulators: A New Approach to Insect Control

by John Siddall

To eliminate harmful and annoying insect pests without endangering other animals and plants, scientists have developed a new class of chemical—insect growth regulator.

Almost since life began insects have gone about their business of eating and reproducing. They have outlived the dinosaurs, mammoths, and other extravaganzas of nature, and today, even under assault from man, far outnumber all other living things. A billion billion are alive at any one time.

Each year about 40% of man's food crops are harvested by insects despite the great array of pesticides used against them. But new lines of research may allow scientists to penetrate the mystery of the most spectacular change of form in the animal kingdom, the metamorphosis of insects from juveniles to adults. Researchers are convinced that knowledge of the unique way in which insects delay metamorphosis will provide a new approach to insect control.

In 1973 California, New York, and Florida made use of a completely new class of chemical, called insect growth regulator, designed to prevent the emergence of adult mosquitoes and flies from their pupal cases. Its chemical structure was modeled on a substance that insects themselves use to maintain their juvenile state until they are ready for their complete change of shape in metamorphosing to adults. The ideas and research that led to the discovery of these substances have involved many fields of the life sciences and can help provide a view of the future of insect control.

Need for selectivity

Man cannot expect to control the voracious insect pests of crops forever by the use of such general poisons as DDT, since these come dangerously close to destroying livestock, fish, and other wildlife, including the predators of insects. The need for more selective methods of control is apparent, but how selective these methods must be is not at all obvious. Of some 1,000,000 known species of insects, fewer than 1,000 are harmful pests carrying diseases or ravaging crops. Many more than 1,000 are predators of those pests, and a special problem of selectivity is to spare these.

A new chemical can form part of a pest-management scheme if it is selective enough to be safe in the environment, but there are two quite different components of the safety of a chemical. First and more easily measured is an inherent property of the chemical itself, something directly associated with its structure, which can be called the inherent level of toxicity. Second, and more important, is the way in which a substance is used. With extreme precautions even the most dangerous substances can be used in relatively safe ways, rather like dynamite, while, conversely, apparently safe substances can be used in very dangerous ways. In planning for selective insect control there should be an idea at the beginning of how the new substance will be used. But since the fine details of this use pattern often can only be worked out much later, a researcher can begin by looking at the first question, the inherent level of toxicity.

Since the idea of a completely nontoxic insecticide is a contradiction in terms, many research workers have tried to develop the idea of selective toxicity. To do this successfully, they must find a unique difference between the insects to be controlled and the many other living things with which they coexist. The concept of evolution can point out such a difference.

Evolution

When Charles Darwin expounded his theory of evolution, one of its most thought-provoking implications was the idea of a common ancestor of all living things. The historical fact of a common ancestry explains why the basic biochemical machinery in all living organisms is the same. The huge molecules that carry genetic information in the chromosomes for the construction of offspring are made up of the same building blocks in all living things. Even the exquisitely complex ways in which different living organisms use this genetic information demonstrate more similarities than differences. Clearly, any insecticide or other chemical that interferes with these basic processes of life will kill many things besides insects. To control organisms selectively scientists must determine the important differences between them. In order to do this, they must look closely at the physiology of different organisms and try to understand the inner workings of the cells and organs.

The mysterious common ancestor probably started life in the

JOHN SIDDALL is vice-president and director of research at Zoecon Corporation.

oceans about 2,000,000,000 years ago. Unfortunately, no fossil records can be found until much much later, because these tiny boneless organisms died and dissolved in the moving water. By the time favorable weather conditions and durable skeletons had evolved so that fossils could form, animal life had already begun to follow two quite different pathways.

From the fossil records that begin in the Cambrian Period about 600 million years ago, it seems fairly certain that a major parting of the ways in the animal kingdom had taken place even earlier. Both the forerunners of the backboned animals (vertebrates) and the earliest known invertebrates, called trilobites, are found together in Cambrian rocks. The significance of this separation of the tree of evolution of animals into two major branches is enormous. The invertebrate branch was destined to lead to the evolution of the insects, which belong to the huge division, or phylum, of Arthropoda (joint-legged, segmented animals). The other branch led to fish, amphibians, reptiles, birds, mammals, and man.

Insects probably evolved as a separate class of the arthropods through constant selection and survival of the more successful forms of their primitive, segmented, wormlike ancestors. There still exist today a number of very primitive wingless insects (Apterygota) such as firebrats, silverfish, and springtails, though even the ancient spring-

Insects can be classified into four major divisions according to the degree of their evolution from the earliest forms on earth. At the top right is the primitive, wingless firebrat, followed in order of development by the fixed-winged mayfly (top left), dragonfly (bottom left), and the flexible-winged green protea beetle (bottom right).

117

tails (Collembola) represent an advanced state of evolution compared with segmented worms.

Other classes besides the insects within the phylum of Arthropoda include crustaceans (crayfish, crabs, lobsters); arachnids (spiders, scorpions, mites); centipedes; millipedes; horseshoe crabs (Merostomata); and the extinct trilobites. Curiously, the horseshoe crab is the only surviving North American member of the ancient class Merostomata and exists today probably exactly as it did 400 million years ago. As a tool for probing the chemistry and physiology of ancient times, this "living fossil" is invaluable. By comparing its physiology and its responses to natural hormones with those of recently evolved insects, scientists may find some important differences between the major classes of arthropods. There appears to be a difference in the structures of the juvenile hormones that control and limit metamorphosis, although it may turn out that all classes of arthropods use similar molting hormones to start the metamorphosis process. Within the insect class the more recently evolved orders, such as moths and flies, but not the older insects, probably use more recently evolved juvenile hormones to hold back metamorphosis until the proper time in their life cycles.

Recalling the major parting of the ways that led to the separate evolution of man on the one hand and of insects on the other, scientists can conclude that the subtle differences between insects and other classes of arthropods also distinguish the insects from man. Because insect juvenile hormones control special events that occur in insects but do not in man and large animals, these hormones provide a sound basis for a chemical that will achieve selective insect control.

Complete metamorphosis

Living insects are classified into about 30 main branches or orders, 4 of which are the primitive wingless insects mentioned earlier. These

118

earliest orders were followed by fixed-wing mayflies, dragonflies, and the flexible-winged Neoptera. All of these undergo "incomplete" metamorphosis, the adults being very similar in shape to the nymphs. "Complete" metamorphosis, in which the juvenile insect is transformed into a different-looking adult after undergoing an immobile pupal stage, did not appear until the Permian Period of geological time, probably 250 million years ago, when lacewings and beetles first came into being.

Complete metamorphosis gives modern insects many advantages for survival. One of the major reasons for this is that it causes their lives to be divided into four separate parts—the egg, the larva or feeding stage, the pupa, and the adult. A safety factor is built into this process in that the four stages are often found in entirely different places: a mosquito larva lives in water where food is abundant, but adults cruise the air where the chances of mating and reproducing are best. Complete four-step metamorphosis occurs in those insect orders that have the largest number of species, such as beetles and flies, and has therefore helped insects to populate virtually every niche in the environment.

Metamorphosis involves great changes in the inner workings of an insect's body tissues and organs at exactly the right time in its life. In the pupal stage, tissues such as the gut and fat body that served special purposes in the fast-feeding larva are broken down and rebuilt to new adult designs. So great are the differences between the four stages that no simple steady growth process can account for such changes in form.

The idea that many small parts of the body are destined to become specialized parts at later times suggests that there must be a very delicate control system within the insect. Otherwise, the rebuilding process might start haphazardly, before the larva had grown to the right size.

Horseshoe crab (above), the only surviving North American member of an ancient class of arthropods, provided researchers with valuable data when they compared its physiology and responses to natural hormones with those of recently evolved insects. Fossil of a horseshoe crab dating from the Jurassic Period (190 million to 136 million years ago) is at left. Opposite page, right, is a fossil of a trilobite, one of the earliest arthropods and now extinct. A young horseshoe crab demonstrates its relationship with trilobites by undergoing a "trilobite stage" in its growth (opposite page, left).

119

(Top left, center left) Dr. Roman Vishniac; (bottom left, below) E. S. Ross

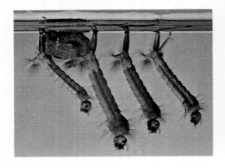

Four stages of an insect's life cycle consist of the egg (top), larva (center), pupa (shown with larvae, bottom), and adult (right), in this case a mosquito.

Hormone control of metamorphosis

Unlike man, insects carry their defensive skeletons on the outside. In order to allow more growth in their feeding stages they must shed them by molting at intervals. At each stage of the insect's progress from egg to larva to pupa to adult, its brain and nearby glands produce specific chemicals that interact in a delicately balanced control system. One such chemical, the juvenile hormone, must be present during the larval stage to prevent the onset of metamorphosis. Many insect cells seem to have an inborn urge to become specialized or differentiated in the absence of any juvenile hormone.

Hormones are substances secreted from one site (endocrine glands) and carried by body fluids to influence receptive tissues at other sites. It is not merely the presence or absence of a hormone in insect blood but the actual amount of it that has a great influence over the target tissues. In the mid-1930s a Cambridge University insect physiologist, Sir Vincent Wigglesworth, discovered the existence of a juvenile hormone in the insect *Rhodnius prolixus* by means of ingenious surgical experiments. He then traced its source to a tiny pair of glands, corpora

120

allata, behind the brain. Strange events followed the surgical placing of extra, active corpora allata glands into a larval insect. Molting continued, but the larva failed to become an adult. More startling were the results of removal of the glands from young larvae; these molted to become miniature nonviable adults. Because of this important property of acting as a brake on development, the juvenile hormone was envisioned in 1956 by a Harvard University professor, Carroll Williams, as a potentially powerful insecticide. However, its chemical nature remained a mystery for 11 more years, despite Williams' 1956 breakthrough in finding the first workable source of juvenile hormone in male giant silkworm moths.

Hormone isolation

The isolation of hormones represents one of the most difficult tasks in the life sciences. Two major problems face the scientist. First, the hormones are present in extremely small quantities. For example, 50,-000 giant silkworm moths yield less than one drop (about 20 mg) of pure liquid juvenile hormone. In the past, this problem has been overcome by brute force as scientists collected hundreds of thousands of insects for chemical extraction. As described below, new methods using test-tube cultures of insect organs are providing a more efficient solution to this problem.

The second major difficulty involves the researcher rather than the insect. Usually, a skillful chemist is not expert in insect physiology or in the breeding of thousands of insects. By the same token the expert physiologist, who can tell the chemist which time of the life cycle is best for the extraction of a hormone, is rarely the best person to operate the complex machinery of a chemical laboratory. Collaboration among chemists, physiologists, and biologists is, therefore, the only practical solution, and the sense of pride in solo achievement must give way to the greater effectiveness afforded by interdisciplinary teamwork.

The arduous task of isolating the insect molting hormone in pure form was begun in Germany in 1939 but was not completed until 1954 by Adolf Butenandt and Peter Karlson. From 500 kg of *Bombyx* silkworm pupae, they recovered less than 0.001 of an ounce of pure crystalline hormone, to which the name "ecdysone" was given in 1956. But this quantity was not sufficient to unravel the hormone's complex chemical structure, and a second attempt was begun using two tons of pupae. This effort yielded approximately 0.01 of an ounce of the hormone, enough to define its chemical structure in detail in 1965 by X-ray analysis of the crystals.

The attempts to isolate the insect juvenile hormone began in 1956. Researchers soon found that it was quite different in chemical properties from the molting hormone, especially that it was not soluble in water. At first, this seemed an advantage because a large part of the contents of silkworm bodies could be rinsed away from the fatty fractions that contained the juvenile hormone activity. Later, however, a

121

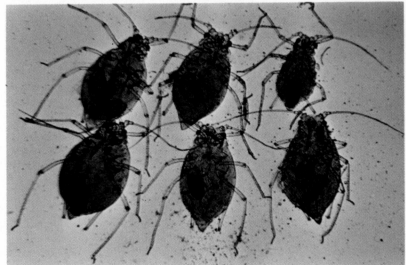

Cecropia moth (top) has a split body connected by a plastic tube, the result of an experiment in which the pupa was halved and joined by the tube. Tissue developed in the tube, but it was not strong enough to support the adult in flight and so the moth fell to its death. Six green peach aphids (center) have abnormally large numbers of embryos per adult; these embryos cannot be born, an effect achieved by treating the female with insect growth regulator. Bottom, the spotted cucumber beetle, an enemy of alfalfa, is treated with insect growth regulator. At the left is a pupa and on the far right a normal, untreated adult. The three adults in the middle show the effects of treatment with the growth regulator: they emerged from their pupae as crippled half-adult insects that will soon die.

major problem arose as to how to separate the relatively minute amount of the oily hormone from the great quantity of fats and waxes in the fatty fractions. After months of tedious experiment, the hormone activity was separated from most of the fats by allowing the latter to fall out as solids at subzero temperatures. But the worst was yet to come; no sooner had the nearly pure hormone been separated from the protective fats than it began to decompose. Understandably, many groups of researchers throughout the world then gave up. However, a combined team of chemists, physiologists, and instrument specialists at the University of Wisconsin persisted and was rewarded with success in late 1966.

The results of an analysis of the isolated juvenile hormone by Herbert Röller and his co-workers at Wisconsin caused a shock in the chemical world. The chemical "skeleton" of the hormone was formed of 12 carbon atoms with regularly spaced branches at 3 points, but the branches were simply too long to be plausible. Though many distinguished scholars voiced their disbelief in the findings no one was moved to publish any criticism, perhaps because of a remarkable feat of prediction that had occurred in 1965. A young insect physiologist, William Bowers, working for the U.S. Department of Agriculture in Maryland, had also started along the trail of juvenile hormone isolation. Despite his considerable chemical skills Bowers could not overcome the purification problems, but he did manage to glean much information about the possible chemical structure of the hormone. With great intuition he combined his ideas with information from Peter Schmialek, working in Berlin, who had found two branched-chain chemicals, farnesol and farnesal, in the feces of yellow mealworm beetles.

Bowers eventually synthesized a chemical compound that demonstrated all the biological properties of the juvenile hormone and declared, "It is believed that this compound will be found to be very similar chemically to the natural hormone when the latter is isolated and characterized." Not only was his prediction correct but his synthetic chemical was found seven years later by a group at Zoecon Corp. to be a natural hormone of the tobacco hornworm moth.

Soon after the first natural juvenile hormone had been found at Wisconsin, a group at Case Western Reserve University in Cleveland isolated a second juvenile hormone from the same oily extracts of cecropia silkworm moths. Again, the chemical structure showed a chain of 12 carbon atoms with 3 branches, one of which was of unusual length. For reasons that remain unknown both the silkworm and hornworm moths make two juvenile hormones, but only the structure determined by the researchers in Cleveland is found in both of these insects. From 1966 to 1974 considerable research activity by chemists and physiologists throughout the world erased any doubts about the correctness of the originally proposed chemical structures. The discovery of these structures has provided the most important information needed for development of a new approach to insect control.

natural hormones

hormone I, R$=CH_3$

hormone II, R$=C_2H_5$

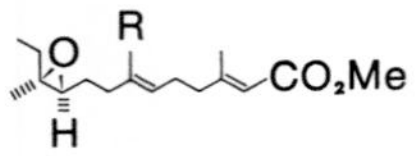

compound	structure	relative potency
A	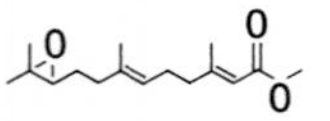	1.0
B	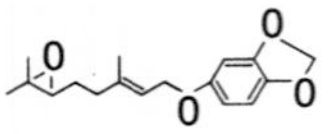	70
C	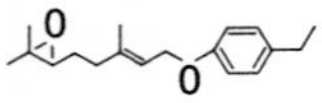	450
D	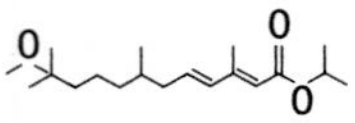	1,900

Chemical structures of insect growth regulator compounds are based on those of the natural juvenile hormones.

Improving on nature

Perhaps the excessive publicity that surrounded the successful isolation of the natural juvenile hormone misled many people into thinking that its commercial use for insect control would arrive before the end of the 1960s. In believing this they were disregarding its fragile nature. Despite its impressive biological effects on insects, the hormone was far too quickly destroyed when sprayed thinly on plants or in swamps for insect control. Considerable improvements on the natural substance were, therefore, called for.

Researchers began their effort to develop a more potent substance by deciding that it must have the following characteristics: (1) high activity against pest insects; (2) low activity against beneficial insects; (3) moderate stability in the environment without being persistent; (4) adequate safety for wildlife, animals, and man; and (5) chemical simplicity for economical manufacture. Bearing in mind that there are about 1,000,000 different species of insects but only about 1,000 pests, scientists had to take care to tailor the hormone structure to ensure its selective action on the pests. For the juvenile hormone this proved to be difficult but not impossible because nature provided much help by timing differently the life cycles of various insects. Many pest insects are preyed upon by other insects that develop out of phase with the pests and are, therefore, not present in the susceptible form when the hormone is applied.

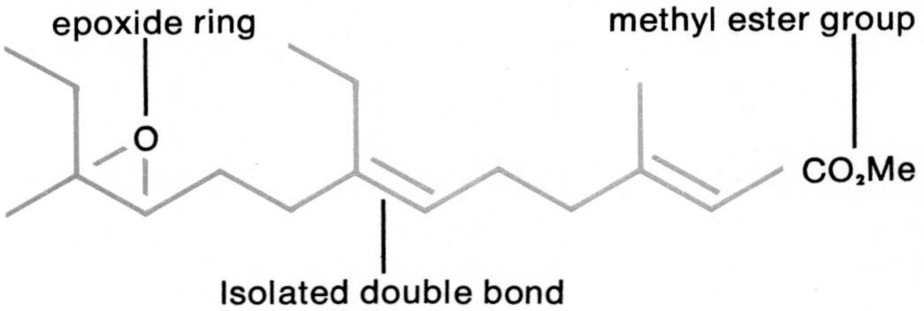

The work to modify the natural juvenile hormone chemically and make it more potent and stable, which began in the mid-1960s, led to the development of insect growth regulators, which were remarkable improvements over the natural hormones. Early chemical changes focused on replacement of the unstable parts of the natural juvenile hormone molecule, particularly the epoxide ring and the isolated double bond. In the case of a chemical called methoprene, the epoxide was replaced by a methoxyl group and the central double bond was "moved" two positions closer to the ester group to form a much more stable system. An important change in the ester group improved both stability and potency simultaneously, when methyl, CH_3, was replaced by isopropyl, $CH(CH_3)_2$.

In the late 1960s insect specialists decided to test the effectiveness of the new growth regulators. Mosquito larvae presented an ideal target, particularly because their rapid larval growth is completed well

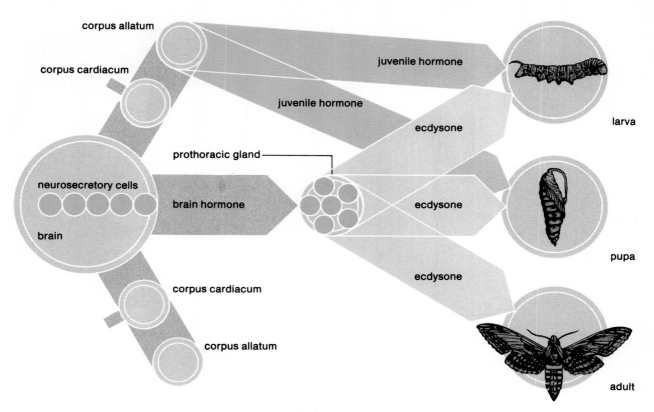

corpus allatum

corpus cardiacum

juvenile hormone

juvenile hormone

ecdysone

larva

prothoracic gland

neurosecretory cells

brain hormone

ecdysone

brain

ecdysone

pupa

corpus cardiacum

corpus allatum

ecdysone

adult

within the few weeks of stability of the chemical. Mosquitoes that transmit yellow fever, and their close relatives, have become adapted so that their eggs hatch when water becomes available as a larval food source. When rice paddies and pasture lands are deliberately flooded, swarms of mosquitoes can be expected about one week later unless controls are applied. Practical demonstration of the control of flood-water mosquitoes became an early goal of the researchers and developers of the new chemicals.

Early field experiments in Panama showed that mosquito control could be achieved for about one day after a spray application but that the unprotected chemical was rapidly destroyed by sunlight and microorganisms in the swamps. To return to the laboratory and redesign the chemical was one solution, but a more attractive alternative was to encase the material within protective capsules that would slowly release their contents. This method allowed great strides in the use of methoprene, which was developed by Zoecon Corp. from modifications of the natural juvenile hormone. Early tests in Panama using the unprotected chemical showed that two ounces of material were needed to control each acre of mosquito breeding waters and that the effects disappeared after one day. By 1974 the use of the encapsulation method allowed five times less material to be required per acre, and the effects remained present for about ten days. The economic advantages of this improvement are obvious, but the increased safety to humans and wildlife because of the introduction of less chemical spray into the environment is equally important.

Pathways by which hormones are secreted to influence insect growth and development are shown schematically. Researchers isolated ecdysone (the insect molting hormone) and the natural juvenile hormone after painstaking and difficult experiments on tons of silkworm pupae.

125

Harold R. Hungerford

Natural predators of insect pests include the praying mantis, attacking a grasshopper (top), and the ladybug, eating aphids (bottom). Protection of these natural predators is a major objective of scientists working on insect control.

Alexander B. Klots

Toxicity

Since the mid-1950s when Williams' idea of using special chemicals made only by insects for selective control of insects was first put forward, considerable research has been devoted to the toxicology of insect hormones. Although the basic premise was that these chemicals would have few if any effects on higher animals, it has only recently been possible to test these predictions.

Most of the tests, which are expensive and time-consuming, have been carried out in private industry, using many different animals. Simple feeding tests of the most promising insect growth regulators showed no signs of acute toxicity, even when abnormally large quantities were fed to laboratory animals. Longer-term studies covering three-month periods have been undertaken on dogs and rodents with favorable results. In the case of methoprene, extensive tests involving many species of wildlife have been set in motion and in most cases completed. As of 1974, with the testing still at a comparatively early stage, the insect growth regulators appeared to be safe chemicals.

126

Helicopter spreads insecticide on the fruit-growing Napa Valley in California. Insect growth regulator is sprayed in such a manner.

The future of insect research

Two fundamental problems remain to be solved, and their secrets are not likely to be yielded easily. One is to understand the process of cellular differentiation, and the other is to determine the mechanism by which apparently simple chemical hormones can trigger, accelerate, or even block this process. Differentiation occurs when cells having no particularly special function divide into groups of cells that are highly specialized for one particular kind of function. Examples include the development of an egg into a many-celled organism and the transformation of the ameboid cells of the slime mold (*Dictyostelium discoideum*) into aggregates that form fruiting bodies. Something remarkable happens when the huge molecules that carry genetic information replicate themselves but in so doing become devoted to using or expressing one small piece of information above all others.

These problems are central to the developmental biology of all living animals and plants, but the insects are an especially attractive vehicle for such studies. Insect juvenile hormones and their chemical variants thus provide a powerful tool for the study of development as well as for pest control.

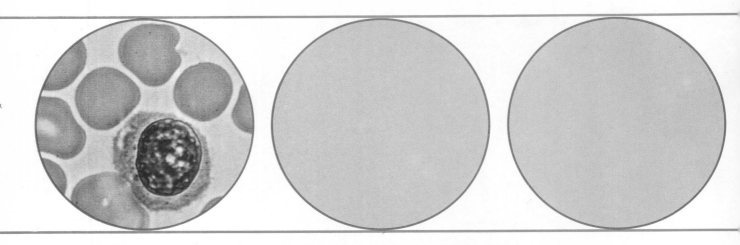

The Lymphatic System: Man's Biological Defense Network

by Clyde F. Barker

By increasing their knowledge and understanding of the tubes, fluid, and cell masses that constitute the human lymphatic system, physicians are becoming able to combat many diseases more effectively.

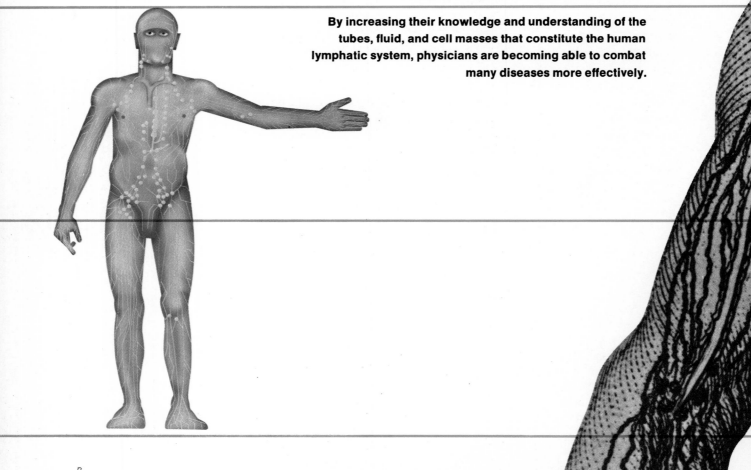

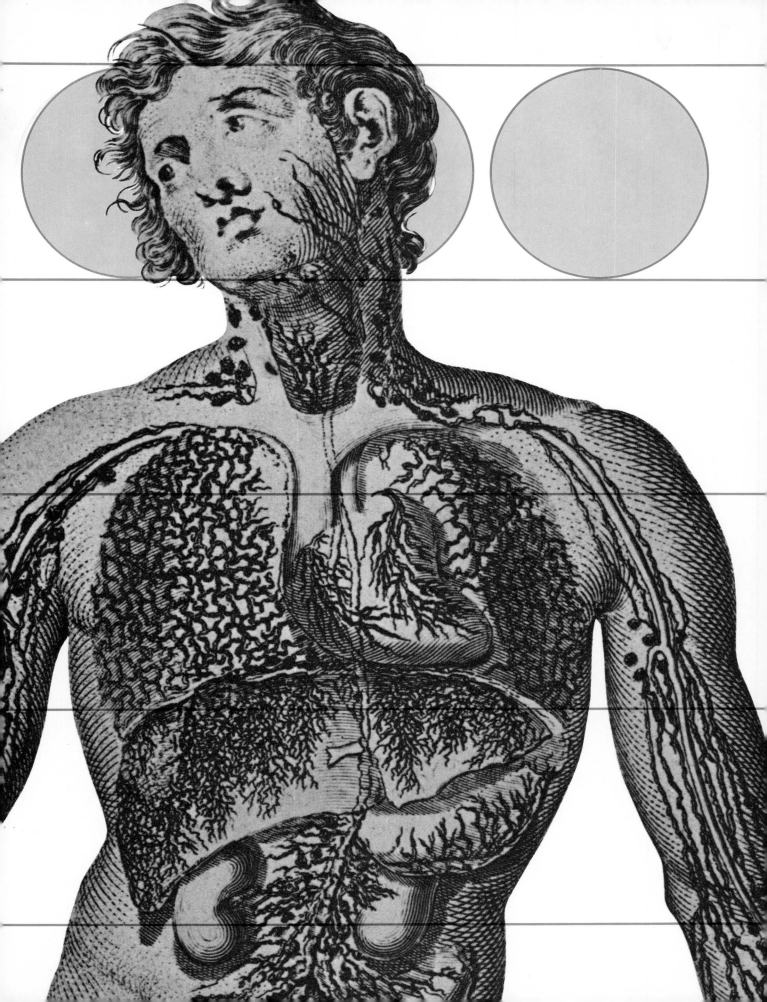

After centuries of virtual neglect the human lymphatic system has received considerable attention in recent years. Scientists have found that this complex network of tubes, fluid, and cellular masses—man's "second" circulatory system—plays a vital role in the maintenance of human health. Research into its constituents and the way in which it functions has provided physicians with information that has helped them combat many of man's most severe diseases and disorders, including those of the lymphatic system itself.

The lymphatics are distensible thin-walled tubes that serve as conduits for clear fluid called lymph. These vessels form an extensive network throughout the body and, together with scattered masses of tissue along their course (primarily the spleen, thymus, and lymph nodes), constitute the lymphatic or lymphoid system. Although these structures constitute 1–1½% of the body's weight and are essential for life, they received little attention from early physicians, probably because lymph is essentially colorless, being made up of water, plasma proteins, electrolytes, and white blood cells (lymphocytes). The red color of blood drew attention to the arterial and venous vessels rather than to the lymphatics. Not until the 18th century did the brothers John and William Hunter, founders of rival schools of anatomy in London, suggest the role of lymphatics in absorption of fluid from the tissues in all parts of the body.

Anatomy and physiology

Living tissue is essentially a collection of cells continually bathed in a fluid. This interstitial fluid, which constitutes the internal environment of the body, brings nutrients to the cells and carries away their waste products. From 4,000 to 5,000 liters of interstitial fluid are formed daily in the human body, by transudation (passage through membranes) or leakage from the arterial ends of the capillaries. Most of it is absorbed into the venous side of the capillaries and thus returned to the body's circulation.

Along with water and salts about 50% of the total circulating plasma protein leaks daily out of the arterial capillaries into the interstitial fluid. It is in the form of molecules that are too large for reabsorption into the venous capillaries and can only be retrieved from the tissue spaces by the lymphatics. This second component of the circulation is therefore as essential to the total metabolism and homeostasis of the body as the transport of the oxygen-carrying red blood cells by the arterial and venous components.

The lymphatics originate as a network of fine capillaries in all tissues except bone, cartilage, the splenic pulp, and certain parts of the eye and brain. It was once thought that this network began as open-ended extensions from the tissue spaces, but it now seems likely that the lymphatic capillaries are closed structures that are permeable to proteins and other macromolecules, perhaps because their walls are only one cell layer thick. Beginning with the capillaries, progressively larger lymph channels are formed as smaller vessels converge. The

CLYDE F. BARKER is professor of surgery and chief of the transplantation section at the University of Pennsylvania.

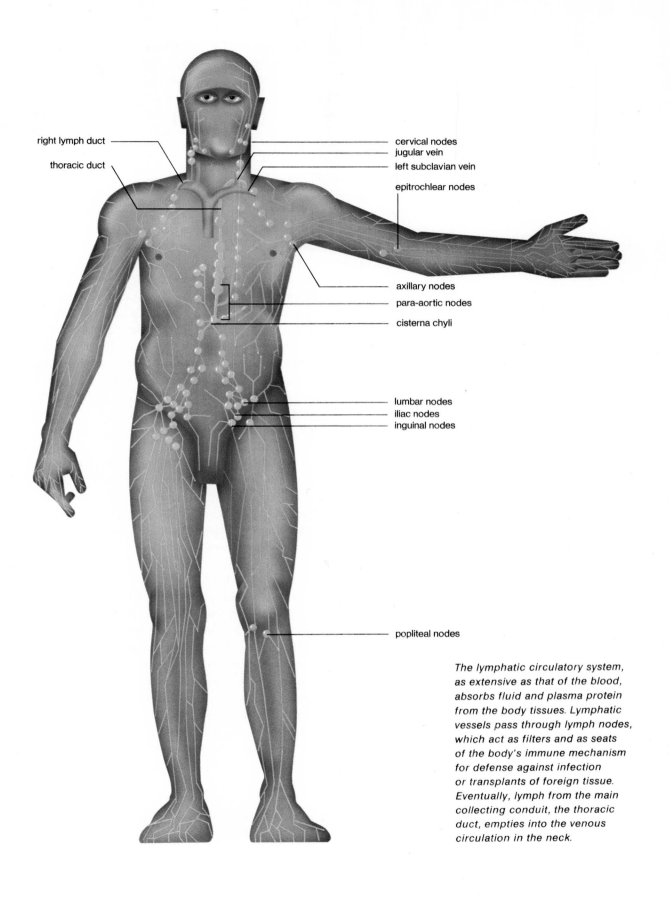

right lymph duct

thoracic duct

cervical nodes
jugular vein
left subclavian vein

epitrochlear nodes

axillary nodes

para-aortic nodes

cisterna chyli

lumbar nodes
iliac nodes
inguinal nodes

popliteal nodes

The lymphatic circulatory system, as extensive as that of the blood, absorbs fluid and plasma protein from the body tissues. Lymphatic vessels pass through lymph nodes, which act as filters and as seats of the body's immune mechanism for defense against infection or transplants of foreign tissue. Eventually, lymph from the main collecting conduit, the thoracic duct, empties into the venous circulation in the neck.

larger lymphatics are joined by smaller tributaries and pass through or around filtration units, called lymph nodes. Those from the legs, pelvis, and abdomen eventually flow into one large lymphatic in the abdomen, the cisterna chyli. This large channel becomes known in the chest as the thoracic duct, and it empties its lymph directly into the venous system in the neck near the junction of the subclavian and jugular veins.

Malfunction of the lymphatic circulation

The importance of the lymphatic system is convincingly demonstrated by individuals in whom it fails to develop normally or becomes obstructed later in life. In 1928 W. F. Milroy described a hereditary, congenital disease that becomes apparent soon after birth, causing striking swelling of the extremities, usually the legs. Radiopaque dye injected into lymphatics demonstrates that the major lymph channels that should drain these legs are malformed or absent. The subcutaneous tissues become spongy and thickened, containing dilated lymph spaces and fibrous tissue. A similar condition (lymphedema praecox) may appear at adolescence in females who have previously seemed normal. Most commonly only one leg is involved, and the microscopic findings are similar to those of Milroy's disease. The cause is unknown but perhaps is related to an increased load thrown on the lymphatics at the time of development of the reproductive organs, with resultant dilatation of lymphatic vessels, incompetence of their valves, and increased intralymphatic pressure.

More common and more clearly understood causes of lymphedema are obstruction of lymphatic channels by tumors, such as cancer of the prostate or malignancies of the lymph nodes (lymphomas). Swelling of the arm can follow surgical removal of the lymphatics and nodes in the armpit and removal of the breast for cancer. This is particularly likely to occur if radiation is used, resulting in further injury to lymphatic channels.

Swelling of extremities can result from repeated bouts of bacterial infection in the skin with extension into the lymphatics. This type of lymphedema is usually less severe than those described above. One type of infection that may lead to striking involvement is caused by the parasitic filarial worm *Wuchereria bancrofti,* an organism transmitted by some species of tropical mosquitoes. This parasite prefers to localize in the lymphatics, and repeated infection with it causes lymphatic destruction and occlusion that leads to such extensive swelling of the lower extremity as to suggest the appearance of an elephant's foot, giving rise to the descriptive name elephantiasis.

In any of these different forms of lymphedema, elevation of the affected area and the use of pressure stockings or bandages on it may be enough to minimize the swelling. In severely involved limbs, operative treatment may be necessary. Procedures have been described for decreasing the edema of lymphatic obstruction by surgically reestablishing lymphatic drainage. These have usually failed, perhaps because of the fragile nature of the fine lymphatic structures or the sluggish rate of lymphatic flow. A more often successful, though cosmetically imperfect, operation involves surgical removal of the boggy subcutaneous tissue through which most of the diseased, distended lymph channels pass. This procedure requires skinning the leg prior to the removal of the subcutaneous tissue and grafting the skin directly onto the normal underlying muscle.

Cellular elements

Thus far the structure and function of the lymphatic circulation have been considered with regard to the fluid and noncellular elements in the lymph. That the lymph also contains "globules" (lymphocytes) was recognized by William Hewson in 1774. During the next 175 years little progress in understanding lymphocytes was made except for better microscopic definition of their structure and the recognition that they are cells with single nuclei, are found in the peripheral blood and bone marrow, and constitute the major cellular elements of the lymph nodes and spleen. Although attention had been called to the presence of lymphocytes in chronic inflammatory lesions such as tuberculosis, essentially nothing was known of their function until the mid-1950s.

The life history of the lymphocyte was determined mostly by J. L. Gowans and his associates, whose experiments in the early 1960s were done by radioactive labeling of lymphoctyes obtained from the thoracic duct of rats. These cells were injected intravenously back into the animals, whose tissues could then be examined to determine the fate of the cells. It was found that within two hours the labeled cells could again be detected in thoracic-duct lymph. The lymphocytes seemed to be recirculated continuously from blood to lymph and back to blood. The entry from blood to lymph was found to be located in the lymph nodes, to which the intravenously infused cells "homed" within 15 minutes.

Rather than the aimless wandering cells they first appeared to be, lymphocytes are now recognized to play a central and complex role in the body's immune mechanism for defense against invasion of bacteria and other foreign agents. Lymphocytes are able to recognize cells and other substances foreign to the individual. This property, together with their continuous recirculation throughout the tissues of the body, qualifies them as ideal agents for a role in surveillance. Furthermore, since some of these cells may live for years, rather than the days to weeks most other circulating cells survive, their immunological "memories" are long.

The work of a number of scientists, notably J. F. A. P. Miller and his associates in Great Britain and Australia, and Robert Good and his associates in Minnesota, has led to the realization that there are two populations of lymphocytes with quite distinct functions: the T lymphocytes (thymus-derived) and the B lymphocytes (bone marrow-derived). The T cells play their primary role in delayed hypersensitivity (such as the allergic response to poison ivy), and in rejecting grafts of

Infections and congenital defects of the lymphatic circulation can lead to massive swelling of the leg, as plasma proteins and fluid accumulate in the tissues.

133

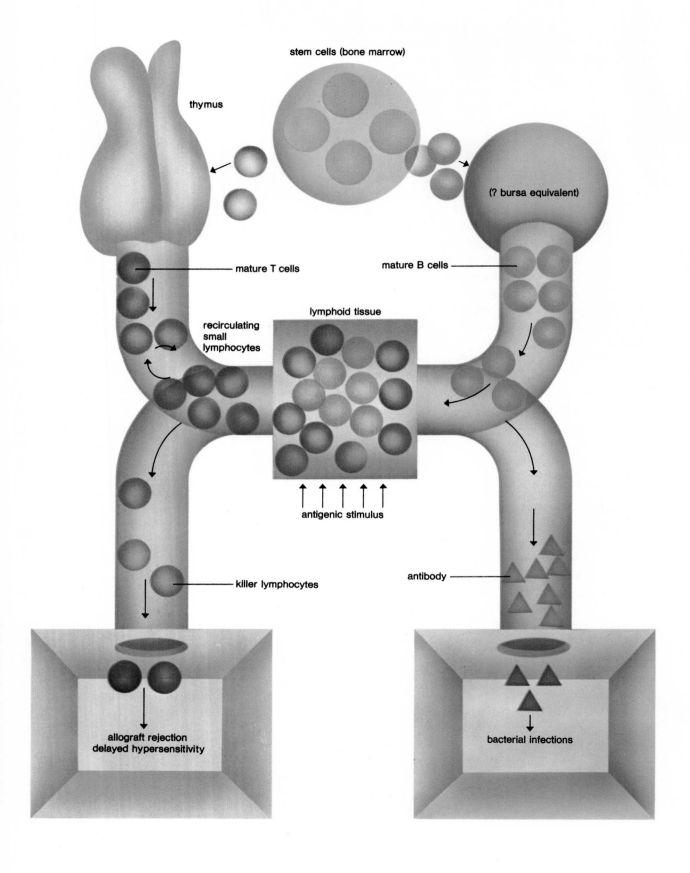

foreign tissue such as kidney transplants. The B cells act predominantly in ridding the body of most infectious bacteria, which they do by secreting immunoglobulins (antibodies).

B and T cells look alike under the light microscope, and both are descendants of stem cells located in the bone marrow. During the embryo stage of a human being some of these cells migrate from the bone marrow to the thymus (a lymphoid organ in the chest just behind the sternum), where they are altered or "educated" in some way to carry out the functions of T cells. When these mature cells leave the thymus, they home into specific anatomical locations in the lymph nodes and spleen. If the thymus is removed from newborn mice before development and migration of the T cells has been completed, these animals will develop normally for a few weeks but then many will die from a wasting disease characterized by weight loss, diarrhea, and atrophy of lymphoid tissues. The mice that survive have only 1–3% of the usual number of recirculating small lymphocytes. Animals that have failed to develop T cells or that have been depleted of them have difficulty in rejecting transplanted tissue from other individuals of their species (allografts) and in expressing delayed-hypersensitivity (allergic) skin reactions to materials such as tuberculin and poison ivy. If they are reequipped with T cells taken from thymuses of normal animals of the same inbred strain, the ability to reject grafts and mount delayed-hypersensitivity reactions is restored.

In birds the bursa of Fabricius, an outpocketing from the gastrointestinal tract, is a lymphoid organ that resembles the thymus. This organ is necessary in chickens for the stem cells in the bone marrow to develop into mature B cells, and if it is removed from unhatched chicks they become immunological cripples of a different sort than if they had lost the thymus. They are not depleted of circulating small lymphocytes, which are mostly T cells, but they lack cells in their germinal centers and medullary areas (areas of the lymphoid tissues associated with the antibody-producing lymphocytes). Such birds are thus abnormally susceptible to infections because of decreased capability to produce antibodies in response to bacteria, but their capacity for graft rejection and delayed hypersensitivity is intact. Some researchers believe that there is a bursa equivalent in mammals, perhaps associated with the Peyer's patches, tonsils, appendix, or other collections of lymphoid tissue in the gastrointestinal tract. In any case, mature B cells have been found in mammalian bone marrow.

The dichotomy found between cellular (T cell) and humoral (B cell) immunity in laboratory animals also exists in man. In 1965 Angelo DiGeorge reported a condition occurring in infants very similar to one found in mice that had their thymuses removed just after birth. X rays of these children do not show a thymus; their lymph nodes have normal germinal centers (B areas) but sparsely populated paracortical (T) areas. They are able to synthesize immunoglobulins (though somewhat less efficiently, because T cells seem to play a role in assisting B

continued on page 138

Derived from a common stem cell in bone marrow, T and B cells then follow different paths and perform different functions. T (thymus-derived) cells migrate during the embryo stage of a human being to the thymus, where they are altered in some way so as to perform their primary role of rejecting transplants of foreign tissue. B (bone marrow-derived) cells develop in birds in the bursa of Fabricius, but no equivalent of this organ has yet been found in mammals. These cells reside in bone marrow and perform their main task of fighting bacterial infections by secreting antibodies.

Photos, courtesy, Clyde F. Barker

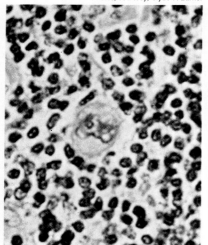

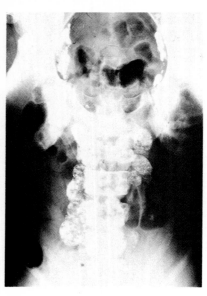

Lymph node of a patient with Hodgkin's disease (top) has a large, malignant Sternberg-Reed cell in the middle of the field. However, most of the cells are benign lymphocytes, probably indicating a potent immune response to the tumor. Lymphangiogram (bottom) of a Hodgkin's patient shows large, diseased lymph nodes alongside the aorta.

Malignancies of the Lymphatic System

The lymphatic system, so intimately tied to the body's anticancer defense, is ironically sometimes the victim of malignant diseases. This group of diseases, the lymphomas, is of great interest to cancer researchers for two reasons: they are among the malignancies for which scientists seem closest to discovering the cause and origin; and they are among the few types of cancer in which complete cures can often be expected.

In 1832 Thomas Hodgkin of London described seven fatal cases of malignant tumors of the lymph nodes. This condition, known as Hodgkin's disease, occurs chiefly in young individuals (20–39 years of age), in whom the risk of developing the disease over a ten-year period is about 0.4 per thousand. Until the early 1960s this disease was assumed always to be fatal. The cause of Hodgkin's disease and other human lymphomas is unknown, but a viral origin seems likely and this has been established in the case of some animal lymphomas.

It has long been recognized that patients with Hodgkin's disease characteristically exhibit the loss of T cell functions while retaining those of B cells. In 1973 Robert Longmire and associates at the Scripps Clinic in La Jolla, Calif., reported that spleen cells (mostly B cells) in patients with Hodgkin's disease produce an antibody that reacts with peripheral blood lymphocytes (mainly T cells). This led to the speculation by Vincent DeVita of the National Cancer Institute that in Hodgkin's disease a virus may cause the patient's B cells to produce antibodies against his own T cells, thus accounting for the disorders of cellular immunity characterizing the disease. Other factors predisposing certain individuals to the development of lymphomas are immunosuppression or immunodeficiency.

Painless enlargement of lymph nodes is usually the initial finding in Hodgkin's disease, although some patients also have fever, night sweats, anemia, and weakness. Lymph nodes throughout the body and in some cases other tissues as well eventually become involved, leading to death in about two years if no treatment is given. Biopsy of the lymph nodes reveals destruction of normal architecture. Multilobed giant cells (Sternberg-Reed cells), presumed to be malignant, are interspersed with normal lymphocytes and plasma cells.

In 1965 R. J. Lukes and associates found a striking correlation between the microscopic appearance of lymph node biopsies and the subsequent course of 377 Hodgkin's disease patients. Their classification was based on the appearance of the nonmalignant cells (lymphocytes) in the nodes, and they noted that when these cells predominated long-term survival of the patient was likely (median survival 9–15 years). If, however, there was a higher percentage of malignant cells, the median survival period was only 6–18 months. Lukes inferred that the nonmalignant lymphocyte population reflected the response of the immune system of the host to the malignancy.

Hodgkin's disease is thought to begin in one group of lymph nodes (frequently in the neck) and then to spread along normal lymphatic pathways to involve nodes in adjacent areas. Less commonly the spread is via the bloodstream, presumably as the malignant cells follow the usual stream of recirculating small lymphocytes. Clinicians have now agreed that in designating the extent of the disease, Stage I indicates involvement of a single node region; Stage II, two or more node groups either above or below the diaphragm; Stage III, nodes on both sides of the diaphragm; and Stage IV, dissemination to tissues other than the nodes or spleen. Because accurate staging can be tricky and determines not only the prognosis but the type and extent of treatment, it has become the function of an interdisciplinary team of specialists: an internist, a radiologist, a pathologist, and a surgeon.

Surgical removal of malignant tissues has been abandoned as a treatment for Hodgkin's disease, and X rays have become the mainstay of therapy. The capacity of this type of radiation to cure many patients with localized Hodgkin's disease has been firmly established during the last decade. As recently as 1960 a 36% five-year survival using conventional 200,000-volt X-ray equipment was highly acceptable and, compared with the 5.8% five-year cure without any therapy, was well worthwhile. By 1970, however, Henry Kaplan at Stanford University reported a 73.3% five-year survival of Hodgkin's disease patients of all stages using a six million-electron-volt linear accelerator for X-ray therapy. With this type of equipment a dose of 4,400 rads can be delivered in four weeks. (One rad is equal to an absorbed energy of 100 ergs per gram of irradiated material.) This dose kills tumors but is safe for the more radioresistant normal tissues, especially if the lungs, heart, kidneys, and ovaries are shielded. Relapse-free intervals of five years can be achieved by X-ray therapy in 80–90% of Stage I and II patients, and 95% of those can be considered permanent cures.

Anticancer drugs, like immunosuppressive agents, work by inhibiting cellular proliferation. Lymphoid cells seem particularly sensitive to this action, perhaps accounting for the fact that drugs are more effective in treating lymphomas than other cancers. The response of Hodgkin's disease patients to MOPP, a combination of nitrogen mustard, vincristine sulfate (Oncovin), prednisone, and procarbazine hydrochloride, has been much better than that to any single drug, and DeVita reported 81% disappearance of tumors at the National Cancer Institute. MOPP as of 1974 was used primarily for Stage III or IV patients or in those who had relapses after X-ray treatment.

The prognosis of non-Hodgkin's lymphomas with treatment similar to that outlined for Hodgkin's disease is not as favorable, with five-year survival ranging from 53% for nodular lymphoma to only 3% for lymphoblastic lymphoma. Origin, classification, staging, and therapy of the non-Hodgkin's lymphomas are less well established than in Hodgkin's disease, but it seems likely that some unifying concept will eventually emerge, quite possibly as a result of continuing advances in understanding the human lymphatic system.

continued from page 135

cell antibody production), but they are unable to express delayed hypersensitivity or reject skin allografts. Most of these patients eventually die of infections. Recently, however, several have been helped by transplants of thymus.

Another immunodeficiency disease was described in 1952 by Ogden Bruton. Bruton's syndrome is a disease of the B cell system. Children suffering from this syndrome have a normal thymus, normal numbers of circulating small lymphocytes, and unimpaired delayed-hypersensitivity and allograft responses. However, the germinal centers of their lymph nodes lack cells, and so these children produce neither B cells nor antibodies. By receiving regular injections of gamma globulin (the fraction of serum that contains antibodies) from normal humans, these children can be helped. Attempts at bone marrow (B cell) transplants have had little success, since these patients have the T cell capacity to recognize the transplanted cells as foreign and to reject them.

Transplant rejection

The above discussion describes how studies of the anatomical changes associated with certain immune deficiencies have led to a better understanding of B and T cell functions, to the direct benefit of those patients with rare immunodeficiencies. It should be pointed out, however, that much of our understanding of lymphocytes and the lymphatic system is the result of work done by those who were interested not in improving the immune response but in avoiding or overcoming it, especially with regard to the rejection of transplants. As early as 1926 J. B. Murphy of the Rockefeller Institute (now Rockefeller University) established that an accumulation of host leukocytes (white or colorless nucleated cells, including lymphocytes, which occur in blood) in the vicinity of a graft was a characteristic feature of rejection. The first solid evidence that host resistance to allografts is mediated by cells was provided in 1954 by N. A. Mitchison with regard to tumor grafts and by R. E. Billingham, R. L. Brent, and P. B. Medawar with regard to grafts of normal tissues. These investigators demonstrated that allograft immunity could not be transferred from one animal to another by means of immune serum (containing antibodies) but was transferable by living lymph node cells obtained from mice that had rejected allografts. These experiments also established that the lymph node was the primary site of transplant rejection.

Lymph nodes studied about six days after transplantation of a skin graft from one individual to another are sites of active cell division. Formed at these sites are large cells called "immunoblasts"; they soon divide, giving rise to a population of cells indistinguishable from small lymphocytes. These activated cells then return to blood circulation, by which route they make their way back to the transplant. There, these cells participate in the destruction of the skin graft probably in two ways: a direct cell-to-cell contact-dependent process, and a release of chemical agents that may recruit uncommitted mononuclear

Through the blood vessels of kidney and skin grafts and then through afferent lymphatics, antigenic material in the form of cellular debris and lymphocytes reaches the lymph nodes. There, it stimulates the production of large cells called immunoblasts. These divide to form killer lymphocytes, which then make their way into the bloodstream through efferent lymphatic vessels and veins in the neck. These killer T cells travel through the blood circulation to the transplants, destroy them, and afterward return through the arteries to the lymph node.

138

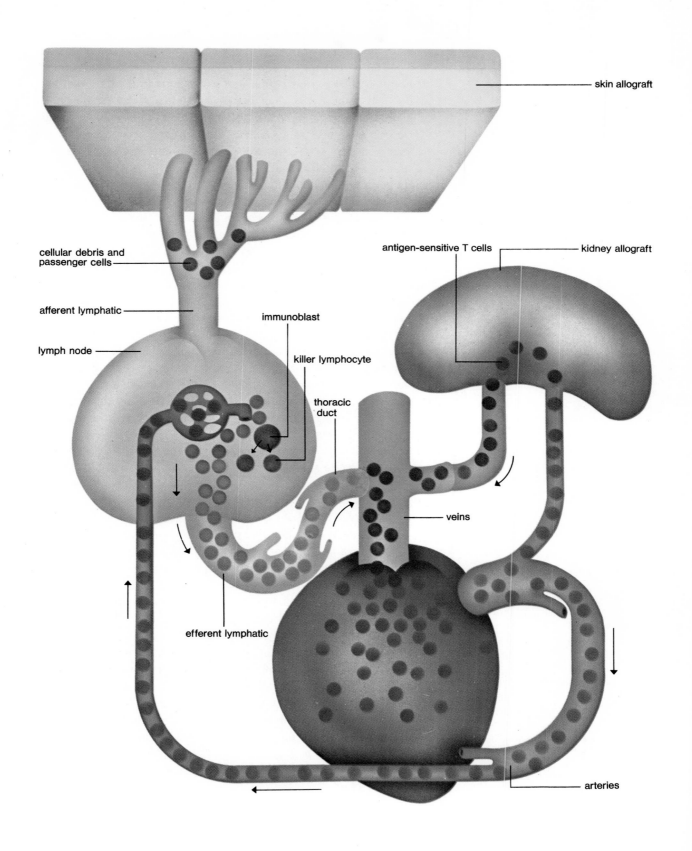

skin allograft

cellular debris and passenger cells

afferent lymphatic

lymph node

immunoblast

killer lymphocyte

thoracic duct

antigen-sensitive T cells

kidney allograft

veins

efferent lymphatic

arteries

cells from the host circulation, cells that are then able to participate in the destruction of the graft.

For the most part the foregoing discussion of the rejection phenomenon applies to kidney and heart transplants in man, as well as the more convenient experimental skin allografts. All of these transplants are destroyed by rejection (usually within 7–12 days) unless the donor and the recipient are identical twins in man or are genetically identical members of the same inbred strain in the case of laboratory rodents.

There are a few exceptions to the inevitability of allograft rejection. Several sites within the human body exist in which grafts of foreign tissue can be placed without provoking rejection. Some of these depend on a lack of a normal blood circulation so that effector (killer) cells cannot reach the implanted foreign tissue to identify or destroy it. Examples of such sites are the anterior chamber of the eye and the cornea, thereby accounting for the success of corneal transplants. Other privileged sites, including the brain, depend on the lack of a lymphatic pathway leading to the regional lymph nodes. Clyde Barker and Billingham created an artificial privileged site by dissecting a skin flap from a guinea pig's flank, thereby maintaining an "umbilical cord" to the flank that ensured arterial and venous blood supply to it but deprived it of lymphatic drainage. Since this prevented the antigens from leaving the isolated flap, skin grafts on the flap were able to survive in some cases for more than 100 days instead of being rejected in the usual 7–12 days.

The privileged-site approach might prove useful clinically in the grafting of hormone-secreting endocrine tissue such as parathyroid glands or isolated pancreatic islets of Langerhans, the insulin-producing part of the pancreas. Only a tiny bit of tissue, which would fit easily into a privileged site, would be needed. An attempt to use this approach was made by isolating transplanted kidneys in plastic capsules so that they would not regenerate lymphatic continuity with their hosts (which normally occurs within a few days). Researchers, however, found that this method did not prevent rejection.

Other methods of manipulating the lymphatic system have been attempted to avoid rejection of organ transplants in man. These include removal of the spleen (which proved ineffective because it leaves the T cells intact) and removal of the thymus (which is not effective because even in young children T cells have already matured and left the thymus before it can be excised). Drainage of the thoracic duct depletes the patient of T cells and will prevent rejection, but it can be used only for days or weeks before the tubing used to drain the duct becomes infected or occluded. Whole-body radiation was used early in the history of kidney transplantation and was effective in preventing rejection, but was also very dangerous because it destroyed the B cells as well as the T cells, thus leaving the patient susceptible to infections that were almost always fatal.

In 1974 the standard antirejection therapy for patients after trans-

140

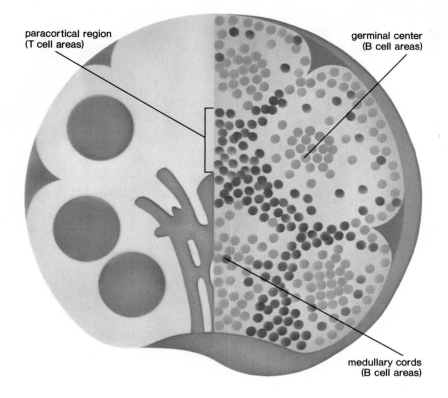

paracortical region
(T cell areas)

germinal center
(B cell areas)

medullary cords
(B cell areas)

plantation was a combination of the following drugs: azathioprine, which retards cell division and thus partially prevents the formation of killer T cells but also somewhat cripples the B cell system; predni-sone, an adrenocortical steroid that can kill lymphocytes but probably acts primarily by decreasing inflammation (which is part of the rejec-tion process but also an important defense mechanism against infec-tion); and antilymphocyte serum, which is an agent directed specifi-cally against T cells but which has been only partially perfected for use in humans. This immunosuppressive regime is far less than ideal since it restrains the functioning of B cells as well as T cells. It is effective enough, however, that its use has allowed 50% of the kidneys trans-planted to patients from related donors to function over a five-year period.

Lymphatic system and cancer

In one sense the immunosuppression techniques in use in the 1970s may be too effective in depressing T cell function. Previous mention has been made that cellular immunity is important in preventing for-mation of tumors. Important evidence for this belief is that immuno-suppressed patients surviving kidney transplantation have about a 200-times-greater risk of developing a malignant tumor than do others of the same age group. Patients with immunodeficiency disease also have very high incidence of malignancies. Many of the tumors in trans-

Diagram of a lymph node shows the locations of the paracortical areas, germinal centers, and medullary cords. T cells are found in the first region, and B cells inhabit the second and third.

141

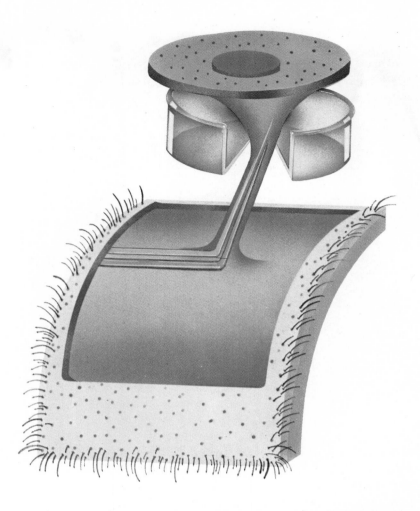

A skin allograft (dark circle) is transplanted to a surgically created skin flap that has no drainage to the lymphatic system. This prevents antigen from the graft from being transmitted directly to a lymph node and, consequently, greatly delays rejection of the graft.

plant recipients are lymphomas of the brain, a rare location for this type of tumor in nonimmunosuppressed patients.

One interpretation of these findings is that malignant cells, continually arising through mutation and differing in their antigenic (antibody-stimulating) characteristics from normal cells, are recognized by the body's surveillance system and destroyed before they can multiply to cause a cancer. This is thought by many to be the major purpose of the T cell system of long-lived lymphocytes, which continuously patrol the tissues of the body from bloodstream to lymph nodes to lymphatics and back to blood.

Although only relatively few are the victims of cancer because of immunosuppressive therapy or immunodeficiencies, the association of malignancy with impaired immunity suggests that improving the immune response in other types of cancer patients might be beneficial. One method being investigated is administration of BCG (a live attenuated organism derived from bovine tuberculosis) as a nonspecific stimulus to heighten immunological responsiveness in cancer patients. When this agent was injected into subcutaneous nodules in patients with malignant melanoma (dark-pigmented skin tumors), approximately 90% of the tumors regressed and several of the patients survived for at least five years.

A time-honored principle of cancer surgery that is being reexamined in the light of the new understanding of the lymphatic system is the desirability of removing regional lymph nodes at the time of tumor surgery. For example, since the early 1900s radical mastectomy (removal of the breast and the lymph nodes of the armpits) has been the accepted treatment for breast cancer. The nodes have been routinely removed on the basis that they may be sites to which the cancer has spread. But because it now seems probable that these nodes are a site of active immunity against the tumor, the possibility that their removal is not always beneficial must be considered. Unfortunately, there is little clinical experience to help decide this issue, though Bernard Fisher, as chairman of the National Surgical Adjuvant Breast Project, funded by the National Cancer Institute, was in 1974 accumulating data that compared the results of treatment of breast tumors with or without removal of the nodes.

FOR ADDITIONAL READING:

Billingham, R. E., and Silvers, W. K., *The Immunobiology of Transplantation* (Prentice–Hall, 1971).

Fairbairn, J. F., Juergens, J. L., and Spittell, J. A., *Peripheral Vascular Disease*, 4th ed. (W. B. Saunders Co., 1972).

Good, R. A., and Fisher, D. W., *Immunobiology* (Sinauer Associates, Inc., 1971).

Rubin, Philip, *et al.*, "Updated Hodgkin's Disease," *Journal of the American Medical Association* (Dec. 4, 1972, pp. 1292–1306 and Jan. 1, 1973, pp. 49–67).

Yoffey, J. M., and Courtice, F. C., *Lymphatics, Lymph and the Lymphomyeloid Complex* (Academic Press, 1970).

Keeping Tabs on Our Turbulent Planet

by Robert Citron

An information clearinghouse for scientists, the Smithsonian Institution's Center for Short-Lived Phenomena dispenses vital data concerning the sudden, unpredictable, and transitory events that take place throughout the world.

The satellite information and communications facility of the Smithsonian Astrophysical Observatory (above) provides the Center for Short-Lived Phenomena an up-to-date record of operating spacecraft. Among the most spectacular events reported by the center was the eruption of Mt. Etna (right) in April 1971.

ROBERT CITRON is director of the Smithsonian Institution Center for Short-Lived Phenomena.

Until the dawning of the nuclear age in the 1940s, the majority of men lived in indifferent harmony with nature. It was man's world, and the land he used to construct his cities and highways belonged to him alone. The wastes and spilled oil he fed into the waters, the choking effluents his factories fed into the air would, he believed, be forever absorbed, for the water and the air were infinite.

With the atomic bomb this belief changed. When man first learned to harness the enormous energy of nuclear fission, the immensity of that achievement caused him to question his responsibilities and recognize his potential for destruction. For the first time, those who sounded the alarm were not a handful of scientists alone—environmentalists, conservationists, biologists—nor even the politicians and statesmen charged with setting the ground rules for the use of this awesome new force. For the first time, mankind as a whole was aroused. Consequently, there has been in this nuclear age a quickened interest in the state of the earth, an enlightened self-interest in

146

the activities of the planet that nourishes and supports all life. The "need to know" has become a top priority.

In January 1968 the Smithsonian Institution, recognizing the urgency of that need to know, established the Center for Short-Lived Phenomena. Earlier, in 1963, scientists from throughout the world had flocked to witness the birth of a new volcanic island, Surtsey, off the southwest coast of Iceland. Impressed by the flow of information provided by the Icelandic geologists, the scientific community began to speculate on the possibility of setting up a permanent center to dispense vital information on the sudden, unpredictable, and transient activities of the earth. One of those men was Sidney Galler, who subsequently became assistant secretary for science of the Smithsonian Institution in Washington, D.C.

Eventually Galler's plans were implemented, and the center entered its planning stage in 1967. Established in a single office at the Smithsonian Astrophysical Observatory in Cambridge, Mass., it had hardly begun to function when it encountered its first major challenge. On Dec. 6, 1967, a new island had erupted in Telefon Bay, Deception Island, Antarctica. The action had begun, according to the center's reports, at 2000 hours Greenwich Mean Time on December 4, when "there occurred an undulatory movement of the surface followed by three major explosions." The volcanic events formed an island 900 m long, 250 m wide, and 120 m high, with three craters and a zone of vents.

The center had been located in Cambridge in order to take advantage of the international communications network that the Astrophysical Observatory had set up for its worldwide optical satellite tracking stations. That network was used by the center to contact people in or near the volcanic eruption area, including members of expeditions from the Argentine Antarctic Institute and two other scientific teams from Chile and the U.K. Information was requested from the United States embassies at Santiago, Chile, and Buenos Aires, Arg., and from the Smithsonian observing station at Comodoro Rivadavia, Arg.

Photographs of the eruption and a preliminary descriptive report of the event were obtained by the center from the Argentine Antarctic Institute and from the naval attaché of the embassy at Santiago. Geological samples of the new island and samples of the gases and the sediments from all the craters and vents were collected by the Argentine Antarctic Institute. In addition, samples of volcanic ash were collected aboard the Argentine vessel "Bahia Aguirre."

In total, samples of eruption products from seven different locations were sent from the site, and the running commentary provided by the center to interested geologists was eventually supplemented by published reports on the entire event, prepared by the men who had conducted the on-site investigations. Thus, coincidentally, the first event to which the center responded was the birth of a new volcanic island, like Surtsey, which had provided the stimulus for the establishment of the center itself.

Rift 100-ft deep on Deception Island, Antarctica, was created by volcanic action in 1970. The first event reported by the center was the creation by volcanic explosions of a new island in Telefon Bay near Deception Island in December 1967.

Mission of the center

From the first, the mission of the center was clear. It was "to obtain and disseminate information on short-lived natural events such as volcanic eruptions, major earthquakes, the birth of new islands, the fall of meteorites and large fireballs, and sudden changes in biological and ecological systems." Its purpose was to encourage research teams to be mobilized and dispatched so that they could take maximum advantage of the opportunities provided by significant short-lived events. Rapid receipt of news of such events would permit such teams, with their instruments and equipment, to travel to the scene of an event in as short a time as possible so as to collect important data that might otherwise be irretrievably lost to science.

148

In the introduction to their book about the center, *The Pulse of the Planet,* James Cornell and John Surowiecki note: "One of the greatest scientific problems of the twentieth century has been finding a means to observe large-scale changes in the earthly ecosystem while they are actually occurring. Many important natural events . . . occur suddenly and unexpectedly in remote corners of the world. Often, too, they end as abruptly as they began. . . . Yet, while the events may be transitory and short-lived, their effects are sometimes long-lasting. Volcanic islands may erupt, subside, and disappear in a matter of days, but volcanic pollutants may linger in the atmosphere for years, the disruption of local biology may persist for centuries, and the alteration in geological substrates may remain forever."

The center's job, then, was to provide an overview, on a continuous basis, of the earth's transient natural phenomena.

Implementing the goals

Once the goals had been established, there began the long, arduous process of introducing the center's services to the scientific community and eliciting its cooperation and participation. Thousands of letters were sent to scientists and institutions throughout the world. To some, perhaps, it appeared to be junk mail, but to the majority it seemed to represent an exciting idea which they would welcome if it proved workable.

It came as a surprise to many when they began to receive from the center a series of reports in their specific fields of interest: Event Notification, for many events in the form of a telephone call or cable, followed within hours by an airmail postcard identifying the scientist or institution making the initial report and quoting that source in detail; Event Information, which included continual updating of the event activity, as reported from the scientists on-site; and Event Status, which continued the information flow throughout the duration of the event—sometimes months—and also included preprints of scientific papers prepared by the investigating scientists that dealt with the results of their research. These preprints presented personal interpretations and opinions in an informal way, often months or years before the formal works on the subject were published in books or scholarly journals.

From the first, scientists and institutions responding to an invitation to become "correspondents" of the center accepted a concomitant responsibility: to provide information to the center about short-lived natural events occurring in their areas and, when called upon, to verify the reports made by other sources. By the end of the first year the center was able to report that it had documented 70 geological, astrophysical, and biological events, including 20 major earthquakes, 12 volcanic eruptions, 11 fireballs, 5 major oil spills, 5 fish kills, 2 rare animal migrations, 2 meteorite falls, 1 red tide, 1 seiche (the sudden oscillation of the water of a lake or bay causing fluctuations in the water level), 1 sea surge, 1 locust swarm, and 1 major drought.

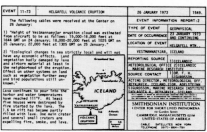

With its 10,000-watt radio (top) the center can maintain communications with all parts of the world. Above, an Event Information report prepared by the center furnishes scientists with updated information on a previously recorded phenomenon.

149

Robert Citron

By the end of that first year, 1968, the center was also able to report, in addition to the wide-scale support of the U.S. scientific community, the participation of 356 correspondents in 74 other countries, including the Soviet Union with 12.

Geophysical events

Between January 1968 and December 1973 the center received and transmitted reports of 282 geophysical events. This by no means included all such events during that time. The 117 reported earthquakes met the selective criteria of being greater than magnitude 7 on the Richter scale; or of occurring in unusual areas; or of being of exceptional interest with regard to crustal movements, faulting and fissuring, major land movements, or landslides. Among the 104 volcanic eruptions reported were those involving underwater eruptions, the birth of new islands, island eruptions and disappearances, caldera

150

collapses, fissure extrusions, *nuées ardentes* (swiftly expanding clouds of hot gases, ashes, and lava fragments), and major mudflows. It is interesting to note that, prior to the establishment of the center, scientists believed that volcanic eruptions occurred at a rate of 10–12 a year.

Geophysical events also include those earthquakes under the sea floor either greater than magnitude 7 on the Richter scale or else affecting the marine geophysical environment considerably, tsunamis (sea waves produced by undersea eruptions), sea surges, and polar and subpolar events. Among the last-mentioned are the formation of ice islands, unusual sea ice break-ups, and surging glaciers. A listing of any year's geophysical events covers a wide range of phenomena, some bearing such exotic names as the Izu-Oshima earthquake swarm (a total of 399 earthquakes recorded between Jan. 14 and 25, 1972, around Izu-Oshima volcano, Japan), the Piton de la Fournaise volcanic eruption at Réunion Island in the Indian Ocean, and the tongue-twisting Grimsvötn Jökulhlaup, a glacial outburst flood in Iceland.

Some events are not without a certain humor. On July 4, 1969, the U.S. Navy's oceanographic office alerted mariners in the Caribbean to a rather bizarre phenomenon: an island off eastern Cuba that was traveling southwest at a speed of 2½ knots. The island was reported to be about 15 yd in diameter and to contain 10–15 trees, each 30–40 ft tall. This unusual event was of special interest to ecologists because floating islands can transport and transplant animal and plant life from one land mass to another. It was suggested that this particular chunk had broken away from the mouth of the Amazon River. Arrangements were made for a botanist and an entomologist from the Smithsonian Institution to fly to the U.S. Naval Base at Guantanamo Bay, Cuba, and from there by helicopter, to the island. On July 16, however, an intensive six-hour search failed to locate the island. By July 19 the island was officially listed as missing, and presumed sunk.

Such episodes are the exception, however, especially among geophysical phenomena. More often, that category includes such violent events as the Managua, Nicaragua, earthquake of December 1972: a 6.25-magnitude shock that produced losses estimated at 6,000 dead, 20,000 injured, and 300,000 homeless. The estimated material loss was about $1 billion. During the period immediately after the disaster, while the city's electrical power was out of service, the center kept in contact with an amateur radio operator. Broadcasting from a portable rig located in his car, this man provided information that was relayed throughout the world.

Astrophysical phenomena

The center recorded 51 major fireballs and 18 meteorite falls from 1968 through 1973. Because the Smithsonian Astrophysical Observatory operates a worldwide network for astronomers and the Central Bureau for Astronomical Telegrams of the International Astronomical Union is based there, the center does not duplicate the observatory's

Lake of molten lava (opposite) is formed in the crater of Nyiragongo volcano in eastern Zaire. The volcano's intense activity in 1971–72 was reported by the center. Fire attacks the Florida Everglades (above) in May 1971, the consequence of a severe drought.

efforts. Nor does it deal with predictable astronomical events, such as eclipses. What the center does deal with, and with maximum speed, are fireballs. With luck, these blazing chunks of rock will fall to earth in places where they can be recovered quickly enough to provide scientists with important data from beyond our own planet.

In August 1972 an exceptionally bright fireball was observed over a large area of the United States and southern Canada. The object was observed from Las Vegas, Nev., to Edmonton, Alta., a distance of approximately 1,200 mi, and there were reports from towns in Montana of sonic phenomena occurring one to three minutes after the sighting followed by an earthquake-like rumble. Observers reported that the fireball was visible for periods ranging from several seconds to nearly a minute, with a smoke trail persisting up to two minutes afterward. In addition to the human eyewitness accounts of the event, the fireball was also observed by satellite for more than 1½ min. The maximum brightness of the object, according to the satellite's sensor, exceeded that of the moon.

Assuming the surface temperature of the meteor to be that of the melting point of an iron-nickel-chromium alloy, astronomers speculated that it had a diameter in excess of 50 m. The brightness of the object indicated a very large body; the long duration suggested a trajectory nearly tangent to the earth. Astronomers consider it conceivable that the meteor skipped through the atmosphere, approaching within 25 mi of the earth's surface, and then skipped out again at about longitude 114° W on a northerly heading. If it had struck the earth, the impact would have been similar to that of a nuclear bomb.

Another remarkable astronomical event recorded by the center was the fantastic meteorite shower that fell on Pueblito de Allende, Mex., in February 1969. Early in the morning of February 8, a blinding blue-white fireball turned night into day over a 1,000-mi path from southern Mexico to El Paso, Tex. Hundreds of observers throughout Mexico reported seeing the brilliant flash of light and hearing a tremendous explosion. Air Force meteorologists, calculating the wind direction and velocities at the time of the event, guided a B-57 airplane through the probable dust train. Special filter traps aboard the plane collected samples of atmospheric dust.

Meanwhile, thousands of individual meteoritic stones rained down over a large area of rural Mexico. They were recovered and identified almost immediately, but it was not until later that it was realized that the area on which they fell (the strewnfield) extended in excess of 300 sq km. Not only was this the largest strewnfield on record but also the most productive; an estimated two tons of meteoritic material were removed from the area, and according to some estimates at least four tons of material actually reached the ground.

The meteorites were carbonaceous chondrites, but among the component minerals were two (grossularite and sodalite) not previously recorded in meteorites. Samples of the meteorites were hand-carried to laboratories at the U.S. National Aeronautics and Space Administra-

152

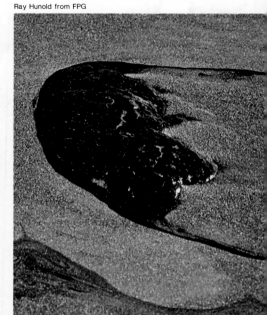

tion and the Smithsonian Astrophysical Observatory, where analysis began less than 100 hours after the fragments fell from space. Within a few weeks of the fall the Smithsonian's Division of Meteorites in Washington, D.C., had distributed meteoritic material to 37 laboratories in 13 countries.

During the late 1960s the Lunar International Observers' Network (LION) was organized by astronomer Barbara Middlehurst. It included 216 astronomical observing stations in 30 countries and had a combined capability of keeping the lunar surface under 24-hour-a-day surveillance for short periods of time. The main objectives of the network were to record the nature, magnitude, frequency, and distribution of transient lunar phenomena such as brightenings of areas on the moon's surface and occasional observations of smoke and/or dust and to obtain independent corroboration of such events from widely separated ground-based observatories.

The network participated in the lunar observing programs of Apollos 8, 10, 11, and 12, and the Center for Short-Lived Phenomena coordinated the communications activities among astronomers throughout the world, the Houston Manned Spacecraft Center, and the astronauts in orbit around the moon. During the latter three missions the center received a total of 169 reports of transient lunar phenomena from 28 observation stations in 19 countries, and on July 19, 1969, during the mission of Apollo 11, observers from six countries on two continents as well as the observers in lunar orbit made either simultaneous or overlapping observations of brightenings in the crater Aristarchus. These brightenings were later determined to be caused by reflected earthshine.

Baby sea lions on San Miguel Island off the coast of southern California are coated with crude oil, the result of a huge spill from an undersea well which in January 1969 created an 800-sq mi oil slick on the surface of Santa Barbara Channel.

Akuri Indians of Surinam, South America, await transportation to new homes with the neighboring Trio tribe in 1971. The Akuri, first brought to the attention of the center in 1968, are primitive stone-ax people who were moved because they faced extinction as the result of a limited food supply.

Archaeological and anthropological events

In the fields of archaeology and anthropology the center confines its activities to reporting newly discovered tribes; rapid changes in human ecological systems; short-lived acculturation; dying languages, customs, and people; major human and animal migrations; and the discovery of archaeological sites threatened with imminent destruction. Such a site was a rural roadway near Chinandega, Nicaragua, worn down by hundreds of years of oxcart traffic. A period of strong rains in October 1968 completed the erosion process to a level where a series of footprints made by prehistoric man was exposed. The prints, smaller than those made by modern men, had been covered by ash flow from the nearby San Cristobal volcano. Immediate investigation and action was undertaken to protect the prints from further erosion.

In January 1969, other footprints were discovered at a cinder block pit quarry near Santa Fe, N.M. These were identified as petrified camel tracks 70,000–150,000 years old. The U.S. federal Bureau of Land Management withdrew the land from public use and constructed a shed over the tracks to preserve them from weather erosion.

Also in 1969 center correspondent Wilhelm Solheim of the University of Hawaii issued an urgent bulletin delineating the "Immediate Archaeological Salvage Needs in the Lower Mekong Valley." His report detailed a 15-year-plan (1971–1986) for the construction of 34 dams in the lower Mekong Valley of Southeast Asia for irrigation and flood control. As Solheim pointed out, discoveries in areas adjacent to those about to undergo flooding by the new dams indicated that

154

evidence of major importance to man's knowledge of Southeast Asian prehistory would certainly be permanently lost. Examples which he cited included the excavations in northern Thailand which produced the oldest plant domesticates yet known (dating between 9000 and 12,000 B.C.), cord-marked pottery dated to about 8600 B.C., the first true Bronze Age site yet discovered in Southeast Asia, and evidence of a sophisticated metal-working technology tentatively dated at 3400 B.C.

Contacted by telephone in February 1974, Solheim provided an updating to his original report. Following a meeting with members of the Southeast Asia Development Advisory Group and the World Bank, and supported by funds from the National Science Foundation, a team of archaeologists had worked for three years to survey the sites in the endangered areas. Although some of the areas had been flooded as scheduled, construction of most of the dams had been delayed indefinitely through a combination of circumstances: the political situation in Southeast Asia, and the recognition by the governments involved that a thorough ecological and archaeological survey needed to be undertaken to ascertain the effects of the dams on the environment.

In 1968 the center began receiving reports from missionaries of the West Indies Mission about an isolated, nomadic, stone-ax tribe in Surinam, South America. During the period between the first contact, in June 1968, and June 1971 eight expeditions were made to the villages of the Akuri people. The Akuri, who were completely nomadic, ranged over an area of 10,000 sq km in a continual search for honey. Their most important tool, the stone ax, they called their "honey gathering thing." They also possessed bamboo knives, tools made from rodents' teeth, and bows and arrows, the latter embellished with curare poison.

By October 1970 the missionaries had made contact with a total of 66 Akuri, and knew positively of 5 more. Convinced that the tribe was rapidly approaching extinction, the missionaries, with the aid of local Indians from other tribes, attempted to establish a permanent village and a garden to overcome the endless wandering of the Akuri in search of a limited food supply. By mid-1971 it was anticipated that the few remaining Akuri would be transplanted to the villages of the Trio, a neighboring, but more sophisticated tribe with whom they had established a relationship of trust. Subsequently, the Akuri were moved.

Biological and ecological phenomena

Perhaps the most intriguing and widely reported of the center's activities are those that fall into the biological and ecological categories. How can anyone fail to be fascinated by events with such tantalizing names as the Sooty Tern Hatch Failure, the St. Louis Spider Invasion, or the Indo-Pacific Starfish Plague?

These colorful titles cover but three of a wide range of phenomena: sudden changes in biological and ecological systems, invasion and colonization of new land by animals and plants, rare rapid migrations,

Monarch butterflies
rest on their annual mass migration
from north to south in the autumn.
The migration of 1970
was particularly heavy.

155

Larry Beaver

Exceptionally bright fireball was reported by the center in August 1972. A meteor estimated to be more than 50 m in diameter, it came within 25 mi of the earth's surface during a 1,200-mi course from southern Nevada to Edmonton, Alta.

unusually abundant reproduction or death of vegetation, ecological aftereffects of short-term human intrusion into an area previously unvisited by man, potentially imminent species extinction, sudden changes in marine and aquatic environments, unusual occurrences of marine vegetation, fish kills, and fires that have a major ecological impact on animals and flora. Between 1968 and 1973, 270 such events were reported to the center.

Among these was the Sooty Tern Hatch Failure, which took place on Bush Key, an island of the Dry Tortugas off the coast of Florida. In the spring of 1969 the center received a report that, for reasons unknown, 98% of the sooty tern eggs in that traditional nesting area had failed to hatch. Strangely, a colony of noddy terns that also nested on the island had a normal hatch. The cause of the failure was never determined, although the investigating scientist speculated that sonic booms may have played a role in this unprecedented and unrepeated occurrence.

The St. Louis Spider Invasion began on the afternoon of Oct. 8, 1969, when masses of fluffy, sticky, threadlike particles floated across St. Louis, Mo. The material, thought at first to be a synthetic precipitate from heavy air pollution, was later identified as spider webbing. The ballooning spider, which crawls to the tops of trees to weave webs into which it discharges its eggs, had done so in unusually large numbers that day, and when the spider mothers set their webs adrift in the wind, weather conditions were such that the webs floated down in large numbers on the puzzled citizens of St. Louis.

In contrast to the spider episode, the Indo-Pacific Starfish Plague was a dangerous invasion by the "crown of thorns" starfish of the coral of Australia's Great Barrier Reef. News of the invasion reached the center in the spring of 1969, and, based on those reports, the Westinghouse Ocean Research Laboratory launched a massive investigation of the coral reefs of the area.

The destruction of the living coral, which serves as an outer layer of the reef itself, was considered especially ominous because the reefs not only shelter a variety of fish and other marine life necessary to sustain the life of the area, but also protect many inhabited islands otherwise open to erosion and flooding. It was determined that the population explosion among the crown of thorns starfish had probably resulted from the disappearance from the area of its major predator, the giant triton or conch. The disappearance may have been caused by shell collectors and tourists. In the early 1970s the starfish plague diminished.

It is disheartening to note that in recent years the "natural" disasters threatening the ecology of the planet, as in the case of the starfish plague, are more and more often initiated or aggravated by the activities of man. For example, in 1968 there were 9 events resulting from oil spills or other forms of water pollution; in 1973 there were 44, ranging from oil and chemical spills to gas escapes and a leak of wastes from a nuclear power plant.

156

The future

By the end of 1973 the center had reported on nearly 700 events to more than 2,300 scientists and institutions in 138 countries throughout the world. Its staff of nine uses nine telephone lines, a short-wave transmitter, a computer bank, six teletype circuits, and U.S. government networks to maintain a round-the-clock capability to respond to the events of this continually active planet. The staff of the center includes a director; an operations manager; two event research specialists, one in ecology and one in the geophysical/astrophysical sciences; a computer specialist; and two communications specialists. The staff occasionally participates in planning transportation and logistic support for Smithsonian-sponsored field research expeditions, and has also taken part in several research expeditions.

Sophisticated new equipment, such as the Earth Resources Technology Satellite (ERTS), is continually adding to that capability. ERTS, a complex earth-orbiting observational system that measures a variety of phenomena on the surface and in the atmosphere of the earth, offers a system similar to that provided by the Apollo astronauts during the lunar observations. In the case of ERTS, the satellite provides observational coverage from space of events occurring on the earth. In some instances, it makes the initial observations and alerts the center, which then disseminates the information to the appropriate scientific authorities.

As an outgrowth of its continuing concern about the impact on humans of natural disasters, the center in 1971 conducted an international survey to determine the number, global distribution, and operational characteristics of existing facilities for warning against and monitoring natural disasters. It also sought to document the extent of current natural disaster research. A published directory of these findings serves as a basic reference guide for the planning and establishment of a comprehensive and effective global natural disaster warning system.

In the same vein, but on a larger scale, the center was instrumental in compiling and publishing in February 1974 a *Directory of National and International Pollution Monitoring Programs.* This was a first step toward the ultimate goal of enabling man to understand and protect his environment.

Road in Managua, Nicaragua, is destroyed, the result of a disastrous earthquake on Dec. 23, 1972, that killed about 6,000 and injured 20,000.

Science Year in Review
Contents

Agriculture

Agricultural scientists continued to attack the problems of providing an ever increasing population with wholesome, inexpensive food and countering onslaughts on crops and livestock by diseases and insects. Also high on the list of priorities was the multifaceted question of energy.

Energy conservation. At no time in recent history had the energy situation been so vitally important to agriculture. Unquestionably, the energy crisis would have a significant effect on food production technology in the U.S. and on the Green Revolution in less developed countries, since both systems of production depended on relatively large energy inputs. Alternative ways to reduce energy consumption were under intensive study by the Agricultural Research Service (ARS) of the U.S. Department of Agriculture (USDA). Only by carefully balancing energy requirements and crop production techniques could both farmers and consumers cope with reduced energy supplies.

One way of reducing energy requirements is for farmers to practice reduced tillage methods. Savings are considerable. ARS engineers calculated that farmers could halve their fuel needs and cut their labor needs 40% by practicing reduced tillage, which skips disking and cultivating whenever possible. No-till systems are even more economical, where soil conditions permit their use. ARS studies also showed that some plants use fertilizer more efficiently with no-till. Some of the fertilizer may carry over from one crop year to the next.

Plants themselves could become a major source of energy in the future. They are the major converters of solar energy into energy that can be used and stored by man, and they are a renewable resource. Scientists were exploring the possibility of chemically treating and then distilling soybean oil in order to produce a substitute for gasoline, liquefied natural gas, and other fossil fuels. However, the practical use of farm crops as an energy source had to await development of technologies that would vastly increase farm production without increasing energy consumption.

Disease and pest control. Another way of reducing energy inputs in crop production is to breed crops resistant to pests, thus reducing the need for conventional insecticide sprays as well as lessening the danger of environmental contamination. Zapalote Chico, an exotic strain of corn brought by the ARS to the U.S. from South America, might provide a source of resistance to corn earworms, probably the most destructive insect pest of corn in the U.S. In corn-breeding research, the strain readily transmitted earworm resistance to its progeny.

New experimental varieties of alfalfa resistant to anthracnose, a disease that damages more than four million acres annually, yielded at least one ton per acre more than several commercial varieties in field tests. This was a major breakthrough in incorporating disease resistance into forage crops. Anthracnose is most prevalent in the South and the southern parts of the Middle Atlantic and North Central states. The disease is caused by the fungus *Colletotrichum trifolii*, which attacks the stems and crown of alfalfa. Affected plants may be killed, or they may become so debilitated that their productivity is severely reduced.

In an effort to develop sugarcane with greater disease resistance and overall higher quality, agronomists at the U.S. Sugarcane Field Station, Houma, La., were launching a new sugarcane-breeding program designed to explore the basic sugarcane species. Those species with inherently desirable traits would be combined with established canes having commercially acceptable features. During 1974 ARS scientists plan to make about 150 biparental crosses that would yield approximately 250,000 viable seeds.

Corn earworm feeds on an ear of corn. An exotic strain of corn introduced into the U.S. from South America might prove to be a source of resistance against these destructive pests.

"Agricultural Research Magazine"

ARS insect virologists were exploring the concept that viruses could be used to control insect pests biologically. A recent experimental finding was that activated charcoal protects sprays containing a virus disease of corn earworms from the destructive effects of sunlight. The virus *Heliothis nucleopolyhedrosis* (NPV) was being used against the cotton bollworm. Normally, one-half of the *Heliothis* NPV, unprotected from sunlight, was inactivated in about one day, but activated carbon extended the half-life to between three and five days. Viral diseases had been found in all major orders of insects. Some 83% of the viruses thus far isolated came from caterpillars of moths and butterflies, perhaps because many economic pests are in that group and considerable work had been devoted to them.

ARS scientists found that resmethrin, a synthetic insecticide related to pyrethrum, was exceptionally effective against both flies and cockroaches when used in the form of a superfine dust. Applied in a confined area, it knocked down stable flies within 10 minutes and houseflies within 30 minutes, and killed the flies within a few hours. Roaches exposed to resmethrin died within 24 hours. Resmethrin was believed to be virtually harmless to warmblooded animals.

The integrated pest management program was another method of coping with insect enemies of man. This approach maximizes the use of beneficial insects, as well as using chemical, cultural, biological, and genetic techniques. In a 20,000-ac pilot experiment in southern Mississippi and adjacent areas of Louisiana and Alabama, several suppression techniques were being used to combat the boll weevil, which had plagued the U.S. cotton industry for 80 years. The techniques included late summer and early fall insecticide applications and cultural measures designed to limit the number of boll weevils entering winter hibernation. In spring, sex attractants were used in traps and in "trap plots" of cotton containing systemic insecticide to kill boll weevils feeding on the plants. After the main cotton crop was planted, sexually sterile male weevils were released weekly to mate with native females. At the end of the second year there was no evidence of boll weevil reproduction in 235 of 236 cotton fields.

Gypsy moths, a serious pest of northeastern forests, were also about to come under an integrated pest management program. Insects that eat gypsy moths, diseases that attack their larvae, and the release of sterile male gypsy moths, among other techniques, were under test in cooperative federal-state research and regulatory efforts. In the sterile male release program, the USDA reared gypsy moths in the laboratory, then sterilized them for release in infested areas. Because adult moths do not eat, sexually sterile moths do not threaten forests. Enough sterilized males must be released to vastly outnumber normal males, thus preventing most female-normal male matings.

Meat supply. Intensive research efforts were under way at the ARS to help farmers and ranchers produce more meat, especially beef. The broad attack on the problem involved breeding, feeding, and management research; more attention to improving pasture, range, and forage production; and increasing the protein content and quality of feed grains.

Gypsy moth larva, a serious pest of forests in the northeastern U.S., dines on a leaf. Efforts to control gypsy moths include sterilization of adult males and the introduction of diseases that attack the larvae.

E. S. Barnard from Bruce Coleman Inc.

A two-part experiment with calves showed that veal with less saturated fat could be produced by feeding calves diets high in polyunsaturated fats. Polyunsaturated milk is produced by cows that have been fed a protected, unsaturated vegetable oil; in the ARS experiment, a mixture of safflower oil and casein, treated with formaldehyde, was used. Normally, unsaturated fats fed to cattle become saturated in the rumen, but this does not happen to the formaldehyde-protected oil, which remains unsaturated as it passes through the calf's digestive system.

In the experiment, calves fed the polyunsaturated milk had four times as much unsaturated fat as control animals, but there was no more total fat in the edible portions of the animal.

Reproductive inefficiency is one of the most costly problem areas in the production of beef. The number of beef calves weaned each year is only about 80% of the number of cows bred. Even this low level of reproduction is achieved at the expense of many repeat matings; 40% of all initial matings are infertile and must be repeated, and 16% must be repeated twice. ARS scientists believed the technology needed to achieve greater reproductive efficiency would include methods to reduce the age at which heifers are able to conceive, to shorten the post-calving anestrus period during which the cow does not have heat periods, to increase the percentage of cows that become pregnant on the first mating, to control sex of offspring, and to induce multiple births.

The discovery of a simple, accurate, and rapid technique for pregnancy testing of sheep promised to save sheepmen millions of dollars otherwise wasted each year in feeding nonpregnant ewes through most of the winter. The ARS-developed test, which was nearly 100% accurate, could be performed 60 to 115 days after breeding. The method also showed promise of being relatively accurate in detecting twin-bearing ewes.

Nutrition. A long-range program to computerize detailed information on the nutrient composition of thousands of foods was begun in 1973. When fully compiled, the Nutrient Data Bank would provide information useful in nutritional labeling of packaged and processed foods and would be especially useful to nutritionists, food scientists, dietitians, physicians, consumers, and the food industries. The USDA had been collecting food composition data for 80 years, but the advent of voluntary nutritional labeling of foods by the processing industry accelerated the need for a more complete central source of such data.

Vanadium was identified as one of the elements that are essential for health. Nutritional studies from three ARS laboratories and the Veterans Ad-

ministration Hospital, Long Beach, Calif., demonstrated that vanadium is an essential trace element for rats and chicks. Past studies had shown that these two animal species are among the more reliable models for studying human dietary needs. According to ARS studies, vanadium deficiency would probably show up in the form of altered blood lipid levels, since it was well established that vanadium alters lipid metabolism. Marginal vanadium deficiencies in humans may be responsible for at least part of the altered blood lipid levels commonly observed among Americans.

An appealing, nutritionally balanced, and protein-rich candy made from corn syrup, nonfat dry milk, and other ingredients drew overwhelmingly favorable responses from both children and adults in taste tests. Synthetic vitamin and mineral supplements could also be added to the candy, the protein quality of which compared quite favorably with casein, one of the best natural sources of protein.

During 1973 an ARS scientist released the first research data showing that blood pressure can be lowered by reducing the level of fat in an otherwise normal diet. The study also showed that the low-fat diet had a pronounced effect on factors involved in blood coagulation. The study, conducted jointly by the ARS and the Georgetown University School of Medicine, Washington, D.C., involved 21 volunteers (10 men and 11 women), all between 40 and 60 years old and in good health.

—Marcella M. Memolo
See also Year in Review: FOODS AND NUTRITION.

Archaeology

The urgent concern of archaeologists about the destruction of sites as a result of the trade in antiquities showed no signs of alleviation, although there were some encouraging trends. Following the signing of an international agreement by Mexico and the U.S., authorities in both countries were clamping down on the trade in major monuments smuggled out of Mexico, and some objects were confiscated and returned. Although not an archaeological object, the Afo-A-Kom statue, sacred to the people of Kom in Cameroon, was returned to that country by a dealer in New York. In India a new antiquities law established new regulations and penalties and required the registration of all antiquities at the district level, including objects in private hands.

As had been predicted in archaeological circles, the trade had both direct and indirect effects on foreign excavations in many countries. Turkey did not close out foreign expeditions, as had been

Stela at Tikal, Guatemala, is covered with Mayan hieroglyphs. New research indicates that some glyphs describe the history of sites and do not refer only to calendrical or theological matters.

feared, but arrangement for such expeditions became much more difficult. Other countries, such as Afghanistan, materially reduced foreign expeditions by requiring that they undertake very expensive restoration projects in order to obtain permits. Foreign expeditions do not contribute to the export of antiquities, since all excavated material is carefully controlled and in most cases remains in the country of origin. Nevertheless, exporting countries place the onus on the importing countries, which are, by and large, the same countries that support most excavations.

America. The controversy over how long man has been in America received a new impetus during the year with the announcement by three geologists of evidence that man lived in the Western Hemisphere 250,000 years ago. Roald Fryxell of Washington State University and Harold E. Malde and Virginia Steen-McIntyre of the U.S. Geological Survey acknowledged that their discovery would lead to a head-on confrontation with existing archaeological data. Sophisticated stone tools found in hard, consolidated deposits at a

162

site called Hueyatlaco near Puebla, Mex., were dated by several techniques at 250,000 years old. An added puzzle was the character of the stone tools, which apparently are more advanced than anything of comparable age in Europe and Asia.

Opinions regarding the tantalizing problem of translating Mayan hieroglyphs appeared to be undergoing a significant shift. Translations of the glyphs on stone monuments recording specific dates in the Mayan calendar had been accepted for many years. Moreover, the work of Eric Thompson and others had led to the recognition of gods' names and symbols relating to other supernatural matters. It was also generally accepted that some glyphs are examples of "rebus" writing, *i.e.*, the use of a picture whose sound suggests another word. Recently, however, Tatiana Proskouriakoff of the Peabody Museum, Cambridge, Mass., and Heinrich Berlin identified certain symbols as names of population centers or of the reigning dynasty, and they proposed that the inscriptions have to do with the history of sites and not only with calendrical and theological matters. Accepting the new point of view, Michael Coe, Yale University, and his brother William, University of Pennsylvania, observed that Mayan writing is now recognized as dealing with actual people, events, history, and everyday affairs.

A famous Mayan temple, "lost" since 1912 in the largely unexplored jungles once occupied by the ancient Maya, was dramatically rediscovered during the year. This unique and well-preserved building, of unknown function and standing 55 ft high, was originally found at Rio Bec in Yucatan by an expedition from the Peabody Museum. It was described as unique in design and the best known example of the Rio Bec style of Mayan art.

A University Museum, Philadelphia, expedition directed by Robert Sharer and David Sedat excavated 10 Pre-Classic Mayan sites and recorded 21 in the Salamá Valley of central Guatemala. An announcement during the year explained that the valley was first settled about 1500–1200 B.C. by small, self-sufficient agricultural units that probably arrived from the south. Before 600 B.C. small hamlets grew up as political and religious centers, and a notational system for predicting natural phenomena had been invented. Between 600 B.C. and A.D. 100, ceremonial and elite residential structures appeared and a codified hieroglyphic writing system was in use. Sharer interpreted the Salamá Valley history as evidence that there was no single mother culture for Mayan civilization, but rather a series of regional centers with many incipient writing systems that eventually coalesced.

Asia. A number of articles and books published during the year emphasized a growing concentration of archaeological research on the history and origins of metallurgy, and new techniques for the analysis and dating of metal objects resulted in some surprising conclusions. For example, in the Near East the use of arsenic to harden copper objects and to facilitate their casting preceded the use of tin. In addition, ancient smelting processes show a sophisticated knowledge of many other metals and their function in various metallurgical techniques.

The great enigma in the study of ancient metallurgy remained the source of the tin used in the bronze industry of the Near East. J. D. Muhly published an account of the earliest Near Eastern inscriptions referring to tin and concluded that it was a precious trade item brought in from the east, although there is no known source in Iran or Pakistan. The discovery of the Ban Chiang site in northern Thailand in 1971–72 provided very good evidence that an extensive tin-bronze industry had developed in that area about 4500 B.C., giving rise to serious questions about the long-standing belief that metallurgy originated in the Near East. During the year a group of archaeologists from the University of Otago in New Zealand found deposits of copper and tin in the same region of northern

Thailand, which would have been accessible to ancient metalworkers. This fact, coupled with the great quantity of tin-bronze objects in the Ban Chiang site and the size of ancient slag deposits in northern Thailand, raised the possibility that tin from Southeast Asia reached the Near East in ancient times.

The remarkably preserved body of a Chinese woman buried 2,100 years ago, which was reported in 1972, was subjected to a full autopsy by a medical staff in Peking. A large bile stone was found in the bile duct and another blocked the hepatic duct. There were signs of tuberculosis in the left lung, and she suffered from serious coronary disease. The doctors concluded that she died of a coronary heart attack as a result of biliary colic. The ova of whipworms, pinworms, and blood flukes were present in the rectum and liver. Some 138 muskmelon seeds were found in her esophagus, stomach, and intestines. She was thought to be the wife of the marquis of T'ai in the principality of Chang-sha.

Near East. The discovery of a new "Rosetta Stone" that might well be the key to the Lycian alphabet was announced in September 1973 by French archaeologists excavating in Xanthos in

Mayan structure at Rio Bec in southeastern Yucatan was rediscovered in 1973 after being lost for about 60 years. Archaeologists believe that the building, perhaps a palace, dates from A.D. 800.

Hugh Johnston

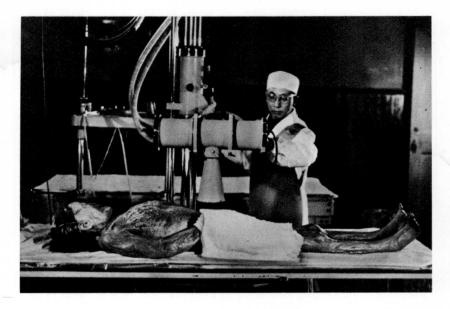

*Chinese physician X-rays
the corpse of a noblewoman
who died at the age of 50
about 2,100 years ago.
Unearthed from a tomb
near Chang-sha,
the body was in an excellent
state of preservation,
a tribute to the skill
of the Han Dynasty embalmers.*

southwestern Turkey. Lycian is thought to be an Indo-European language related to Greek, but even the most basic characteristics of the language have remained obscure. The tablet from Xanthos is a quadrangular prism about four feet long with a trilingual inscription in Greek, Aramaic, and Lycian. There are 4 lines in Lycian, 35 in Greek, and 27 in Aramaic. It was known that the tablet predates the arrival of Alexander the Great in Asia Minor in the late 4th century B.C.

Great emphasis was being placed on the restoration of ancient monuments throughout the Near and Middle East. One outstanding example was the Iranian-Italian project at Persepolis, founded by Darius I in 518 B.C. and considered by some specialists to be the grandest and most artistic architectural monument of the ancient world. Most interesting to Near Eastern archaeologists was the discovery by the restoration team of a large palace, three miles from the main Persepolis complex, that may have been built before the founding of Persepolis proper. Foundations and 28 columns were unearthed.

Syrian teams, aided by 14 foreign expeditions, were excavating a series of sites to be inundated by the dam at Medinat al Thawra on the Euphrates. Perhaps the most significant discovery to date was that of a 2000 B.C. town, occupied by Syrians under Hittite rule, which was believed to be Emar. Some 50 clay tablets written in Akkadian cuneiform indicated that the town was one of a chain of trading centers connecting Mesopotamia with Phoenician centers on the Mediterranean.

What was reported to be the "oldest song in the world" was described in news articles in March 1974. Anne Kilmer, an assyriologist at the University of California at Berkeley, succeeded in translating a Hurrian text of 1800 B.C., found near Ras Shamra in Syria, containing the song and instructions for playing the music. The song was subsequently played on a lyre designed after the remains of one of the instruments discovered in the Royal Tombs at Ur in Iraq. The song was thought to be in the diatonic scale of contemporary Western music and was thus said to prove that Western music was some 1,400 years older than previously believed (actually the famous silver flute from the Royal Tombs had been recognized as proving this fact many years earlier). Some sumeriologists doubted that a Hurrian or Sumerian of 1800 B.C. would recognize the music played in Berkeley, since nothing in the tablet indicates the pitches or rhythms.

In Egypt excavations led by David O'Connor of the University Museum, Philadelphia, delineated a huge Nile harbor with an area of nearly 26.9 million sq ft. Now completely silted over, it was originally linked to the river by a canal 1.5 mi in length and was in use only during the annual period of high water. The harbor, in an area across the river from Luxor now known as Malkata, was created in the reign of Amenhotep III (1417–1379 B.C.).

Europe. The discovery at Chania (ancient Kydonia) in western Crete of a large cache of tablets inscribed in Minoan Linear A was described as the most important find of the year in that area. The tablets were significant because the principal difficulty in the decipherment of Linear A was the paucity of inscriptions of this kind.

From Italy came the announcement of the discovery of a sophisticated civilization that flourished about 27 centuries ago in the region of

Rome. The site at Castel di Decimia, ten miles south of Rome's center, had been unearthed some 20 years earlier during road construction, but the current excavation of an extraordinarily rich tomb of a woman with ornaments of gold, silver, pearls, amber beads, and a funeral chariot, as well as three other rich tombs of men, showed an unexpected wealth and sophistication in central Italy shortly after the traditional date of the founding of Rome.

Another dramatic discovery was announced by Alfonso de Franciscis of the National Museum at Naples: the ancient city of Oplonti, ten miles north of Pompeii at the modern town of Torre Annunziata. Excavations, which had just begun during the year, already uncovered an almost perfectly preserved villa of the Roman period with magnificent wall paintings, well preserved and in place. The villa was buried by the eruption of Vesuvius in A.D. 79.

In a footnote to classical archaeology, Ioannis Sakas, a mechanical and electrical engineer, demonstrated how Archimedes utilized sun reflectors to set fire to the Roman fleet at Syracuse sometime between 215 and 212 B.C. Sakas theorized that the Greeks did not have curved mirrors at that time but could have used a series of highly polished copper shields. In Piraeus 70 sailors from the Greek Navy arranged 200 flat shields (5 × 2 ft and covered with a polished copper spray) so that upon command they could be turned to focus the sun's rays on a wooden boat 60 yd from shore. Within two minutes the boat was ablaze.

What was described as the first Neolithic settlement in Britain known to date as far back as about 3000 B.C. was found at Carn Brea in Cornwall. It is a hilltop site surrounded by a wall of massive granite blocks, some weighing up to half a ton. Roger Mercer of the Cornwall Archaeological Society described the people as farmers living in a sizable fortified village.

Techniques. The application of neutron activation analysis to soapstone artifacts in the eastern U.S. showed great promise for the identification of the quarries from which they came. With known sources, trade routes and cultural contacts over a wide area could be worked out. Ralph O. Allen and Charlton G. Holland at the University of Virginia were working primarily on material from Virginia but also on some samples ranging from Pennsylvania to Georgia. The most useful trace element so far was europium, but other elements under investigation were chromium, cobalt, scandium, zirconium, and iron. Soapstone was used extensively to make vessels, particularly in the late Archaic period (c. 2000–500 B.C.), and soapstone artifacts have been found far from any known sources.

The research of Gary Carriveau at the Applied Science Center for Archaeology (MASCA), University Museum, Philadelphia, promised a new method of dating ancient metal artifacts. It involves the application of the thermoluminescent (TL) technique to the dating of metal slag, crucibles with metal drops attached, and ceramic tools encrusted with metal and slag. Anne L. Berry, Department of Geology, Stanford University, reported some success with TL dating of Hawaiian basalt flows ranging from 12,000 to 17,000 years ago.

Another application of TL dating, published in *Archaeometry* in February 1974, showed that the controversial bronze horse in the Metropolitan Museum of Art, New York City, is not a modern forgery, as had been claimed some years earlier. The core of the sculpture consists of sand, calcite, clay, and other minerals, which presumably could be dated by TL except for the fact that the horse had been subjected to X rays and had accumulated a considerable artificial radiation dose. However, in the core were grains of zircon with uranium concentrations of hundreds of parts per million. Zircon grains would accumulate a dose of tens of thousands of rads in 2,000 years and would not be seriously affected by recent X-ray doses. TL analysis of zircon grains in the core by the Oxford Laboratories proved that the sculpture was made in antiquity, although dating was not accurate enough to determine the exact period.

MASCA published carefully calibrated charts for correcting carbon-14 dates. The correction factor was based on several years' study of bristlecone pine tree rings and covered the period from about 5000 B.C. to the present. Major corrections of up to several centuries must be made in the period from 5000 to 1000 B.C.

—Froelich Rainey

Architecture and building engineering

Major events in architecture and building engineering during the year included the near completion of The Netherlands' huge project to prevent the tidal flooding of its southern coast; the opening of London's New Covent Garden Market; and the topping out of Chicago's Sears Tower, the world's tallest full-sized building. Much attention was devoted to conserving energy through changes in building design and new methods of waste recovery, and to saving and restoring Italy's flood- and pollution-threatened historic cities.

Dutch Delta Plan. After two decades of arduous effort, the vast Dutch Delta Plan, one of the most imaginative engineering schemes of modern times, entered its final stage: the main construction works in the Oosterschelde (East Scheldt) estuary. By 1979, if the project proceeds on schedule, the turbulent tidal complex of The Netherlands' southwest coast will become a tame freshwater reservoir with immense economic benefits to the country.

The Delta Plan was begun after a catastrophic 1953 storm that inundated about 400,000 acres in the southern Netherlands, killing approximately 1,830 persons and tens of thousands of livestock. In addition to putting an end to such periodic calamities, the scheme was designed to stop the saltwater poisoning of farmland by the tidal flow, create more reclaimed land for the growing Dutch population (expected to reach 17 million by the year 2000), and halt or at least substantially reduce the erosion and soil subsidence that have caused The Netherlands to "sink" inch by inch throughout the centuries. This last goal would be achieved by radically shortening the coastline.

The three large rivers that flow through The Netherlands—the Rhine, the Maas (Meuse), and the Scheldt—create a sandy tidal archipelago of islands and peninsulas, extremely vulnerable to storms. The problem of the Delta planners (headed by a hydraulic engineer, Johan van Veen) was to dam the estuaries while permitting the river waters to reach the sea and to keep open the ports of Rotterdam to the north and Antwerp, Belg., to the south. Four primary dams, fronting the sea, were designed to block the mouths of the four great estuaries between Rotterdam's New Waterway and Antwerp's Westerschelde (West Scheldt). Inland, three secondary dams were designed to protect the work sites on the primary dams and to serve, when the project is complete, as part of the total system for controlling the flow of the rivers. Many other secondary works—dikes, dams, bridges, roads—were involved.

The four primary dams were built in order of increasing difficulty, beginning with the relatively small Veerse Gat dam, completed in 1961, and continuing with the Haringvliet and Brouwershavensche Gat dams, completed in 1970 and 1971, respectively. Much valuable experience was gained on these works, particularly in the development of two techniques. One of these was the overhead cableway, employed with a turning device that permitted cable cars to follow a continuous circuit of loading and dumping in the construction of the dam embankments. The second was the even more important use of caissons. Dutch engineers had first experimented with caissons in restoring the ruined dams of Walcheren Island immediately after World War II. The caissons used in the Delta Plan are "culvert caissons," huge boxes that can be lowered to the sea floor. They have gates on either side that can be left open to permit the ebb and flow of the tide. When all the caissons of a dam are in place, the gates are closed simultaneously.

One of the large dams, that sited in the Haringvliet estuary, is equipped with giant sluice gates. These serve a double function: to help control the water level in the northern basin and to permit the masses of ice from the upper Rhine and Maas to escape. The 17 massive gates in the sluice structure are each 186 ft wide and are separated by 18-ft-thick piers. To assemble so large but, until complete, so vulnerable a structure it was necessary to build a giant rectangular cofferdam, the largest in engineering history, 1,500 yd long and 600 yd wide.

The Oosterschelde Dam, under construction in the largest (5-mi-wide) estuary, had to be given a very broad, gradual slope to resist the pounding of the immense volume of water. The design called for a width of 200 ft at the waterline, slanting down to a full kilometer—0.6 mi—at the base. The first stage of construction was to dump fill in order to create three artificial islands. These provided anchors for the other components of the barrier and also accommodated service harbors for construction craft. One island also formed the cofferdam for building another outlet sluice system. The second stage was the building of a central embankment dam connecting two of the artificial islands and forcing the tidal current into three remaining gaps, one to the south and the other two, separated by the third artificial island, to the north. Closing these three deep channels was the final and most difficult task of the project.

As the scheme neared completion, Dutch planners saw benefits even larger than those hoped for when it was begun. It seemed likely that the location, resources, and transportation network of the reclaimed region would enable it to become one of the world's great commercial-industrial-agricultural complexes.

Engineering and energy. Much engineering and scientific effort in 1974 went into attacks on the energy problem. Among the several different approaches were conservation of energy through building design, conservation through waste recovery and recycling, and a novel heat-storage technique.

In the U.S. the American Society of Heating, Refrigerating, and Air-Conditioning Engineers was enlisted to help write a set of energy conservation standards that could be adopted by the states. Some of the targets of the new standards included: (1) restriction of heat flow through a building's exterior envelope (walls, roof, and windows); (2) the installation of shading devices (draperies, venetian blinds, awnings, fins) or of insulating glass for the windows of buildings; (3) the development of standards for combined heating, ventilating, and air-conditioning systems to improve efficiency; (4) the installation of energy-recovery equipment for air exhaust systems; and (5) improved control of lighting to reduce waste.

In St. Louis solid-waste shredders proved successful in processing residential trash. The small-scale experimental plant was designed to produce 300 tons of refuse fuel per day, recovering steel for recycling and producing a supply of fuel for a power plant. The shredded trash is then mixed with coal for burning. Since in 1974 more than three-fourths of residential trash in the U.S. ended up in open dumps, constituting a totally wasted, not to mention unsightly, potential resource, the St. Louis experiment and similar, much larger projects planned for that city and Chicago may prove of immense long-range significance. Environmental impact and energy waste aside, many cities and towns are running out of places to dump. New York City in 1974 produced about 55,000 tons of solid waste per week, and was landfilling the last remaining site on Staten Island with partly burned trash.

The U.S. Environmental Protection Agency in 1974 was funding research in several cities on developing techniques for utilizing solid waste. These included using it to produce steam, an oil-like fuel, and compost material. France, West Germany, and Japan have incinerated trash for several years to produce steam and electric power. A giant plant in Rotterdam incinerates about 770,000 tons of waste a year, producing power and operating a desalting system.

The blackout of Christmas light displays on buildings in 1973 proved only the beginning of a cutdown in outdoor illumination. It was decided that several new buildings specifically designed for exterior lighting would forgo it, at least for the time being. The 56-story First International Building of Dallas, with X-type wind bracing, was equipped with high-intensity quartz lights running vertically down each corner of the rectangular tower and outlining the double-Xs (one above the other) on each facade. Lights were all installed but left dark until the crisis abated. Maryland's 30-story trade center in Baltimore was also given its lights, designed to provide a silhouette effect, but

Refuse Processing Facilities in St. Louis, Mo., shreds solid wastes and separates them into such reusable materials as steel, glass, and copper. Light fraction, including paper, cloth, and other combustible materials, is burned with coal to generate electric power.

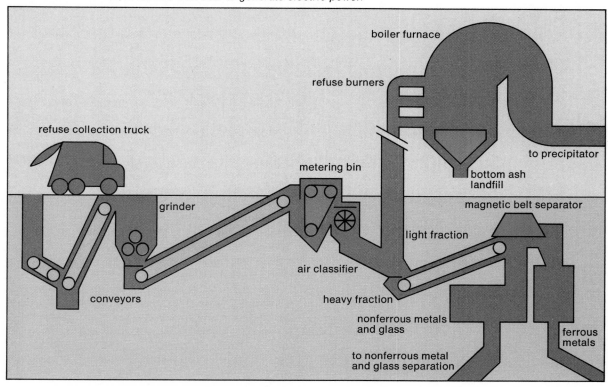

it was not given the power to operate them. A number of other new buildings cut back on exterior lighting so that it provided no more than necessary security.

Solar power was the subject of much research and a number of applications. On the 12th-story setback of the 70-story RCA building in New York City a conference center was designed to make use of solar energy for part of its heating requirements. Part of the center is black-walled and windowless to absorb sunlight; part is enclosed with insulating glass. (*See* Feature Article: POWER FROM THE SUN.)

Exploitation of geothermal energy also continued in many places in 1974. The Geysers field in northern California attained a capacity of 400 Mw of power to pass Italy's Larderello field as the largest in the world. The ultimate capacity of Geysers may reach 5,000 Mw. In addition to such conventional geothermal sources, which exploit natural sources of steam and hot water, a new technique was under development to attempt to harness the immense amounts of heat in the earth's rocks. By

Canadian National Tower in Toronto rises to 1,464 ft to become the tallest structure in the world excluding broadcast antennas. With the addition of a television mast the structure achieves a height of 1,805 ft.

B. Brooks from Bruce Coleman Inc.

drilling into dense rock formations and pumping in water at high pressure, the rock formations are fractured. The water is later recovered at several hundred degrees higher temperature. The method was devised at Los Alamos Scientific Laboratory. (*See* Feature Article: POWER FROM THE EARTH.)

Taking note of the considerable differential in the cost of electricity during peak daytime hours and slack periods during the night, a Connecticut corporation developed and patented a system that stores heat underneath buildings at night and releases it during the day. The heat is captured nightly by resistance heating mats (sets of coils heated by electrical resistance) buried in sand under the concrete floor slab and is radiated upward during the day.

London's new marketplace. In the summer of 1974 the New Covent Garden Market opened after three years of construction, replacing London's historic but crowded and obsolete Covent Garden Market as the wholesale center for the city's fruits, vegetables, and flowers. New Covent Garden consists principally of two immense single-story structures on either side of a major rail line and a multistory administration headquarters. To connect the two single-story buildings, one of which contains the fruit and vegetable market and the other the flower market, tunnels were driven under railway tracks.

Rejected was a design for a multistory building, which studies showed would be less efficient in handling the huge flow of truck traffic than the one-story structures. The fruit and vegetable building is actually two 12,000-ft-long rectangular sheds with roofed-over links. The flower market is a square one-story building with temperature and humidity control throughout.

Saving Italy's cities. Ever since the flood of the Arno River that devastated Florence in 1966, scholars and art connoisseurs have been conscious of the multiple hazards that endanger the irreplaceable treasures of Italy's cities. The treacherous soil mechanics of the Piazza del Duomo in Pisa threaten to topple the Leaning Tower, while architectural and sculptural masterpieces in Venice, Florence, Milan, and many other cities are eroding under the assault of air pollution. Engineering studies are under way to find means of saving the cities that were long the vanguard of Western civilization.

One proposal enacted in 1974 promises to rescue Venice at least from the perennial danger of Adriatic storm waters that flood through the three openings in the lagoon's sandy barrier to wash into the Rialto Bridge and the Piazza San Marco. Tides sometimes rise six feet above normal, more than enough to cover the entire city. In response to a government competition, several engineering firms offered designs for floodgates to be installed in the three channels, Lido, Malamocco, and Chioggia. The winning design calls for "floating" gates, hollow steel cylinders that will normally lie flat in concrete boxes on the seabed but can in case of threatening tides be raised on hinges to a sloping position, providing a flexible but firm barrier to the waves. (*See* Feature Article: CAN VENICE BE SAVED?)

Highest tower. The latest entry in the continuing skyscraper race, the Canadian National Tower in Toronto, rises to 1,464 ft, beating the Sears Tower in Chicago by 10 ft. The Toronto structure may, however, require an asterisk in the record books, since it is only 182 ft in diameter at its base and slims down toward its summit to a 50-ft-diameter restaurant and observatory and then to a 36-ft-diameter tube. A television mast makes the ultimate height 1,805 ft, not a record for TV-radio towers.

The Sears Tower, meantime, was topped out as the world's tallest full-sized building at 1,454 ft. Its completion gave it a record perhaps more surprising than its height. At $160 million, the building actually came in under its budget, a rarity in any kind of construction in recent times. The record, however, in cost competition still belongs to the Empire State Building, which was completed in 1931 for $41 million, against an original estimate (in 1929) of $50 million.

"House of a Thousand Holes." The tubular design concept pioneered by Fazlur Khan and other architects in such noteworthy high-rise buildings as the John Hancock Center and Sears Tower in Chicago and the World Trade Center in New York City found a striking new expression in a 50-story office building in Hong Kong. In tubular design the exterior walls of the building are made strong enough to resist lateral (wind) forces without the aid of the "steel skeleton" bracing of the old-fashioned Empire State-type of skyscraper. By spacing exterior columns close together, as in the World Trade Center, or by adding diagonal bracing, as in the John Hancock Center, adequate strength is achieved against lateral forces while retaining the conventional rectangular window architecture.

The Hong Kong building, Connaught Centre, takes a new tack; its round, porthole-like windows leave more mass in the walls, creating in effect a large, solid, rectangular tube pierced with holes, an extraordinarily powerful object to resist Hong Kong's notorious typhoon winds. The result is a highly unusual facade, which quickly won from Hong Kong residents the title "House of a Thousand Holes."

—Joseph Gies

Astronautics and space exploration

Unmanned probes to Mercury, Venus, Mars, and Jupiter were among the most dramatic events of the year in space. In the United States the Skylab manned program was completed successfully, and the first domestic satellite communications system was inaugurated.

Earth satellites

In the three general classes of earth satellites, communications, earth survey, and navigation, the spacecraft launched during the year generally continued to demonstrate improved performance. Through mid-1974 the United States, the Soviet Union, and Canada were the only countries that launched and operated such satellites.

Communications satellites. The U.S. Communications Satellite Corp. (Comsat) dominated the category of communications satellites during the year. Comsat operated a global space communications network for an 85-nation consortium, Intelsat. All Comsat satellites were launched into synchronous orbits about the equator. This meant that at an orbital altitude of 22,300 mi the satellite velocity matched the angular rate of the earth's rotation and thus remained fixed over one portion of the earth's surface.

In 1974 Comsat had five Intelsat 4 satellites in operation, three over the Atlantic Ocean and one each over the Pacific and Indian oceans. The first Comsat satellite, Early Bird, launched in 1965, had a capacity of 240 telephone circuits. By 1974 capacity had increased to 25,000 two-way telephone circuits or 60 color television channels. Comsat planned to launch three more Intelsat 4s and then follow them with more advanced designs. Ground facilities to service these satellites increased to 93 antennas at 80 ground stations in 59 member countries.

In 1974 Comsat inaugurated a service that increased the speed of transmission of digital data. Basic data language is expressed in "bits," with eight bits roughly equal to one letter in the alphabet. The new service makes possible transmission rates as high as 9,600 bits (or 280 words) per second. The cost of the new service, called Digisat, is substantially lower because the amount of time required to use a satellite is reduced.

In the fall of 1973 the U.S. Federal Communications Commission (FCC) opened the way for domestic U.S. satellite communications systems. Six separate requests of companies (or consortiums of companies) to establish satellite communications systems were approved — Western Union, Comsat/American Telephone and Telegraph Co. (AT&T), General Telephone and Electronics Corp., RCA, and American Satellite Corp. (which later withdrew because of financing problems). Several hundred millions of dollars of risk capital are involved. Ground stations and antennas were being built; advanced designs of communications satellites developed; and customer service contracts established. All satellites were to be launched by the U.S. National Aeronautics and Space Administration (NASA), which would be reimbursed by the private concerns.

RCA became the first private firm to provide domestic satellite service. It used the Canadian Anik II satellite system and ground relay stations in New York, San Francisco, and the Alaskan cities of Anchorage and Juneau. The state of Alaska benefited greatly, receiving improved telephone service at lower cost and live television programs for the first time. Within the U.S., RCA rented coast-to-coast telephone satellite circuits for $1,700 per month compared to $2,298 per month for private land-based lines. Thus, a three-hour football game via satellite transmission cost $3,500, much less than if it were transmitted by land lines. RCA planned to launch two satellites of its own in 1976.

Western Union expected to launch three domestic satellites in 1974; Comsat/AT&T one in 1974 and two in 1975. All of these companies were constructing ground stations to be in operation by 1975, the deadline set by the FCC. Whether there would be sufficient demand for this great increase in U.S. communications service remained a major question.

Other nations watched these developments and began considering the establishment of their own domestic satellite systems. The slow-moving Franco-West German Symphonie satellite was scheduled for launch late in 1974. Japan awarded contracts to Philco-Ford and General Electric to design and build experimental domestic communications satellites. Although China built three 98-ft-diameter antennas at Peking and Shanghai to link with communications satellites over the Indian and Pacific oceans, no plans for the development of a domestic system were known.

On the military side, the U.S. Navy was contracting for a global multi-purpose synchronous satellite system for the 1975–76 period. Called FleetSatCom, it would provide secure military communications to surface ships, patrol aircraft, and submarines. A portion of the system would service U.S. Air Force requirements for USAF strategic aircraft and airborne command posts. The U.S. Air Force separately planned a global satellite communications system.

The Soviet Union continued to operate a communications satellite system. Unlike the U.S. satellites, the Soviet craft flew in a 65° (inclined to the equator), highly elliptical orbit with a perigee (place in orbit closest to the earth) in the Southern Hemisphere. These satellites, called Molniyas, could be viewed by major Soviet ground terminals for about eight hours a day. This orbital path provided good reception at near north-polar latitudes, harder to reach from the U.S. equatorial orbit system. By spacing three Molniyas in 12-hour orbits, it was possible to obtain full 24-hour coverage. Late in 1973 the U.S.S.R. demonstrated a portable 22-ft-diameter antenna ground station, designed to transmit television broadcasts of Communist Party leader Leonid I. Brezhnev during his official visit to India.

In experimental work, NASA scheduled a launch of the Applications Technology Satellite (ATS-F), which would broadcast from synchronous orbit. To be used first in Alaska, the Rocky Mountain states, and Appalachia, its purpose was to investigate the feasibility of inexpensive community-type antennas in remote areas. Afterward, it was to be moved to Asia for experimental broadcasts to community antennas in several thousand villages in India. The educational programs for these broadcasts were to be supplied by the Indian government.

The European Space Research Organization (ESRO) was negotiating with European aerospace firms to develop a synchronous-orbit, maritime communications satellite, Marots. Scheduled to be launched in 1977, it was designed to provide ship-to-shore and shore-to-ship communications and data exchange. Belgium, France, Italy, Spain, the United Kingdom, and West Germany indicated their intent to participate.

Earth-survey satellites. Included in the category of earth-survey satellites are spacecraft employed for meteorological, earth resources, geodetic, and reconnaissance applications. Such satellites are designed to survey the earth with sensors, both photographic and electronic, in order to obtain and transmit data that can not be gained by other means.

Weather satellites. The U.S. and the Soviet Union continued daily global operations to obtain atmospheric and environmental data. In the U.S., the National Environmental Satellite Service (NESS) was responsible for conducting weather satellite operations. NESS is a major component of the National Oceanic and Atmospheric Administration (NOAA) of the U.S. Department of Commerce. All NOAA satellites are launched into polar orbit, providing global coverage as the earth rotates. Developmental work on advanced-design

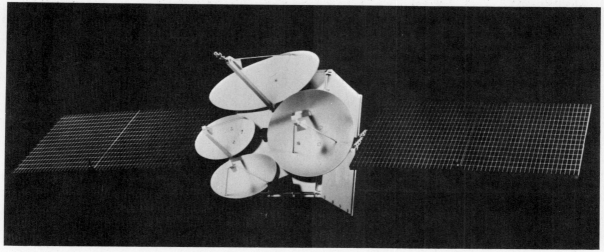

Communications spacecraft to be used by RCA for its U.S. domestic satellite system is scheduled to become operational by 1976. The high-capacity satellite will contain 24 transponders.

satellites (such as Nimbus) and all launchings are conducted by NASA.

On Nov. 6, 1973, NOAA-3, the second of a new class of environmental satellites, was placed in orbit. Along with the previously launched NOAA-2, it provided daily measurements of various phenomena on an almost global basis. Radiometers powered by solar cells provided high-resolution images of cloud cover, snow and ice, and the surface of the oceans, as well as information on temperature and moisture in the atmosphere. A solar proton monitor detected the arrival of energetic protons from the sun and measured their rates and distribution in polar regions. These latter data were transmitted to NOAA's Environmental Research Laboratories at Boulder, Colo. From there warnings were issued on impending solar storms that could influence weather and radio communications.

Later in 1974 NASA planned to launch two synchronous meteorological satellites (SMS). In synchronous orbit at longitude 70° and 135° W, these satellites were designed to "see" continuously all of North and South America and adjacent ocean areas. They are prototypes for a subsequent 1974 NOAA satellite, geostationary operational environmental satellite (GOES). This spacecraft will obtain data on the evolution of weather patterns and cloud motion by day and night, information previously obtainable only with radar. In addition, images will be provided over oceans and remote areas where no radar exists. Finally, GOES is expected to observe weather systems which radar cannot see, such as fog and stratus (nonprecipitating) clouds. Scanning coverage of nearly one-quarter of the earth's surface will be obtained

every 20 minutes. Including data processing, information can be transmitted to weather stations every 30 minutes.

GOES will transmit its data to a central processing facility near Washington, D.C., from which relevant portions of the total image will be extracted and sent by telephone lines to regional weather stations in San Francisco, Kansas City, Mo., and Miami. The system provides also for collection of significant environmental data from up to 10,000 sensing platforms in remotely located sites on the earth's surface.

NASA planned to launch Nimbus F, with advanced environmental sensors, in the summer of 1974. Among other capabilities, the satellite was to measure the winds in the lower stratosphere above the tropics. The purpose of this study is to improve knowledge of how atmospheric energy moves from the tropics to temperate zones.

Japan contracted for and began development of a geostationary meteorological satellite (GMS), scheduled to be launched in late 1976. The GMS was to utilize the visible infrared spin-scan radiometer (VISSR) developed for NASA's SMS. Images based on the data obtained by the radiometer would be transmitted to Japan every 20 minutes and alert meteorologists to potential storms and typhoons in the Far East. The GMS was to be stationed on the equator at longitude 140° E.

Looking ahead, scientists believed that continued observation, study, and understanding of global weather evolution and movement would enable forecasters to predict weather accurately two weeks in advance. Such knowledge would aid many elements of a nation's economy. A study in the state of Wisconsin, for example, shows that

171

such forecasts would be worth $20 to $35 million a year to farmers in relation to the hay crop alone.

Earth resources satellites. NASA's earth resources satellite, ERTS, after two years of operation has produced an immense quantity of data. In the more than 300 separate investigations utilizing these data, practical uses have most notably occurred in mineralogy, land use inventory, water impoundment assessment, and photo-map compilation.

After 10,000 orbits of the earth at an altitude of 565 mi ERTS had obtained more than 100,000 black-and-white pictures and photographs in strange artificial colors produced from the multispectral scanning system. All had been made available to investigators and the public. Overlapping photos are taken of areas covering 13,225 sq mi.

At NASA-sponsored meetings of investigators and potential users of ERTS data, more than 300 papers were presented. They described the value of such data for such purposes as monitoring urban development and planning future land use; estimating crop yields and taking inventories of

timber; discovering areas of air and water pollution; mapping strip-mine and forest-fire scars; exploring for minerals and petroleum; detecting landscape features that may someday help in predicting earthquakes; keeping watch on volcanoes; surveying the breeding ground of migratory waterfowl; studying flood hazards; and determining the distribution of marine life. Investigations by ERTS under way in the U.S. included the development of a national geological and soil feature map; a study of vegetation damage from highway construction in Maine; the obtaining of information on the dynamics of Lake Pontchartrain, La.; a study of land use in the great urban megalopolis stretching from Boston, Mass., to Richmond, Va.; and an inventory of timber resources in selected U.S. forests.

Among the many unexpected results of ERTS are that its data have helped scientists develop a better understanding of earth processes and weather dynamics. For example, photographs from the satellite showed smoke particles streaming from Illinois and Indiana steel mills along the southern shore of Lake Michigan. The smoke, mixed with moist air rising from the lake, provided nuclei for snow formation, which caused heavy snowfall along the shores of the lake. Acid waste dumps near the harbor of New York City were shown to be too extensive for normal dilution processes and were consequently polluting shore waters. ERTS photos of massive pollution in Lake Champlain by a New York paper mill provided key evidence in an antipollution suit instituted by the state of Vermont.

Based on ERTS data, land use maps have been updated by such states as Connecticut, Massachusetts, Maryland, Michigan, Minnesota, New York, Rhode Island, and Wisconsin. During the 1973 floods of the Mississippi River, ERTS data were used to estimate the amount of flooded acreage and evaluate flood control efforts. Another successful application was the estimation of crop acreage. Estimates of wheat acreage for the states of Kansas and Nebraska proved to be approximately 99% accurate when compared with "ground truth."

In other nations ERTS experiments proposed by foreign scientists included detection of potential locust breeding sites in Saudi Arabia, snow surveys to assess the risks of spring flooding in Norway, assessment of land use and soil erosion in Guatemala, determination of the hydrologic cycle of the Santa River basin in Peru, and surveys of winter monsoon clouds and snow cover in Japan. Brazil found ERTS photography of the Amazon River basin enormously valuable in charting impenetrable jungles never before accurately mapped.

Land use photograph of the earth was taken by ERTS satellite at an altitude of 568 miles. Gulf of California at the mouth of the Colorado River is black area at lower right, urban area is at upper left, farmland at upper right, and mountains at lower left.

Courtesy, NASA

Not only is ERTS data providing new information, but it also reduces greatly the cost of some routine surveys. For example, 800,000 ac of cotton-producing land in California are inventoried each year at critical times in an effort to control the infestation of pink bollworms. By using ERTS data surveyors were able to perform the task in 16 man-hours compared to 128 man-hours before ERTS, at the same 90% accuracy level. Rhode Island was mapped in eight days. The cost of the map, which showed land use, was estimated at less than one-tenth that of a similar map produced by other means. It is estimated that a survey map of a state the size of Iowa or Illinois made by medium-altitude, specially equipped aircraft would cost $1 million; if made by high-altitude aircraft, it would cost $200,000. But ERTS allows the task to be done for about $80,000. Because ERTS-1 continued to perform well, a scheduled launch of ERTS-2 was delayed until 1976.

Geodetic satellites. By simultaneous observation of a satellite from two or more points on the earth's surface, scientists can determine with great accuracy the distance between those points. Both laser and Doppler techniques are employed.

By utilizing a U.S. Navy navigation satellite scientists determined the precise locations of oil-drilling platforms in the Gulf of Mexico. These platforms, 40 and 100 mi from the Louisiana coast and thus out of sight, were located within an accuracy of 6 ft. A Navy-built geodetic receiver, weighing less than 100 lb, used Doppler transmissions from the satellite. These data, plus precise observation time measurements and precise orbital data of satellite path, permitted calculation of the location of the observation point at each platform.

Reconnaissance satellites. The U.S. and the Soviet Union continued their reconnaissance satellite programs, but also continued to release no details. It was presumed, however, that the spacecraft involved in these programs utilized extremely high-resolution photographic and narrow-band electromagnetic radiation surveillance techniques. Since the early 1960s both countries have regularly launched satellites in low orbits for military observation. One type of Soviet surveillance satellite apparently has a measure of maneuverability, enabling a small change in orbital period. During the Arab-Israeli conflict in 1973 there was indication of specific Soviet satellite coverage of the Middle East combat zone.

In addition to the probable continual satellite inspection by the U.S. and the Soviet Union of ballistic missile launch facilities in the other country, there also appeared to be an increased emphasis on ocean surveillance. Tracking of ships is feasible not only by photography but by infrared detectors. The relative temperature difference of a ship compared with the surrounding water and engine combustion exhaust gases is presumed visible even at night. Furthermore ship wakes tend to be distinctive in pattern. Such satellite intelligence data may be supplemented by long-range surveillance aircraft.

Another surveillance satellite used by the U.S. is the Midas. This satellite system, in synchronous orbit over the Indian Ocean, provides early warning of ballistic missile launches by monitoring infrared radiation that they generate. It is presumed that a ballistic missile attack would thus be discovered at the time of launch.

Navigation satellites. The U.S. Navy Transit system continued to be fully operational and in use by the U.S. fleet throughout the world. Although commercial ships were permitted to use this system, it required a shipborne computer to calculate position, based on the Doppler shift of the satellite's signal. Thus, it was too expensive for widespread civilian use. Future commercial navigation seemed more likely to depend upon a system wherein a signal is transmitted from a ship via satellite to a shore-based computer; there, position data can be determined and broadcast back to the ship.

Quite separately, the U.S. Air Force proposed a global system of four synchronous satellites to provide position-fixing and communications for military and commercial aircraft. The U.S. Navy favors a different, greatly improved, Transit satellite system known as Timation. It would require 27 medium-altitude satellites in polar orbit for global coverage. The proposed Navy satellites would incorporate extremely precise timing systems utilizing crystal oscillators. An even more advanced satellite design would use an atomic clock.

—F. C. Durant III

Manned space exploration

During the last year, manned space exploration was confined to earth orbital missions. The United States completed the Skylab program with manned missions of 28, 59, and 84 days duration. The Soviet Union launched Soyuz 12 for a short two-day flight to test spacecraft modifications that had been dictated by malfunctions occurring in previous missions. The U.S. and the Soviet Union continued the necessary research and development efforts for the joint Apollo/Soyuz Test Project. The U.S. also proceeded with the early development stages of its space shuttle program. A joint Western European team, the European Space Research Organization (ESRO), assumed

Astronaut Charles "Pete" Conrad walks in space to retrieve the solar telescope film canisters on Skylab 2. Conrad and two other U.S. astronauts spent 28 days in space on the mission.

responsibility for the design, testing, and fabrication of a large spacelab that was to be carried into orbit by the space shuttle orbiter spacecraft. The development of this spacelab appeared to presage a new era of international cooperation in the exploration and utilization of the space environment for useful purposes to mankind.

Skylab. The fourth major U.S. manned space program, Skylab was designed to provide a manned orbiting laboratory that would perform scientific investigations in such areas as earth resources, space technology, materials processing, solar astronomy, and medical sciences. The project utilized modified spacecraft, launch vehicles, and equipment developed for the Apollo lunar landing program. A large orbital workshop was the crew's living quarters and main experiment laboratory. This workshop was constructed from the Saturn IVB, the upper stage of the Saturn V launch vehicle, which was 21.7 ft in diameter, 48.5 ft long, and 78,000 lb in weight. The workshop contained two floors and all the necessary consumable goods to sustain the crew. The crew living quarters, food galley, toilet, collapsible shower, entertainment equipment, and medical experiments hardware were located on one floor. The upper level contained some experiment equipment, food storage lockers, drinking water

supplies, and other mission equipment stowage. Large external panels containing solar cells were deployed in orbit to convert solar energy into electrical power to operate various systems.

Apollo command and service modules were used to transport the Skylab crewmen to the orbital workshop cluster and return them back to earth. The command modules were launched by Saturn IB booster rockets.

Launched on July 28, 1973, the second manned Skylab mission, Skylab 3, was commanded by Alan L. Bean with Jack R. Lousma as the pilot and Owen K. Garriott as the science pilot. Approximately eight hours after launch, the crew piloted the command module to a rendezvous and docking with the orbital Skylab workshop. During the flight problems were encountered with the attitude control propulsion system of the command module. If this system had failed, the crew would not have been able to use the spacecraft for return to earth. The ground team immediately prepared a backup command module and launch vehicle for a possible rescue of the Skylab 3 crewmen; however, the problem was isolated and it was determined that a rescue would not be required.

During the mission, three space walks were completed to install and retrieve experiment film, to deploy another thermal shield in order to provide better insulation to the damaged workshop, and to replace faulty rate gyroscopes used in the workshop attitude control system. The crew operated all planned experiments, and a large quantity of scientific information was acquired. On the 59th day of the mission, the crew entered the command module, separated it from the workshop, and reentered the earth's atmosphere to splash down in the Pacific Ocean 225 mi SW of San Diego, Calif. The crewmen were recovered in excellent physical condition. The recommendation of the previous Skylab crew to increase the amount of exercise aided in maintaining crew health in Skylab 3.

Skylab 4, the third and last manned Skylab mission, was commanded by Gerald P. Carr, with William R. Pogue as the pilot and Edward G. Gibson as the science pilot. The command module was launched into orbit on Nov. 16, 1973, following a six-day delay because of problems with the launch vehicle. Additional provisions were launched with the crewmen to allow them to extend the planned mission of 56 days for an additional 28 days. The crew carried out an extensive scientific program and accomplished the additional task of observing and photographing Comet Kohoutek. Although the comet did not appear bright to observers on earth, the crew was able to observe it and obtain valuable scientific knowledge with instruments and cameras on board the workshop.

During the flight the crew completed four space walks to perform tasks essential to the mission. At the end of the flight and before the crew left the orbital workshop, material samples and other equipment were assembled into a collection bag and left for possible retrieval by future space flight crews. The crew also used the command module propulsion system to fly the orbital workshop to a higher orbital altitude (280 mi) in order to extend the expected orbital life of the unmanned orbiting workshop approximately ten years.

On the 84th day of the mission, the crew flew the command module back to earth to land in the Pacific about 170 mi SW of San Diego. The crew returned in excellent physical condition and showed no apparent medical problems following this longest manned space flight.

Skylab science program. The Skylab experiment program encompassed the fields of medicine, earth resources, science, technology, and engineering. Approximately 75 major experiments were flown in this program for participating investigators in the U.S. and other countries. In addition, a number of experiments devised by high school students were selected and flown.

The basic objectives of the Skylab medical experiments were the study of man during extended space flight and after his return to earth. A mineral balance study was conducted to measure precisely the input and output of electrolytes and other body chemicals in order to quantify their rate of change during weightless flight. To accomplish this experiment, the crew members consumed a fixed diet and samples of their urine and feces were collected and returned to earth for analysis. An experiment was also conducted to analyze body fluids (urine and blood) to determine if hormonal or other changes occurred in the space environment. The cardiovascular system was studied to determine if the functions of the heart, arteries, and veins were altered during the mission or following return to earth. A bicycle ergometer was used as one device to study changes in heart rate, blood pressure, metabolic gas exchange, and the electrical activity of the heart muscles. Equipment was flown to study the functions of the vestibular system to determine if alterations occurred in the susceptibility to motion sickness or in a crewman's ability to make spatial judgments. The sleep characteristics of the crew were also studied to determine if the quantity and quality of sleep changed during space flight. Studies were also completed to determine if changes in bone mineral levels occurred as a result of extended space flight.

The preliminary results of this comprehensive series of medical tests and measurements showed

Vasily Lazarev (left), commander of the mission, and design engineer Oleg Makarov orbit the earth in September 1973 in Soyuz 12, the first Soviet manned flight since three crewmen died in Soyuz 11 in 1971.

that man can work and live in the space environment for up to approximately three months without any major problems. Based on the data obtained from Skylab, it would appear that man can fly on much longer missions.

The earth resources experiments were designed to use the Skylab workshop as a manned orbital laboratory in order to obtain information from various instruments and photographs to study the earth. A team of investigators from 32 states of the U.S. and 19 other countries took part in these studies. The instruments included multispectral and earth terrain cameras, an infrared spectrometer to measure the effects of atmospheric attenuation and radiation from the earth's features, a multispectral scanner to measure and record energy emitted and reflected by earth features in the visible and infrared bands, a radiometer/scatterometer and altimeter to gain information on features of the earth's terrain and water bodies, and an L-Band radiometer to map the earth's thermal radiation and surface temperatures. To demonstrate the usefulness of earth resources measurements from space, the information obtained from these sensors was to be evaluated and compared with information collected on the earth at specific test sites. The investigations included agriculture and crop inventories, geology mapping and ex-

175

Skylab space station is seen against the cloud-covered earth in photograph taken by astronauts in the Skylab 4 command module. The missing left solar panel was torn off during the launch of the station.

ploration, water resources mapping and utilization, sea currents and water temperatures, and atmospheric studies.

Vast quantities of data on the earth were obtained by the Skylab crewmen for evaluation by the science team. The results from these and other resources studies may provide new methods and techniques for the management and control of earth resources and provide a better understanding of the earth's total ecology.

A solar telescope was used by the Skylab crewmen to obtain photographs and to measure the sun without the interference of the earth's atmosphere. A white coronagraph was used to measure brightness, form, and polarization of the solar corona. An X-ray spectrograph obtained X-ray photographs of solar flares and other activities. Instruments also were used to obtain ultraviolet photographs and information on solar activity.

Apollo/Soyuz Test Project. In July 1975 the United States and the Soviet Union planned to conduct the first international manned space flight. This program resulted from studies started in 1970 on the design of docking devices that would allow spacecraft from the two countries to be coupled in space. The ultimate objective is the development of an international space rescue capability. Through the use of this common docking device, crewmen in a disabled spacecraft could be retrieved by a rescue spacecraft. Formal approval for the joint flight program was achieved in 1972 when U.S. Pres. Richard Nixon and Soviet Premier Aleksey N. Kosygin signed a joint agreement for cooperation in peaceful exploration of space. Since that time, engineers and scientists have held numerous technical meetings in the Soviet Union and the U.S. They reached agreement on the design of an airlock module, which the U.S. began building. This module is to be used by the crewmen to dock the Soyuz and Apollo spacecraft and will permit the crewmen to transfer to either spacecraft. During the last year, detailed operational plans including the launch date and time, spacecraft trajectories, control center procedures, and other details for this mission were approved. In addition, the Soviet and U.S. astronauts started training for the flight.

The mission was to consist of the rendezvous and docking of the two spacecraft and a period of joint operations during which crewmen would visit in the other nation's spacecraft. In addition, a limited series of experiments was to be conducted both jointly and separately by the two crews. Scientists from the two nations established joint science studies that would be carried out by the astronaut teams. The U.S. crew was to be commanded by Thomas P. Stafford, a veteran of both Gemini and Apollo flights. Slated as the docking module pilot was Donald K. Slayton, making his first flight into space, and the pilot for the mission was to be another space rookie, Vance D. Brand. The two-man Soviet crew was to be commanded by Aleksey A. Leonov and would also include Valery N. Kubasov, both space veterans.

The Soviet Union was scheduled to launch its spacecraft on July 15, 1975, from Tyuratam launch site. During the first day, the Soyuz spacecraft would be flown into an orbit 230 mi in altitude. Approximately 7½ hr after the Soyuz launch, the Apollo command module would be launched from the Kennedy Space Center in Florida. It would then be maneuvered by the crew into an orbit to rendezvous with the Soyuz spacecraft. About two days after the launches, the two spacecraft were to dock. The docking module would be activated, and the U.S. astronauts would first visit the Soviet spacecraft. Docked operations would then proceed for two days during which time a Soviet crew member would visit the Apollo command module. Joint experiments and radio/television coverage were to be provided during the docked operations. Upon completion of the joint operations, the two spacecraft would separate for their return journeys. During the entire operation ground control teams in both countries were to be in constant communication so as to direct the mission jointly.

Space shuttle. During the year the U.S. continued development work on the space shuttle program. This project aimed at the development and operation of a relatively low-cost reusable space transportation system. The major components in the program were to be the orbiter, solid rocket boosters, and spacelab. The orbiter would be a winged vehicle designed to transport the crew and passengers into space and return them to earth. It was to be about 125 ft long and have a wing span of 78 ft and a tail height of 57 ft. The solid rocket boosters would launch the orbiter to an altitude of approximately 27 mi above the earth, after which they would fall away into the ocean for recovery and reuse. At this point in the trajectory the orbiter propulsion system would propel the spacecraft into earth orbit for missions ranging from 7 to 30 days. After a mission, the orbiter would be flown by the crew to reenter the earth's atmosphere and would land on a long airplane runway. The first manned flight of the orbiter was scheduled for 1979.

In the last year, a cooperative agreement was reached between the U.S. and ESRO for the design and construction of a pressurized manned spacelab and for orbiting unmanned platforms to use with the orbiter. The spacelab was to be a pressurized module capable of accommodating typical laboratory and observatory equipment. Along with the spacelab, the unmanned platforms would be carried into orbit in the orbiter payload bay and when deployed would support such items as telescopes, instruments, cameras, and other equipment that might be needed to complete studies in the space environment. Under the agreement reached, nine European countries would provide the resources for the joint development of these

systems. The first operational flight of spacelab was scheduled for 1980.

Soviet manned space flight. On Sept. 27, 1973, the U.S.S.R. launched a two-man crew in the Soyuz 12 spacecraft. This was the first manned mission since June 1971 when the flight of Soyuz 11 ended in the tragic death of the three crewmen because of an air leak in one of the spacecraft's pressure relief valves. The Soyuz 12 flight crew consisted of Vasily G. Lazarev and Oleg G. Makarov. The mission lasted approximately two days and was an apparent test flight to evaluate spacecraft design changes made as a result of Soyuz 11 problems. Television pictures of the crew indicated that space suits were worn by the crew during certain mission phases of the mission as opposed to the unprotected flight suits worn by previous crewmen. The space suits would provide protection to the crew in the event that spacecraft pressure was lost as in Soyuz 11. The mission ended on Sept. 29, 1973, after completing 31 orbits at orbital altitudes of 202–214 mi above the earth. The spacecraft landed in the Kazakhstan Steppe, about 250 mi from the launching site at Baikonur.

Soyuz 13 was launched by the Soviet Union on Dec. 18, 1973. Pyotr I. Klimuk and Valentin V. Lebedev orbited the earth for eight days at altitudes ranging from about 140 to 170 mi. During the flight the cosmonauts grew nutritive protein in a special device designed to provide space-grown nutrients for space travelers. Other experiments undertaken during the mission included extensive tests of the spacecraft's flight control and automatic navigation systems, surveys of the earth's surface, and studies of the stars in the ultraviolet range of the spectrum. Soyuz 13 returned to the earth and landed in Kazakhstan on December 26.

—Richard S. Johnston

Space probes

As 1973 drew to an end, interplanetary space seemed literally filled with scientific probes on their way to one or another of the sun's satellites. Pioneers 10 and 11 were en route from the United States to Jupiter, while four Soviet Mars vehicles were bound toward the planet for which they were named. The U.S. Mariner 10 was headed first for Venus and then for Mercury. Data on the sun's activity came from two probes in highly eccentric orbits around the earth: the Soviet Prognoz 3 and the U.S. Explorer 50, the latter the tenth and last in a series of probes known as IMP (Interplanetary Monitoring Platform).

Probing Mars. The four Soviet probes, Mars 4, 5, 6, and 7, were launched in July and August 1973.

Their ambitious mission included the landing on the planet's surface of two instrumented packages that would transmit information to two satellites in orbit about the planet. These, in turn, would relay the data to the earth. By Jan. 9, 1974, the probes were 7.5 million mi from the earth and transmitting data about the cosmic rays and the solar wind in the region between the earth and Mars.

Mars 4 and 5 reached their target on February 10 and 12, but misfortune struck at once. The braking motor of Mars 4 did not fire on command, and the probe shot past the planet after approaching within 1,320 mi of its surface. Two days later, however, Mars 5 successfully entered Martian orbit, circling it with an apocenter of 19,500 mi and a pericenter of 1,056 mi. It had a period of 25 hours, and the orbit was inclined at an angle of 35°. Mars 4 was reprogrammed to send back data on interplanetary space as it sped away from Mars.

On March 12 Mars 6 ejected a landing capsule at an altitude of approximately 3,400 mi above Mars. The capsule entered the tenuous atmosphere at a very low angle to minimize drag on the descending vehicle. Its two parachutes functioned properly, and the probe telemetered information to the orbiting Mars 6. As the capsule plunged through the thin atmosphere of Mars, it radioed that there was several times more water vapor in the atmosphere, in that region, than had been previously

thought. However, the probe ceased transmitting before it reached the surface of Mars. Scientists in the U.S.S.R. theorized that high winds near the surface may have caused the failure. Mars 7, the sister probe, also had bad luck. On March 9 its landing capsule separated from the main body, but it missed the planet by about 780 mi.

Despite the failures, several things were learned about Mars. It is surrounded by an envelope of hydrogen reaching to a distance of about 12,000 mi and having a temperature of approximately 350° K. There are "broad oscillations" in both the pressure and water content of the atmosphere on the surface of the planet, and traces of ozone were found in the atmosphere.

The Soviet setbacks with their Martian landers caused U.S. designers to reconsider the Viking probe, which was scheduled to be launched in 1975. The Viking faced problems of its own during the year. Engineers had trouble with the probe's biology instrumentation and the all-important on-board computer. Both resulted in cost overruns at a time when money for space was scarce.

Flights to Jupiter. Pioneer 10, launched on March 2, 1972, from Cape Kennedy, rendezvoused with Jupiter on Dec. 3, 1973. The 570-lb probe, carrying only 65 lb of scientific instruments, survived its first major threat in mid-1973 when it passed through the potentially hazardous asteroid

Trajectory of Pioneer 11 is plotted so that the probe will pass on the sunrise side of Jupiter and use the planet's gravitational field to aid in setting a course for Saturn.

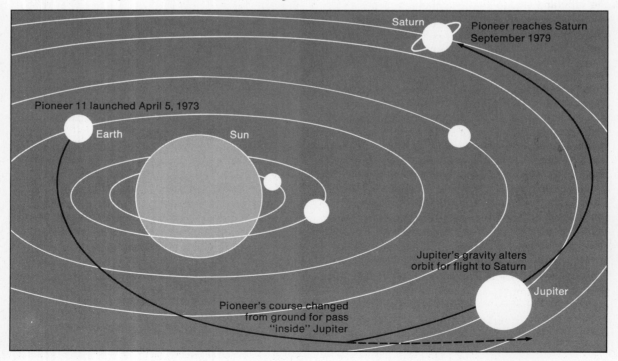

belt circling the sun. Pioneer 10 overcame its second major threat only a few hours before its nearest approach to Jupiter. Engineers at NASA's Ames Research Center, Mountain View, Calif., which managed the project, were not sure whether the probe could withstand the intense radiation in the belt surrounding Jupiter. But aside from a 5° rise in the temperature of the radioisotope thermoelectric generators, which supplied electrical power for the probe, Pioneer 10 performed well. It survived a radiation dose a thousand times greater than that required to kill a human being. At the point of closest approach to the planet, 2.86 Jupiter radii, electrical instruments onboard Pioneer 10 were 95% saturated by the intense radiation flux, and the probe's asteroid-meteoroid detector was seriously damaged. Engineers at the Ames center speculated that if the probe had approached only 0.5 Jupiter radius closer to the planet it would not have survived.

Pioneer 10 first encountered the Jovian environment on Nov. 26, 1973, when it crossed the bow shock wave formed where the planet's magnetosphere meets the interplanetary magnetic field. This occurred at a distance of about 108 Jupiter radii from the planet. Since the earth's magnetosphere (on the side facing the sun) extends to a distance of only 13 earth radii, that of Jupiter is great indeed. The Jovian magnetosphere proved to be extremely complex when compared with that of the earth. It apparently has two distinct regions. Out to a distance of 20 Jupiter radii the strength of the dipolar field is about 4 gauss, or some eight times greater than that of earth. Beyond that lies a non-dipolar field. Pioneer 10's magnetometer also confirmed that the polarity of Jupiter's field is opposite to that of the earth's, with respect to rotation.

With respect to the atmosphere of Jupiter, Pioneer 10 made the significant discovery that helium is a constituent, along with the previously noted hydrogen, ammonia, methane, and, in lesser amounts, deuterium and acetylene. Helium cannot be detected by earth-based instrumentation. Astronomers speculated that because of its low temperature, Jupiter's present atmosphere may well be almost primordial and that closer study of it may reveal to us more about the formation and evolution of the solar system.

The probe's infrared radiometer for the first time measured the temperature of the dark side of the planet, again something that cannot be done from earth. It found that there was no appreciable difference in temperature between it and the sunlit side. The same instrument also proved that the planet emits approximately twice the amount of energy it receives from the sun. This knowledge indicates that the additional heat may be derived from the planet's interior.

Pioneer 10 also provided photographs of the planet with a resolution five times better than any that could be taken from the earth. In addition, the probe provided the first pictures of the phases of Jupiter. Since the planet always presents the same face to the earth, its phases cannot be seen. The imaging photo-polarimeter of the craft measured the red and blue light reflected by the planet. Preliminary pictures showed more browns and grays in Jupiter's turbulent atmosphere than seen in previous years. The yellow seen in 1971 in the South Equatorial Belt had diminished considerably. Other pictures showed that the North Tropical zone was becoming narrower and fainter, while the South Tropical zone, in which is found the mysterious Great Red Spot, was becoming wider and more clearly defined.

The pictures did little to solve the mystery of the Great Red Spot. While it is prominent in many of the photographs, little new information or details were revealed other than a somewhat darker core and a darker border than seen from the earth.

The versatile scientific probe also sent back new data on some of Jupiter's 12 moons. Io appears to have a density of 3.5 g per cc, a value some 20% greater than previously estimated. The mass of the satellite is about that of the earth's moon. Pioneer 10 also sent data indicating that Jupiter's moons decrease in density the farther they are from the mother planet.

The satellite Io also has a tenuous atmosphere. Earthbound astronomers had long believed this from observing Io when it was eclipsed by Jupiter. The S-band radio transmitter on Pioneer 10 sent signals that were refracted slightly as the craft occulted Io, proving the existence of the atmosphere.

Assuming that the probe's electric power supply does not deteriorate at a greater rate than it has over the past two years, Pioneer 10 should continue sending information on the interplanetary medium until it reaches the orbit of Uranus in 1979. In 1987 it will cross the orbit of Pluto, more than 3,600,000,000 mi from the sun, and leave the solar system, the first man-made object to do so.

Also on a trajectory for Jupiter was Pioneer 11. Differing from Pioneer 10 only in the inclusion of a second magnetometer that can measure magnetic fields as intense as one million gammas, it successfully traversed the asteroid belt on March 20, 1974. Four days later it received commands that would alter its course drastically. The new path would take Pioneer 11 around the leading edge (sunrise side) of Jupiter, back across the solar system at an angle of about 15° to the ecliptic plane (the region within which almost all the planets and asteroids orbit the sun), and swing it past Saturn about Sept. 5, 1979. On its new path the probe would pass within about 325 million mi of the sun and arch over the asteroid belt it had earlier threaded. Pioneer 11 was expected to arrive at Jupiter on Dec. 5, 1974. Information provided by the probe was expected to be of great value in planning the 1977 Mariner Jupiter-Saturn mission.

Venus and Mercury missions. Mariner 10 proved itself the equal of Pioneer 10. Launched on Nov. 3, 1973, the 1,108-lb spacecraft undertook the most ambitious missions of any U.S. probe to date. It was to investigate both the planet Venus and then the innermost satellite of the sun, Mercury, about which little was known.

The probe contained seven major scientific experiments, which accounted for 175 lb of its weight. These included twin television cameras, an infrared radiometer, two ultraviolet spectrometers, two magnetometers, a charged particle telescope, a scanning electron spectrometer, a scanning electrostatic analyzer, and an X-band radio transmitter. The total was twice the number of instruments carried by Mariner 9, and the probe was built in part from components left over from that vehicle.

On February 5 Mariner 10 encountered Venus. As the probe passed above the planet's surface at an altitude of about 3,600 mi, its twin television cameras took the first close-up, high-resolution pictures of Venus' thick, turbulent cloud cover. These pictures, more than three thousand in number, revealed that the cloud cover is highly structured and is constantly agitated by convective cells and divergent streams from the planet's equatorial zone. They also disclosed a large high-pressure region along the planet's equator, where the sun's heat causes convection currents that divide the primary stream and send portions spiraling outward to the planet's poles.

Pictures taken at intervals of 24 hours indicated that the clouds over Venus' equator are moving at almost the same speed as the jet streams in the earth's atmosphere: 220 mph. Thus, the atmosphere of Venus has a rotational velocity of about four days, approximately 60 times greater than that of the planet itself. With Mariner 10's pictures in hand, astronomers found that the atmosphere above lat 50°–60° N and S rotates much faster than that above the equatorial region. The pictures also showed that there are at least three distinct layers in the atmosphere. A multiple haze layer is evident about 44 mi above the surface. A more clearly defined layer exists at about 37-mi altitude, and a third layer ranges between 18 and 30 mi above the surface. The indications are that the Venusian atmosphere is far denser than that of the earth. At an altitude of 37 mi, it is thought to be as dense as is the earth's at only 8 mi.

Other instruments aboard Mariner 10 telemetered back equally significant data. The magnetometers and the charged particle telescope proved that Venus has no detectable magnetic field surrounding it. Unlike the earth, which has a strong magnetic field that creates a bow shock wave toward the sun and diverts the solar wind around the planet, Venus experiences the full force of the solar wind, undoubtedly adding energy to the planetary disturbances that churn up its atmosphere.

The two radio transmitters aboard beamed their signals toward Venus to measure its mass, gravitational field, and atmospheric density. They showed that the planet was considerably more spherical than the earth. One of the ultraviolet spectrometers detected a cloud of atomic hydrogen surrounding the planet, and helium, atomic oxygen, and atomic carbon were found along with large amounts of carbon monoxide in the upper regions of the atmosphere. The charged particle telescope confirmed that there is little or no radiation emanating from Venus.

After sweeping past Venus, Mariner 10 headed for Mercury. On March 16, its engine was fired for 51 sec to correct the probe's course for the innermost of the sun's satellites. As it approached Mercury, the spacecraft began taking its first pictures. It reached its closest point to the planet on March 29, only 460 mi from the surface. The television cameras and non-pictorial instruments began their observation of Mercury six days before that date and continued it until five days afterward.

Superficially, Mercury looked amazingly like the earth's moon. Conspicuous in a picture taken from a distance of 124,000 mi was a large crater containing a smaller and obviously more recently made crater, with rays extending from it like those emanating from craters on the moon. Close-up pictures of the far side of the planet showed it to be incredibly rough and cratered. Other photographs indicated features such as maria, rilles, and a variety of craters, some with sharp edges, others with eroded rims, and still others with central peaks similar to those on the moon. Indeed, the photographs revealed a greater variety of craters on Mercury than appears on the moon.

The dual magnetometers of Mariner 10 recorded a weak magnetic field around the planet, 90 to 100 gammas at the point of closest approach. This value compares to the 30,000 gammas at the earth's equator. The same instruments also detected a detached bow shock wave where the magnetic field deflects the solar wind around the planet.

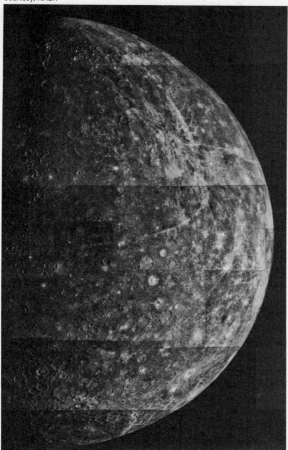

*Photomosaic of the cratered surface of Mercury
was constructed of 18 photographs taken at 42-second
intervals by Mariner 10 six hours after it flew
past the planet on March 29.*

By combining data from the radio science experiments with some celestial mechanics, scientists were able to compute a more accurate value for the mass of Mercury. It is 1/6,023,600 that of the sun. By comparison the earth's mass is 1/332,958 that of the sun; thus, the earth is some 20 times greater in mass than Mercury. The radio science experiments also indicated that there was no atmosphere or ionosphere on either the night side or the day side of Mercury.

Temperatures ranging between 370° F on the day side of the planet and −280° F on the night side were reported by the probe's infrared radiometer. However, scientists speculated that the temperature on the day side could range between 560° and 800° F, depending on the distance of the planet from the sun.

On April 2 Mariner 10's cameras were shut off to save power. Earlier, the temperature in the equipment compartment of the probe had risen to 108°

F, only 12° from the maximum upper limit, and the craft's main electrical boost regulator had begun drawing increased power from the solar cells. Despite these difficulties, mission controllers were hopeful that the probe would "stay alive" for another six months, when after circling the sun it would return to Mercury. On the return voyage it was targeted to pass over the planet's south pole.

Future probes. The amazing successes of Pioneer 10 and Mariner 10 heartened planners of future U.S. probes. At a conference in Washington, D.C., in May, NASA officials discussed the scientific and technical feasibility of launching an unmanned probe to Mars in cooperation with the Soviet Union. To be launched in 1981, it would travel to that planet, scoop up samples of the soil, and return to the earth with them. But the estimated cost of $1 billion made it unlikely that the probe would take place.

As for future explorations of Venus, indications during the year seemed to be that the Venera probe series, which the Soviets had used to explore the planet in the past, had come to an end. Western observers seemed inclined to agree that the U.S.S.R. probably was in the process of designing a second-generation probe for Venus.

As 1974 began, NASA revealed that it was seeking the cooperation of Europe and the U.S.S.R. between that year and 1991 to realize its program of launching 57 scientific probes during the period. Robert S. Kraemer, director of NASA's planetary programs, said that such cooperation would be needed to undertake one-fourth to one-third of the missions so planned. Kraemer and six other NASA officials toured Europe in February looking for such assistance. Despite enthusiasm by some European space scientists for such projects, neither the European Space Research Organization nor the individual countries had the funds to commit for such joint ventures.

—Mitchell R. Sharpe

Astronomy

During the past year, astronomers' interest in the solar system remained at a high level, stimulated by the increasingly sophisticated use of interplanetary probes. Their primary concerns, however, continued to be stellar astronomy, high-energy astrophysics, and cosmology.

Comet Kohoutek. Probably the most intensively studied comet in history, Comet Kohoutek was rewarding to professional astronomers, despite the disappointment of many amateur viewers who had expected to see a far more spectacular sight. The comet was discovered by the Czechoslovakian-

Comet Kohoutek is photographed by the 42-in. Schmidt telescope at Catalina Observatory on (top to bottom) November 28, November 29, and December 6. Top and center photographs show how the division in the tail has moved away from the head, possibly the result of interaction with the solar wind.

born astronomer Lubos Kohoutek at Hamburg Observatory in West Germany on a photograph taken March 7, 1973. Soon afterward, he found a prediscovery image on a plate taken January 28. As the sixth cometary discovery of that year, it was officially designated 1973f. At the time of these first observations, made with the 31.5-in. Schmidt telescope at Hamburg, Comet Kohoutek was a very faint (16th-magnitude) tailless object, moving very slowly in the constellation Hydra.

The first calculations of its orbit showed that the comet was still more than four astronomical units distant from the sun (1 astronomical unit = approximately 93 million mi) and would not come closest to the sun (perihelion) until more than ten months later, when it would be a much brighter object. Therefore, astronomers had time to prepare for cooperative observations and to build equipment for studying the comet. The U.S. National Aeronautics and Space Administration (NASA) set up a special office to coordinate investigations of Kohoutek, and erected a cometary observatory on South Baldy Mountain, New Mexico. This station, operated jointly with the New Mexico Institute of Mining and Technology, received a 14-in. Schmidt camera, very effective equipment for studying comet structure.

The path of Comet Kohoutek across the sky caused it to vanish into the evening twilight from early May until late September 1973, when it emerged into the predawn morning sky, having brightened to about magnitude 10 or 11. Naked-eye visibility was attained in late November, and by December 18 favorably situated observers could see a tail 15° long.

For several days around the date of perihelion passage, Dec. 28, 1973, Comet Kohoutek was again hidden in the sun's glare. Theoretically, the comet should have been most brilliant at this time, some informed predictions of peak magnitude being about −3 (intermediate between Venus and Jupiter in brightness). However, the comet was about 2° from the solar disk as it passed the sun on the far side, at a distance of 13 million mi, and so was not visible from the earth's surface.

During this interval, the only successful sightings were made from spacecraft outside the earth's atmosphere. Only 11 hours before perihelion, Comet Kohoutek was photographed as a brilliant blob superimposed on the solar corona, by the OSO-7 (Orbiting Solar Observatory) satellite. On December 29, the day after perihelion, the comet was seen by Skylab astronauts Edward Gibson and Gerald Carr during a space walk 272 mi above the earth's surface.

Beginning about Jan. 1, 1974, the comet again became visible to ground observers, this time in the western evening sky as a bright naked-eye object that very quickly faded during the next weeks. In mid-January, the tail was as long as 25° on photographs taken with the South Baldy Schmidt telescope. Meanwhile, 1973f was traveling rapidly outward from the sun, as it crossed the orbits of Mercury, Venus, and the earth on January 4, 18, and 29, respectively. By mid-1974, the comet was still observable in large telescopes as an inconspicuous patch in the northern sky.

Comet Kohoutek, when near the sun, was surrounded by an enormous halo of glowing hydrogen gas, detectable only in ultraviolet light. This feature was photographed on Jan. 7, 1974, with an image converter flown on an Aerobee rocket from White Sands Proving Grounds, New Mexico. The bright inner part of this halo, out to 4° from the comet's nucleus, appeared to be spherically symmetrical, while the faint outer part was elongated in a direction away from the sun. On January 17 this same hydrogen envelope was observed by sensors aboard the Mariner 10 spacecraft, then on its way to Venus. The hydrogen glow could be traced as far as about 12° from the nucleus in the direction opposite the sun.

Similar huge, tenuous hydrogen clouds had been observed by satellites around Comet Tago-Sato-Kosaka (1969g) and around Comet Bennett (1969i). Apparently, these appendages are normal features of large well-developed comets.

Radio astronomers eagerly applied to Kohoutek the microwave techniques so successfully used in recent years for discovering interstellar molecules. French observers, employing the large radio telescope at Nançay, detected evidence indicating the presence of the OH (hydroxyl) radical. This same evidence was studied by Barry Turner with the 140-ft antenna of the National Radio Astronomy Observatory at Green Bank, W.Va. He concluded that the OH lines arise from an extended halo around the nucleus. The CH radical was observed as an emission line at the 9-cm wavelength by radio astronomers at Harvard and Smithsonian observatories with an 84-ft antenna. Particularly significant was the detection of two emission lines of methyl cyanide (CH_3CN)—the most complex molecule yet found in a comet—by B. L. Ulich and Edward Conklin with the 36-ft radio telescope on Kitt Peak in Arizona.

Measurements of the comet's infrared radiation were conducted during December and January at the University of Minnesota, with a 30-in. reflecting telescope, at many wavelengths up to 18.5 microns (1 micron = 0.0001 cm). Since the daytime sky is quite dark at infrared wavelengths, the comet (though optically unseen) could be mapped in broad daylight. By this means it was found that,

in addition to a broad fan-shaped dust tail pointing away from the sun, there was a narrow antitail directed sunward. (The antitail was seen by the Skylab astronauts and later was widely photographed.) The spectral-intensity curves for both the coma and the tail show a conspicuous peak near the ten-micron wavelength, which is believed to indicate silicate dust grains. (The same peak had been detected in Comet Bennett, 1969i.) On the other hand, this silicate peak was not shown by the antitail. If the dust throughout the comet is of uniform composition, these findings indicate that the particles in the coma and tail are micron-size or smaller, whereas those in the antitail are large as compared with ten microns. These larger grains are matter expelled from the comet and concentrated in the plane of its orbit.

A significant development was the identification of a number of red emission lines in the comet's tail, recorded at Lick Observatory in California, Asiago Observatory in Italy, and Wise Observatory in Israel. Canadian scientists Gerhard Herzberg and Hin Lew found that these previously unexplained lines are produced by positive ions of water vapor. This discovery tends to confirm the "dirty snowball" model of a comet proposed by Fred L. Whipple in 1950, in which coma and tail consist of dust and gases released by surface evaporation of a small nucleus of dust-contaminated ices and solidified gases.

The orbit followed by Comet Kohoutek since its discovery was derived chiefly from calculations by Brian Marsden at the Smithsonian Astrophysical Observatory. On the comet's initial approach to the sun, when it was still so distant that perturbations by the planets were insignificant, the orbit was a very elongate ellipse, with a period of several million years. Near the date of perihelion, the effect of the planets' gravitational attractions was to speed up the comet enough for its orbit to become slightly hyperbolic. Later, after the comet had receded far enough from the solar system for planetary perturbations to become inappreciable again, the orbit would again be an ellipse with a period of approximately 75,000 years.

Planetary probes. Radar observations from the earth and the Mariner 10 flyby dramatically upgraded astronomers' knowledge of the cloud-covered planet Venus. Radar mapping of Venus' solid surface was carried on mainly at Arecibo Observatory in Puerto Rico and the Jet Propulsion Laboratory (JPL) in California. Arecibo produced its first radar map of Venus in 1968, which showed the surface to be somewhat less rough than that of the moon. In August 1973, JPL radar astronomers announced that detailed charting of a 900-mi-wide area in the equatorial region of Venus had revealed a dozen large, shallow craters. The largest, 100 mi in diameter, is only about a quarter of a mile deep. The others range from 20 to 60 mi across. This area of the planet appeared to be as crater-infested as the moon or Mercury.

The JPL observations were obtained at the Goldstone Tracking Station in the Mojave Desert with two radio telescopes, one 210 ft in diameter and the other, 85 ft. The two were located 14 mi apart and were operated jointly as an interferometer. The outgoing radio signals (wavelength 12.6 cm) were transmitted from the larger dish, and the returning echoes were received by both. In this way, resolution of surface features 6 mi in diameter could be achieved and depth differences of 600 ft could be measured.

The Mariner 10 spacecraft, which was launched Nov. 3, 1973, flew past Venus on Feb. 5, 1974, at a distance of only 3,600 mi above its solid surface. Perhaps the most significant results came from Mariner's two television cameras, which took 3,400 frames of ultraviolet photographs. These pictures resolved the Venus cloud patterns into fine detail and revealed for the first time the general pattern of atmospheric circulation. In the equatorial zone large areas of light and dark material were seen to move westward at about 300 ft /sec relative to the planet's surface. In the middle latitudes, north and south, conspicuous streaks move from the equatorial region obliquely poleward, merging into a bright polar ring at latitudes 50° N and S. This pattern is symmetrical with regard to the two hemispheres. No seasonal effects are expected because the planet's equator and orbital plane nearly coincide.

High-resolution photographs of the planet's limb (outer edge of its apparent disk) show clearly the presence of several haze layers lying above the smooth top of the cloud deck. A highly stratified structure of the upper atmosphere was also revealed by Mariner 10's occultation experiment. The craft carried radio transmitters whose signals were monitored on the earth as Mariner passed behind Venus. This observation provided a vertical temperature profile of the Venus atmosphere, indicating four distinct temperature inversions at levels between 35 and 39 mi above the surface. The top inversion could mark the transition between the stable atmosphere below and the convective region above. The temperature at the top of the cloud deck was found to be about −18° C, according to scans made by Mariner's infrared radiometer, which was working at a wavelength of 45 microns.

If Venus has any magnetic field, it is less than about 0.0005 that of the earth. This result from Mariner 10's magnetometer confirmed the measurement made by the Soviet spacecraft Venera 4 in 1967. Therefore, the surface of Venus is not magnetically protected against the solar wind of charged particles rushing at high speed outward from the sun. This lack of shielding is in accord with the narrowness of the wake reaching outward from Venus through the interplanetary plasma, as monitored by Mariner's plasma experiment. Remarkably, this wake could be traced for about 700 Venus radii downstream.

Analysis of the Doppler shifts in the radio-tracking data transmitted by Mariner gives an accurate determination of the mass of Venus as one part in 408,523.9 (±1.2) of the sun's mass. The gravitational field of the planet appears to be highly symmetrical, suggesting that Venus is more nearly spherical than the earth.

On March 29, 1974, Mariner 10 flew past the planet Mercury at a distance of only about 750 km (460 mi). Between March 23 and April 3, more than 2,000 pictures of Mercury were obtained with the craft's twin television cameras, which recorded half the planet's surface.

These pictures showed for the first time a heavily cratered surface, much like the moon's. The craters range from fresh, sharp-rimmed specimens to battered, low-walled rings of great age. Some craters have smooth floors, as if once flooded by lava. No evidence of atmospheric erosion was detected, though a tenuous atmosphere consisting mainly of helium, neon, and argon was observed.

Preliminary analysis of the television pictures hinted that Mercury's outer layers are probably low-density silicate rocks. If so, the mean density of the planet (5.5 times that of water) indicates that it has a very dense interior, perhaps a large core of iron.

Gamma-ray bursts. In June 1973, scientists at the Los Alamos Scientific Laboratory in New Mexico announced that the earth's atmosphere is occasionally sprayed by intense bursts of soft gamma radiation, lasting a few seconds and coming from outside the solar system. A series of these bursts, which consisted of high-energy photons with wavelengths of less than 0.01 Å, was recorded in 1969–72 by four Vela satellites, launched by the U.S. Department of Defense to monitor violations of the nuclear test-ban treaty of 1963. The four spacecraft fly in a circular orbit of about 75,000 mi radius, and each carries six cesium iodide scintillation counters that form a nondirectional array for detecting energetic photons.

The spacecraft detected 16 bursts, each of which was recorded practically simultaneously by two or more of the satellites. The small differences in arrival times at the several satellites indicated

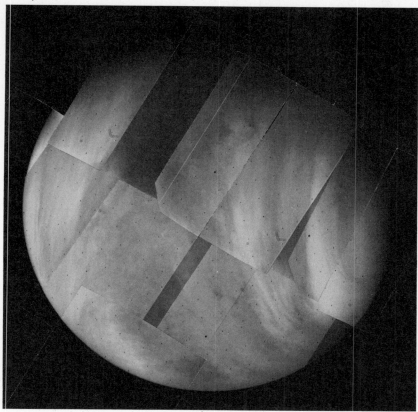

The nearly full Venus, covered by a thick layer of clouds, is seen in a mosaic of television pictures taken by Mariner 10 from a distance of about 440,000 miles. The cloud patterns indicate the general circulation pattern of the planet's upper atmosphere.

roughly the direction from which the gamma-ray burst came. In this way it was demonstrated that these bursts did not originate from the earth, sun, or planets, but from outside the solar system.

By late 1973, the total number of gamma-ray events recognized by the Los Alamos group had risen to 23. Astrophysicists took a deep interest in this remarkable phenomenon and advanced several possible explanations of its cause. In 1968, Stirling Colgate had predicted that a star exploding as a supernova should emit a flash of X rays and gamma rays. However, this did not seem to be the same phenomenon detected by the Vela satellites, for the dates and directions of the observed bursts did not seem to be correlated with known supernovas.

Floyd Stecker and Kenneth Frost advanced the suggestion that the gamma-ray bursts came from stars undergoing "superflares," analogous to but enormously more energetic than the familiar short-lived localized flares on the sun. A serious objection to this idea is that a superflare would cause a concurrent conspicuous flash of the star at microwave and optical frequencies, and this has not been observed.

In an alternative theory, Jonathan Grindlay and Giovanni Fazio proposed that such a gamma-ray

burst is what should be expected from an interstellar dust grain speeding toward the sun with nearly the speed of light. If such a particle is made of iron and has a radius of one millimeter, electrostatic forces would break it up into a spray of partially ionized iron atoms when it comes within 300 astronomical units of the sun. These fast-moving atoms, on interacting with photons of sunlight, should produce an intense burst of gamma radiation.

Grindlay and Fazio speculated that pulsars (now generally believed to be rapidly rotating neutron stars) may be the origin of the high-speed interstellar iron grains. In particular, a pair of gamma-ray bursts were observed by Vela satellites on Oct. 17, 1969, just 18 days after a significant change in the period of the radio pulses from the Crab Nebula pulsar was noted. Both bursts came roughly from the direction of the Crab Nebula, and 18 days is plausible for the difference in travel time to the sun between an interstellar dust particle and the radio waves.

Planets of Barnard's star? The nearest star to the earth except for the triple system Alpha Centauri is Barnard's star, only 5.9 light-years distant from the sun. This 10th-magnitude red dwarf has a large proper motion, 10.3 sec. of arc per annum,

185

which has been known since 1916. (Proper motion is that component of a star's motion in space that is at right angles to the line of sight, so that it constitutes the apparent change of position of the star on the celestial sphere.)

Interest in Barnard's star was heightened by Peter van de Kamp's discovery in 1963 that its proper motion shows a very slight periodic waviness, as if influenced by an unseen orbiting companion. In 1968, van de Kamp announced, on the basis of measurements of more than 3,000 photographs taken with the 24-in. Sproul Observatory refractor, that this unseen companion moved around the star in an elongated 25-year orbit. Since the deduced mass of the companion was only about 1.7 times that of Jupiter, it could properly be called a planet. Alternately, he noted, the observations could be reconciled with two planets having periods of 26 and 12 years. These findings were widely interpreted as a strong indication that planetary systems must be numerous in our galaxy, if not only the sun but also the second nearest star to it has one.

By 1974, however, it appeared that van de Kamp misinterpreted his original observations. A careful independent study of the motion of Barnard's star was made by George Gatewood and Heinrich Eichhorn, using 241 photographic plates taken over a 55-year period with the 30-in. refractor of Allegheny Observatory and the 20-in. refractor of Van Vleck Observatory. Though the measurements were conducted with extreme refinement, they failed to show any indication of a waviness in the proper motion while confirming the Sproul Observatory results in every other particular. If Gatewood and Eichhorn are correct, it would follow that Barnard's star is single after all and that the slight waviness originated in minute systematic errors of some kind.

Interstellar communications. In 1973, a number of radio astronomers in the U.S. and the Soviet Union continued their efforts to detect possible radio transmissions from extraterrestrial civilizations at stellar distances. The advocates of this search maintain that many dwarf stars in our galaxy presumably have planetary systems resembling ours and that many of these planets harbor life. They further argue that such abodes of extraterrestrial life are so numerous that some of them must have developed technologically advanced civilizations. These astronomers conjecture that attempts of such a civilization to communicate with others would be by radio signals at a wavelength near 21 cm (a spectral region of prime interest to radio astronomers). Project Ozma in 1960 was a first, unsuccessful effort by U.S. astronomers to detect signals of this kind.

Since then, enormous improvements in the sensitivity of radio telescopes have encouraged further researches. In 1973, Gerrit Verschuur announced that two years earlier he had monitored ten nearby dwarf stars, including Tau Ceti, 61 Cygni, and Barnard's star. He observed with the 300-ft and 140-ft steerable antennas of the U.S. National Radio Astronomy Observatory, listening near the 21-cm wavelength. His results were negative. Verschuur stated that he would have been able to detect a transmitter on a planet belonging to Tau Ceti or Epsilon Eridani, if six megawatts of power or more were being beamed earthward.

A much more extensive search for artificial signals of extraterrestrial origin was mounted by Benjamin Zuckerman and P. E. Palmer in 1973, using the same two radio telescopes as had Verschuur. They scanned approximately 500 stars, believed to include all the likely candidates for life-bearing planetary systems within 80 light-years of the sun. A preliminary analysis of the observations yielded no hint of artificial signals. As of 1974, the simplest explanations of the negative results by both U.S. and Soviet investigators are that either technologically advanced civilizations are relatively rare in space or that some of the hypotheses guiding the searches are invalid.

—Joseph Ashbrook

Atmospheric sciences

A growing realization among scientists and the general public is that weather and climate play crucial roles in the most pressing problems facing people throughout the world. The availability of adequate food depends on the character of the weather during the growing season. Widespread droughts over the grain belts lead to submarginal diets and starvation for large segments of the world's population. Flooding in some instances can have equally serious effects. The consumption of energy for heating and cooling buildings depends to a significant degree on the temperature and wind. An abnormally warm 1973–74 winter over most of the U.S. played an important role in alleviating the consequences of a fuel shortage.

Despite the increasing recognition of the societal influences of weather and climate, there has not been an appropriate increase in investment in the relevant research and development.

Global weather programs. During the summer of 1974, the Global Atmospheric Research Program (GARP), a massive international research effort, took a major step toward realization of its goals—to develop a mathematical model of the earth's atmosphere that would make it possible to predict the weather up to two weeks in advance and to study factors governing climate. In order to attain these objectives much more needs to be learned about the effects on the earth's atmosphere of air-sea interactions and the role played by tropical clouds in transporting energy and water vapor. For this reason, during the period of June 15–Sept. 30, 1974, about 30 nations participated in the GARP Atlantic Tropical Experiment (GATE). It involved a detailed study of the tropical region between latitudes 10° S and 20° N and extending across the Atlantic Ocean from the west coast of the Americas to the east coast of Africa. The aim of the experiment was to study the development of all types of tropical storms and their interactions with the atmosphere as a whole.

During the course of GATE many different devices were used to observe the atmosphere and the oceans. Instruments were carried on buoys, ships, airplanes, and balloons. The region was scanned by a number of weather satellites. The sheer magnitude of the effort, which was under the overall direction of Joachim P. Kuettner of the U.S. and an international group of scientists working out of Bracknell, Eng., caused the experiment to be regarded as a major achievement. It was the culmination of approximately five years of planning, and it laid the groundwork for the first global experiment, scheduled for the late 1970s.

Research on climate. One of the most significant developments in the atmospheric sciences over the last few decades has been the realization that the climate of the earth changes appreciably over periods of a few decades or less. This recognition has led to studies of the physical and dynamical characteristics of climate and, more recently, to research on mathematical models of

global climate. Such models attempt to take into account, in a realistic fashion, all of the crucial factors, such as incoming solar radiation and outgoing infrared radiation, the water and land distributions over the planet, the gaseous and particulate constituents in the atmosphere, the character and extent of clouds, the snow and ice cover, and the interactions of air and sea.

One of the reasons for developing a mathematical model of climate is that it would make possible the prediction of future changes in climate. Such information is of vital importance in planning programs concerned with food, water, and energy. Thus, in late November 1973, the U.S. National Weather Service, recognizing the need for all available information on expected weather during the winter season, issued publicly for the first time an experimental forecast of average temperatures over the subsequent three months. The forecast, which later proved substantially correct, indicated above-normal temperatures for the central, northeastern, and southwestern parts of the U.S. and below-normal readings for the west coast and northwestern states.

Once a satisfactory model of the earth's climate has been developed, it should be possible to evaluate the effects of human activity. Of particular interest is the role played by emissions into the atmosphere—carbon dioxide, and the particles and heat released by energy-producing and -consuming processes. M. I. Budyko in the Soviet Union and others have been concerned that if the amounts of heat released into the atmosphere continue to increase at the expected rates, they may reach levels where they can significantly modify the climate. The questions of the possible effects of emissions into the atmosphere have

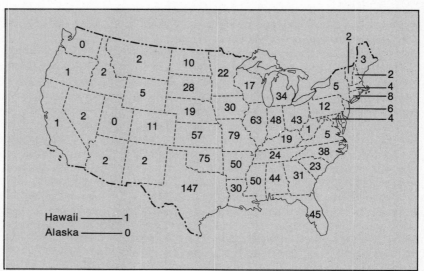

A record number of tornadoes, 1,107, broke out across the U.S. in 1973, the previous high being 929 in 1967.

Courtesy, NOAA

become increasingly urgent because of steps being taken in response to fuel shortages and the drastic increases in fuel costs. Controls on industry and motor-vehicle emissions have been relaxed to allow the more efficient use of available fuels. The consequences of such actions are greater quantities of atmospheric pollution.

In a recent review Lester Machta of the National Oceanic and Atmospheric Administration (NOAA) noted that the carbon dioxide concentration in the atmosphere continues to increase at a rate of about one part per million (ppm) per year and should reach 330 ppm by the end of 1975. The increase can be attributed to the combustion of fossil fuels. According to Machta's summary the data on atmospheric turbidity are mixed, but he expresses the view that there has been "a very widespread increase of dust north of about 30° N; second, forecasts of future dust loadings are, at best, educated guesses; and third we must know more about the optical properties of the particles before the scientific community will ever agree on the sign of the temperature change resulting from an increase in tropospheric dustiness." It has been shown by others that whether dust leads to warming or cooling also depends on the reflection properties of the earth's surface.

Violent weather. During the calendar year 1973 there were more tornadoes reported in the United States than ever before. The record number of 1,107 caused at least 87 deaths. According to reports by the National Weather Service there have been about 640 tornadoes and 125 deaths in an average year during the period from about 1954 to 1969. The relatively small number of fatalities in 1973 can be attributed, at least in part, to better weather services, public awareness, and community preparedness. However, on April 3–4, 1974, a series of tornadoes that ripped through the Middle West and South caused more than 300 deaths and property damage in excess of $1 billion. Ohio, Kentucky, and Indiana were the hardest hit states in the nation's worst tornado disaster since 1925.

When considering the frequencies of tornadoes, it should be recognized that tornado observations and recording procedures leave a great deal to be desired. Tornadoes are small, short-lived, violent storms that are difficult to predict and observe. They usually are less than a few hundred meters in diameter, last no more than a few minutes, and are very destructive and frightening. Few are observed by instruments, their occurrences and locations usually being reported by people who happen to be in the vicinity of the storm. As the population has increased, the likelihood of storm detection has increased, and as communication systems have improved, it would seem reasonable that the number of reported storms should also increase. Therefore, a comparison of tornado frequencies from one period to another should be made with caution. Nevertheless, the 1973 statistics probably do indicate that there were many more tornadoes than in previous recent years. Important questions that remain to be answered are why so many occurred in 1973 and is the number likely to remain high in years to come.

In August 1973 Rep. Larry Winn, Jr. (Rep., Kan.), convinced that it is possible to reduce the hazards of tornadoes by an increased federal investment in research and development on tornado detection, prediction, and warning, introduced a bill to that effect in the House of Representatives. Hearings were held in Washington, D.C., and Kansas City, Mo., but little hope was held for passage in 1974. In the meantime, various groups, particularly in NOAA and several universities, were working on the development of improved techniques for tornado detection. Emphasis was placed on the use of Doppler radar, which should be able to measure the high wind speeds associated with tornadoes. More widespread use was also being made of instruments called spherics receivers, which detect the unusual electrical activity in tornadic storms.

Weather satellites. On Nov. 6, 1973, a weather satellite, NOAA-3, was successfully launched into a polar orbit around the earth. It carried a variety of radiometers for observing the character of clouds, the extent of snow cover, and the temperature of the atmosphere. A device called the Vertical Temperature Profile Radiometer made continuous broadcasts of its measurements. Ground stations under the satellite's orbit were able to obtain the data as the vehicle flew over them. Several countries, including France and Norway, constructed stations for receiving the data.

Cloud and temperature measurements from the NOAA weather satellites by 1974 were routinely incorporated into analyses of weather conditions over the Northern Hemisphere. Plans for extending this procedure over the Southern Hemisphere were under way. Such satellite observations were expected to be particularly valuable because of the vast ocean expanses and the small density of observing stations over the land masses.

The geostationary Applications Technology Satellites 1 and 3, which for a number of years were stationed over the mid-Pacific Ocean and Brazil, respectively, have shown that such spacecraft yield much valuable meteorological data. (A geostationary, or synchronous, satellite is one placed in such an orbit that it will always remain above the same area on the earth's surface.) In about 20 minutes, the spin-scan cameras on these satellites have observed cloud patterns over half the earth.

Solar forecaster keeps a watch on the sun and studies its effect on the earth by monitoring a roomful of electronic equipment at the Space Environment Services Center in Boulder, Colo. Particular attention is paid to solar flares.

Smaller regions, for example the tropical Atlantic, could be examined at much shorter time intervals when a hurricane was developing. (*See also* Year in Review: ASTRONAUTICS AND SPACE EXPLORATION, *Earth Satellites.*)

Weather modification. Weather modification research in the U.S. during 1974 was substantially reduced from that in earlier years. In fiscal 1973 federal expenditures for weather modification amounted to about $20 million, while in fiscal 1974 they were about $17 million. This reduction occurred in spite of a report from the National Academy of Sciences in 1973 calling for increases in expenditures. Particularly hard hit by the reduction in funding was research on precipitation modification and research on the mitigation of damage by violent storms.

The National Hail Research Experiment during the spring and summer of 1973 completed its second field season of testing a much-heralded Soviet scheme for suppressing the fall of damaging hail. Three more years of observation and testing were planned, after which the researchers hoped a decision could be reached on the effectiveness of suppressing hail damage by seeding clouds with ice particles. The program, which in 1974 operated at a greatly reduced budget, was expected to shed a great deal of light on the physical nature of hailstorms as well as on the efficacy of hailstorm modification.

The effect of cloud seeding on snowfall was being tested over the San Juan Mountains of southwestern Colorado in a five-year project being supported by the U.S. Bureau of Reclamation. The program was designed to be a comprehensive evaluation of earlier procedures developed by scientists at Colorado State University. An analysis of the techniques used in the project was expected to be completed by about 1975.

The Bureau of Reclamation was also assigned responsibility for carrying out a major research effort over the high plains along the eastern slopes of the Rocky Mountains in order to ascertain the degree to which rainfall can be increased from summer clouds. Archie M. Kahan, chief of the bureau's Division of Atmospheric Water Resources Management, expected that the project would be conducted over a period of five to seven years at a cost of $15 million to $20 million. Joanne Simpson of NOAA completed a series of cloud seeding experiments in Florida during the summer of 1973. These tests, part of a program started in 1970, were expected to yield a great deal of information on the physical properties of tropical, convective clouds and the effects of seeding with heavy doses of silver iodide particles on the clouds themselves and on the rain they produce.

Mathematical modeling of clouds. Significant advances have been made in recent years in the development of mathematical models of convective clouds. Notable contributions have come from groups led by Y. Ogura at the University of Illinois, Harold D. Orville at the South Dakota School of Mines and Technology, and E. F. Danielsen at the National Center for Atmospheric Research. The last two groups developed models of a hailstorm that take into account many of the dynamical and microphysical aspects of these complicated disturbances. There are still major deficiencies in elucidating the conversion of water to ice within the

189

storm, but the researchers hoped to overcome them in the next few years.

Relatively simple models of convective clouds have been used successfully by various investigators in the study of techniques to increase rainfall. Realistic hailstorm models should make it possible to evaluate procedures designed to reduce the fall of damaging hail.

Air quality. During 1974 the energy crisis pushed concern about air pollution into the background, but it seemed likely that this would be only a temporary state of affairs. Given prolonged, stable atmospheric conditions and high levels of emissions from motor vehicles and industry, problems of maintaining air quality could be expected to arise.

Scientists at NOAA established a network of stations in the U.S. and various island locations in the Pacific to monitor important atmospheric constituents, such as carbon dioxide, ozone, nitrogen oxides, and sulfur oxides. William E. Cobb reported the results of indirect measurements of particles in the air over the ocean areas. He found that over the North Pacific, downwind of Japan and Asia, and over the North Atlantic, downwind from the U.S., there are extensive plumes containing high concentrations of particles. They can be attributed to heavy industrial activity. A third major plume of particles extending southward over the Indian Ocean is thought to be caused by natural dust sources. Cobb concluded that pollutants generated by man have not affected the atmosphere over most of the global oceans. These results are quite important in evaluations of the effects of environmental pollution on the earth's climate.

Debates over development of supersonic transport airplanes focused attention on the chemistry of the stratosphere and of the ozone layer, which is centered at an altitude of about 20–25 km (12–15.5 mi). Michael B. McElroy and his associates at Harvard University confirmed earlier results showing that the introduction of nitric oxide in sufficient concentrations could lead to significant decreases in the ozone concentration. It was predicted that a supersonic fleet of 320 Concordes (or presumably Soviet Tu-144s) operating seven hours a day at an altitude of 17 km (10.5 mi) would lead to a decrease of 1% in the column density of ozone.

Ozone is a very important constituent of the atmosphere because it is a strong absorber of ultraviolet radiation from the sun. A reduction of ozone is expected to result in increased ultraviolet radiation reaching the ground and a consequent increase in sunburns and skin cancers. Recently, the NOAA Air Resources Laboratory established a network of stations to measure the incoming ul-

traviolet radiation, ozone, and atmospheric particulates. This data will be correlated with the incidence of skin cancer.

The Climate Impact Assessment Program of the U.S. Department of Transportation, NOAA, and various university groups were involved in the study of the stratosphere. Measurements of the concentration of nitric oxide at altitudes between 15 and 30 km (between 9 and 18.5 mi) were being made by means of equipment carried aloft by high-altitude balloons. A flight measuring the maximum natural concentrations found them to be lower than had been expected on theoretical grounds. General conclusions must await the completion of more extensive analyses.

—Louis J. Battan

Behavioral sciences

Considerable emphasis in the behavioral sciences continued to be placed on their practical application to social problems. The insights of anthropology found increasing use in such fields as politics and economics. At the same time, debate continued over the use of behavior modification techniques and the relevance of psychological findings to education.

Anthropology

An event conferring special honor upon herself and her profession was the election of Margaret Mead to the presidency (for 1975) of the American Association for the Advancement of Science. She was one of the very few anthropologists to hold the post in the 126-year history of the association, which spans all scientific disciplines.

Mead was also chosen to write the lead article in the second *Annual Review of Anthropology* (1973), in which she addressed herself to "Changing Styles of Anthropological Work." One of the article's major concerns was the unity of the field in the face of an increasing diversity of subfields. Mead noted that when 80 colleagues, variously selected from many anthropological specialties, were asked to name the five most important books produced in the last five years, only four books were mentioned more than twice. She disagreed with the suggestion that this indicated that "nothing of very great importance has happened in the last five years"; rather, she argued that a "great deal has happened, but no consensus can be reached because of the extraordinary diversity within the subject."

One could not help sharing Mead's nostalgia for the days of Boas and Kroeber, when it was possible to maintain a sense of the full scope of anthropology without being lost under a torrent of publications. It was encouraging, however, in contemplating these "Babe Ruths" of the discipline, to be able to hail the publication of *The Interpretation of Cultures* (1973) by Clifford Geertz, who probably came closest to being American anthropology's "Hank Aaron."

Geertz's place among anthropologists of his generation was unique; he was one of the most highly admired and esteemed scholars not only within his own profession but outside it. In the 17-year period during which the 15 articles in *The Interpretation of Cultures* were written, he produced over 40 notable publications, 5 of them seminal books. Geertz's approach to the study of culture was well set forth in the opening essay in *The Interpretation of Cultures,* "Thick Description: Toward an Interpretive Theory of Culture" (the only article written especially for the volume), in which he attempted to set forth his views on the matters discussed in the rest of the book. For Geertz, "culture is not a power, something to which social events, behavior, institutions, or processes can be causally attributed; it is context, something within which they can be intelligibly—that is, thickly—described." While not uninterested in theory, Geertz believed that the best of anthropology comes from its intimate relation with data: "What generality it contrives to achieve grows out of the delicacy of its distinctions, not the sweep of its generalizations."

For the most part, the other 14 essays in the volume reflected the rich context of Geertz's fieldwork in Indonesia (1952–54, 1957–58, and 1971), though some space was devoted to recent research in Morocco (1964, 1965–66, 1968–69, and 1972). Geertz managed with rare skill to achieve the aim he held out for anthropology, *i.e.,* "to draw large conclusions from small, but densely textured facts; to support broad assertions about the role of culture in the construction of collective life by engraving them exactly with complex specifics." Though Geertz was by no means representative of the only theoretical trend within anthropology, his emphasis on fieldwork undoubtedly was central to the discipline as a whole.

Growth. The continued expansion of the field of anthropology was certainly one of the important facts of the year. According to the *Guide to Departments of Anthropology 1973–74,* 25 new departments of anthropology and 15 museums were added during the period. Three hundred and eleven new professionals (96 women and 215 men) from 68 universities were indicated as having completed dissertations in 1972 and 1973.

A majority of these were presumably on the job market, but there was at least some superficial evidence that they might have been more fortunate than those new anthropologists who sought employment in 1972–73. At the 1973 annual meeting of the American Anthropological Association (AAA) in New Orleans, La., 502 job applicants sought 198 listed positions, a 16% increase in positions and only a 9% increase in applicants over the previous year. However, this might merely reflect the fact that more positions were openly listed at the 1973 meetings than previously, following an open employment referendum initiated by more than 400 fellows and voting members and accepted by 85% of those voting in May 1973. Compared with the last time the AAA met in New Orleans, in 1969, attendance (2,907) was up 62% and formal participants had increased by 82%.

Extrapolations of current statistics indicated a diminishing demand for anthropologists as well as other professionals through the 1980s. Because anthropology had grown very rapidly in recent years, its academic population was younger than that of most other disciplines. As a result, the anthropological job market would benefit less from retirement in the next 20 years than would be true of most other disciplines.

The continued shrinkage of academic positions and the increasing production of Ph.D.s in anthropology led to increasing interest in employment opportunities for those anthropologists in "applied" fields outside academia. Despite the fact that a recent survey of anthropologists revealed that only 5% were employed outside the academic community and museums, applied anthropology was given status equal to that of social anthropology, archaeology, physical anthropology, and linguistic anthropology in the 1973 reorganization of the AAA's major publication, the *American Anthropologist.* Clearly it was anticipated that the number of anthropologists working with interdisciplinary teams, administrators, and politicians would increase.

Applied anthropology had languished in the two decades following World War II, when there was a rapid expansion in the number of academic positions in the field. However, the movement of anthropologists from applied fields at that time was also accompanied by disillusionment with government, whence many of the "applied" projects emerged. In particular, many anthropologists felt that their skills would be or had been used as a cosmetic cover for oppressive dictatorships or to subvert in other ways the best interests of the people among whom they worked.

Political and economic anthropology. The importance of the anthropological contribution to the study of development received strong endorsement from the political scientist T. V. Sathyamurthy (*Current Anthropology,* December 1973). Sathyamurthy wrote the article following fieldwork in Uganda, during which he found that an awareness of anthropological studies both assisted his work with politics and led him to question the applicability of many assumptions and theories from orthodox political science. He became aware that "political scientists tend to treat political behaviour and political values of the people of developing societies without reference to what is peculiar" in those societies. In contrast, he asserted that good research in political science requires "an increased sensitivity to what is traditional, regional, local, tribal, provincial." This greater sensitivity can "be achieved by a discriminating use of anthropological approaches."

Sathyamurthy's article was subjected to close critical review by 18 anthropologists, to whom Sathyamurthy made a careful rejoinder. The process exemplified one of the features of anthropology he most admired: "animated debate leading to reinterpretation of existing work, refashioning of prevailing tools, and attempts to achieve quantum jumps in our understanding of society." One of the important conclusions emerging from the discus-

Adult male mandrill demonstrates large canine teeth, rivaling those of the great cats. These teeth are not needed for obtaining food and seem to have developed to support the role of the male as the aggressor and protector in primate societies.

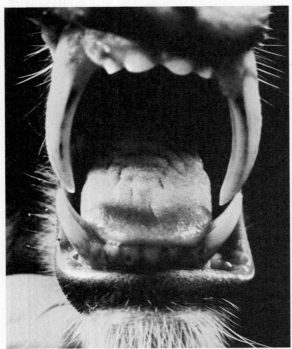

American Scientist

sion was the view that anthropology needs to expand its horizons beyond microstudies to deal with the national and international considerations that can impinge upon even the most apparently isolated village.

A similar relationship existed between anthropology and economics, although in this case it was an anthropologist, Harold K. Schneider (*Economic Man: The Economics of Anthropology,* 1974), who reached across the boundary between the two disciplines. Schneider's book was the first by an anthropologist to deal systematically with economics, deliberately moving into territory that many anthropologists had considered off limits for several decades.

In essence, Schneider's effort closely paralleled the central concern of the institutional economist, Gunnar Myrdal, in his earlier massive inquiry into the poverty of nations. Myrdal stated his goal as the demonstration of "the inadequacy of our inherited economic theories and concepts," while at the same time "replacing conventional theories and concepts by other, new ones better fitted to the reality of these countries." He believed that economists needed not only to establish mechanisms that could explain the unique properties of these economies, "but also to build an analytical structure fitted to the dynamic problems of development" (*Asian Drama,* 1968, p. 27).

Myrdal has been criticized by anthropologists for taking the potential contribution of anthropology too lightly. Myrdal himself anticipated that criticism by observing, in effect, that other social sciences often take potshots at economic formulations but never in a way that is systematic enough to be meaningful. Schneider made an important beginning toward that task, for he drew on the concepts and theories of microeconomics but with the selectivity of an anthropologist. No doubt he would be criticized by economists, but the issue had been joined.

Physical anthropology. Somewhat analogous developments were taking place in other subfields of anthropology as anthropologists reacted creatively with adjacent disciplines. The appearance of *Methods and Theories of Anthropological Genetics* (1973), edited by M. H. Crawford and P. L. Workman, represented one such interaction. In 22 chapters, including an overview by James Spuhler, geneticists and anthropologists discussed the methods and theories arising from their common concern with the study of variation within and among human populations. The book might be regarded as a summing up of two decades of research marked by several significant advances in method and theory. These derived both from sophisticated computer applications and from the isolation of a large number of genetic markers in the blood that are useful in characterizing gene pools.

The continuing debates over the interpretation of the fossil record of human evolution were influenced not only by newly discovered fossils but also by recent biochemical comparisons between man and primates. In two publications aimed at a general audience (S. L. Washburn and E. R. McCown, THE NEW SCIENCE OF HUMAN EVOLUTION, Feature Article in the *1974 Britannica Yearbook of Science and the Future,* and S. L. Washburn and R. Moore, *Ape into Man: A Study of Human Evolution,* 1973), Sherwood Washburn supported the hypothesis of Allan Wilson and Vincent Sarich that man and the chimpanzee are very closely related and possibly shared a common ancestor as recently as five million to ten million years ago. In support of his case, Washburn cited the extensive biochemical similarities between apes and man.

On the other hand, a continually mounting pile of fossil evidence from Lake Rudolf, Kenya, and from the lower Omo basin of Ethiopia demonstrated the existence four million to five million years ago of hominids with a very progressive cranial structure. This seemed to suggest that a separation of man from the great apes must have occurred much earlier than Washburn hypothesized. Just how much earlier remained cloudy, however, for the argument was not only one of the fossil record versus the biochemical evidence, it was also a debate about the methodologies involved in the interpretation of both types of evidence.

—Raymond Lee Owens

Psychology

Progress in psychology is more readily identified by broad trends than by discrete events. Although it may be true that discrete events are ultimately responsible for the advance of the science, their identification at or shortly after the time of their occurrence would require a great deal more in the way of prescience than is possessed by most observers. This review accordingly focuses on some of the more prominent trends in psychology that have developed over the past few years, and that seem to offer the most promise for future progress in the discipline.

Simulation of mental disease. Psychologists in recent years have become appreciably more active in research on pressing social problems. For example, Philip Zimbardo of Stanford University paid unselected volunteers to serve as "captives" or "guards" in a simulated prison situation; even within the relatively short time span of this experi-

ment remarkably severe behavioral aberrations, mainly relating to "depersonalization," were produced in the subjects, suggesting the potent behavioral effects that real imprisonment must have.

Much interest was generated by the report of another Stanford psychologist who carried out a different kind of simulation. D. L. Rosenhan reported the results of the deliberate self-commitment by eight individuals into 12 mental hospitals. In each case the pseudopatient stated that he had been suffering from auditory hallucinations and requested admission. After the pseudopatients were admitted, they behaved "normally," insisting that they were no longer disordered. The major finding of this research was the remarkable tenacity with which the hospital personnel clung to their original diagnosis (typically schizophrenia) in spite of the complete absence of any symptomatology. Other patients detected the normality of the experimental "patients" much more readily than the psychiatrists, with whom the pseudopatients had little contact.

This research provoked a great deal of discussion in the mental-health field. The results seemed to confirm the most serious suspicions of many critics concerning the undue rigidity of psychiatric diagnosis and to raise major questions about the value of psychiatric diagnosis generally.

Behavior modification. The behavior modification technique continued to win friends and influence psychologists concerned with an effective means of replacing undesirable with desirable behaviors by a direct ("reinforcement") attack upon the behaviors themselves. However, storm clouds were gathering. The most prominent of these during the year was the closing of programs at some federal prisons under which chronic of-

"I believe I have a new approach to psychotherapy, but, like everything else, the FDA tells me it first has to be tested on mice."

Sidney Harris

fenders had been deprived of certain privileges in order that the privileges could be reinstated as reinforcement for improved behavior. Also, a federal court, in a potentially far-reaching decision, decreed that state hospital patients who work must be paid in accordance with their productivity. Apart from that issue, legal authorities pointed out that token-economy programs in mental hospitals, in which desirable behavior is rewarded by tokens that can be exchanged for certain privileges, run the risk of abridging patients' civil rights. Thus, at a time when the furor over B. F. Skinner's defense of behavior modification in his controversial *Beyond Freedom and Dignity* seemed to be settling down, a whole new set of problems for operant-conditioning procedures as applied to human engineering technology promised to add legal expertise to the spectrum of skills the practicing psychologist must have.

Biofeedback. For many years it was commonly believed that visceral (smooth-musculature) processes are not subject to instrumental learning. But as a result of new experimental work over the past few years, researchers learned that visceral processes can indeed be instrumentally influenced—that they are, in other words, amenable to reinforcement in the same fundamental manner that skeletal responses are.

One of the most important by-products of this research was the biofeedback technique: training people to control visceral processes that are normally unconscious by means of feedback from devices that monitor and display those processes. Biofeedback had proved useful in improving certain disorders (*e.g.*, asthma, migraine headache) not readily helped by conventional medical treatment. Nevertheless, there were still serious practical problems that needed to be resolved.

Although research on biofeedback was just beginning and there were many unanswered questions, a large number of new biofeedback-training devices suddenly appeared on the market. The problem of providing useful guidelines to protect the public from unreliable or unscrupulous producers of these devices while at the same time avoiding undue restriction of legitimate research and application was pondered during the year by the American Psychological Association. In the meantime, until adequate guidelines could be developed, interested persons were well advised to proceed with caution.

Intelligence testing. The controversy initiated in 1969 by Arthur Jensen's scholarly review and genetically inclined interpretation of black/white differences in test intelligence showed no sign of slacking off. New attacks on intelligence testing continued to appear. For example, psychologist

Adapted from Medical World News

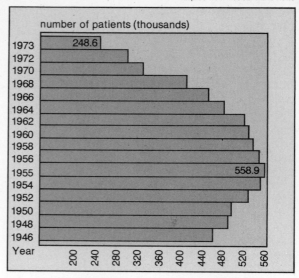

The number of patients in U.S. public mental hospitals declined sharply after the mid-1950s, in part a result of the effort to return patients to community settings.

David McClelland delivered a broadside attack on orthodox intelligence testing as essentially unrelated to real-life job requirements.

With regard to the role of "nature" and "nurture" in such complex behavioral functions as intelligence, the only reasonable position at the moment seemed to be some kind of "interactionism." Moreover, the data available simply did not permit any kind of definitive conclusions as to which of these factors is more important. In fact, some of the more extensive data suggested interrelationships quite different from those commonly assumed by either of the opposing points of view. For example, Sandra Scarr-Salapatek obtained intellectual capacity scores on all of the twins enrolled in the Philadelphia public school system during a given year and subjected these scores to an elaborate statistical analysis, including correlations with socioeconomic measures. Her conclusion was that the genetic factor (inferred from estimates of identical twins as compared with fraternal twins) becomes significant only under favorable environmental conditions; in other words, genetic differences have a greater influence on intellectual performance when children enjoy superior environmental opportunities.

Instructional innovations. In the academic world, psychologists were actively engaged in many and varied attempts to improve the caliber of college instruction. The most prominent such innovation was first formally proposed by Fred Keller in 1968 in a now-famous essay entitled "Goodbye, Teacher . . . ." Commonly referred to as the Keller Plan, the method was also called the unit-mastery procedure or, more generally, Personalized System of Instruction (PSI). The Keller Plan as applied to science instruction in college was the subject of a comprehensive survey during the year. Evaluation of the results of a large number of experimental and questionnaire-type investigations revealed a surprisingly consistent advantage of PSI over the orthodox lecture-discussion procedure. Not only was it favored by students but it also proved to be at least as effective as and often much more effective than the older method of providing them with the kind of knowledge tested for in final examinations.

The most attractive feature of PSI for students was self-pacing; the students take a short examination on a unit of instruction whenever they feel they are ready and are allowed to repeat the examination without penalty or prejudice if they do not pass. It was especially interesting that students found increased freedom in PSI courses, in light of the fact that the system was developed within the operant (Skinnerian) framework that is not ordinarily regarded as conducive to freedom. Interaction with undergraduate peers serving as tutorial assistants, usually students who have done well in the same course the previous year, was the second most cited attraction.

In PSI each unit of instruction (usually a part of a chapter in an orthodox textbook) must be mastered (examination passed at a very high level, typically 90% accuracy) before the next examination is permitted. The role of the instructor is mainly administrative. The dependence of PSI on the written rather than the spoken word accentuates the need for good textbooks. The results reported from approximately one thousand applications of PSI at the college level in psychology courses alone were so encouraging as to make this technique one of the most promising to appear on the educational scene in many years.

Language in chimpanzees. The major theoretical interest in training of chimpanzees to use language is the ultimate investigation of how language is learned and utilized by organisms complex enough to master a syntax but not so complex as to disguise the fundamental principles by learning too much too fast—as human children typically do. Among chimpanzees being so trained, David Premack's Sarah, who was taught to place metal-backed pieces of plastic on a magnetized slate in order to communicate with her human associates, was joined by Duane Rumbaugh's Lana at the Yerkes Regional Primate Research Center at Atlanta, Ga. At age three, Lana used a computer console as a means of communication, pushing panels of 75 varied geometric symbols in

1900-1975

A landmark event in the study of human behavior occurred three-quarters of a century ago when Viennese physician Sigmund Freud published *The Interpretation of Dreams.*

a systematic manner to express her wants. Like Sarah, Lana mastered a kind of syntax and could be said to have learned a rudimentary form of language. The computer was programmed to reject any sentence that was not properly arranged (*e.g.*, all questions had to be initiated by the question symbol; all requests had to be started with the symbol for "please"; all sentences had to be ended by a period).

The other classic chimpanzee of language-learning research was Allen and Beatrice Gardner's Washoe, trained to communicate by American Sign Language (the hand signals commonly used by the deaf). Washoe had been moved from the University of Nevada to the University of Oklahoma, where Roger Fouts was engaged in a variety of sign-learning research with a small group of young chimpanzees. When Washoe was first placed in the Oklahoma colony of untrained chimpanzees, she continued to use her well-learned hand signs, even though they were useless as means of communication with the other animals. Chimpanzee–chimpanzee signing was among the many interesting problems under investigation; such signing does occur between two chimpanzees who have been previously trained individually, and is said to be relatively rapid in the social situation as compared with the individual (one chimpanzee signing to a human associate).

Neomentalism. Perhaps the most striking trend within contemporary psychology was the resumption of interest in "mental" phenomena, after the many decades during which Watsonian behaviorism held almost undisputed control of methodological and substantive issues, at least in the field of experimental psychology. Mind now appeared to be back in psychology, and in experimental psychology, to stay. Many of the most intensively researched problems continued to involve distinctly mental operations—the use of imagery, for example, as a variable in learning and memory.

Information processing, which consists of both perceptual and memorial functions, hinges on numerous mental operations, such as rehearsal. With mental phenomena back in the good graces of experimental psychologists, the key methodological problem was how to identify—by overt behavioral measures, which are all that can be observed—the crucial covert processes. More explicit attention to this general problem was anticipated and was already evident in some ways (*e.g.*, the recent advent of the first issues of a new journal, *Behaviorism*, designed to focus on the ways in which behavioral methodology and theory can tackle psychological and philosophical problems).

—Melvin H. Marx

Botany

Research in botany during the year continued to expand knowledge of the structure, physiology, ecology, genetics, development, and economics of plants. A major concern was to obtain more food, fiber, and other materials from plants, and some examples of current developments relating to this are described below.

Food plants from the sea. Continuing concern over feeding a growing world population increased attention on the sea as a source of food for both livestock and human beings. This is not a new idea, since fish and other marine animals and plants have been used for food throughout much of man's history. While most food from the sea has been in animal form, certain peoples have long utilized algae in their diet. More attention may be paid to large-scale harvesting of plants from the sea since, as a general rule, each link in the food chain produces a biomass (weight of organisms per unit area or volume of habitat) of only 10% of the previous link. Direct consumption of plants, therefore, is ten times more efficient than consumption of animals that have eaten the same plants. Some interesting recent findings have shed new light on the potential productivity of the sea.

Phytoplankton, the microscopic green plants that form the beginning of the food chains upon which virtually all animals of the sea depend, are known to require certain nutrients from the water in which they live. Only parts of the ocean, such as certain coastal waters of continents, are rich in these nutrients and thus support phytoplankton and the complex communities of animals dependent on them.

Work recently reported by S. Apollonio shows that glaciers enrich the nutrient concentration in the waters of the fjords into which they empty. Two nutrients in particular, nitrates and silicates, are greatly depleted each summer because they are utilized by rapidly expanding phytoplankton populations during the growing season. It is these two nutrients that appear to be augmented by glacial activity; nutrient-bearing rock is entrapped by the scouring action of moving ice and is eventually deposited in the fjord. Evidence for this comes from a comparative study of two similar fjords of Ellesmere Island in Canada's Northwest Territories. One fjord is fed by a glacier, and the other fjord is not.

A study of the only known case of harvesting grain from the sea was reported during the year by Richard Felger and Mary Beck Moser. The Seri Indians of Sonora, Mex., wade into the waters of the Gulf of Mexico in April or May to harvest the floating masses of eelgrass, *Zostera marina*, which

Three glaciers enter the northern end of South Cape Fjord on Ellesmere Island. Glaciers increase the concentration of nutrients, especially nitrates and silicates, in the waters into which they empty.

they then dry on the beach. The grain is separated, winnowed, toasted, ground into flour, and cooked in water. While the Seri no longer harvest as much *Zostera* seed as their ancestors, Felger and Moser proposed the grass as a potential food source for man because of its widespread distribution, because it has protein and carbohydrate content comparable to that of terrestrial grains but lower fat content, and because its cultivation demands no artificial fertilization and pesticides.

Sea turtles also feed on *Zostera*. This observation, along with similar observations of sea animals feeding on natural marine vegetation, leads to the conclusion that the vegetation of the sea may provide feed for livestock. The *Fishery Bulletin* of the U.S. Department of Commerce reported that turtle grass, *Thalassia testudinum*, is a likely ingredient of livestock feed. Its 13% protein makes it richer than wheat grain. Sheep showed significant weight gain and feed utilization when fed a ration replacing 10% of the usual alfalfa with turtle-grass pellets. *Thalassia* grows abundantly in an area of about four million acres off the coast of Florida and over a much greater area around Caribbean and South Pacific islands. It can be harvested twice yearly with no apparent damage to the plant's ability to grow back.

Crop disease. Increasing productivity by attacking plant disease is an important aspect of research in botany. A recent example relates to a serious outbreak of southern corn leaf blight. This disease spread rapidly in 1970, in nearly epidemic proportions, from regions in the southern U.S. to some corn-growing areas in the north. The resulting decrease in yield was of major concern to growers and consumers alike, while identification of the disease-producing agent and its effect on corn plants became of concern to researchers.

The rapid spread of the blight was found to be caused by the crossing of certain strains of corn used to produce hybrid seed in 1969 and 1970. The cytoplasm of the corn grown from this seed contained genes that made the plants male-sterile. (Hybridization is controlled more easily when the ear-producing plant is male-sterile, because the plant can not pollinate itself.) For some reason this cytoplasm, designated *cms-I*, also made these corn strains more susceptible to infection by the southern corn leaf blight fungus, *Helminthosporium maydis*. Control of the blight epidemic was achieved by producing seed from resistant strains, but research on the mechanisms producing the blight continued.

At the University of Illinois, it was found that *H. maydis* produced a pathotoxin (disease-causing poison) that disrupted membranes, particularly those of the mitochondria. Such disruption decreased the production of adenosine triphosphate, the principal compound involved in a cell's energy-yielding and energy-requiring reactions; this would cause growth of the plants to be retarded and thus explain the blight at least partially.

197

C. A. Beasley, University of California at Riverside; research supported by Cotton Incorporated

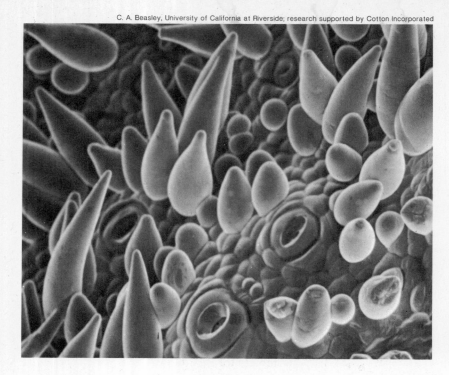

Unfertilized cotton ovules being grown in vitro produced elongated fibers when treated with phytohormones.

During the past year it was reported that *H. maydis* phytotoxin acts even more quickly against photosynthesis than it does against mitochondrial function. Evidence indicates that rapid closing of stomata (tiny openings in leaves or stems through which occurs the gaseous interchange between the atmosphere and the intercellular spaces of the plant) greatly lowers the amount of carbon dioxide available for fixation (metabolic conversion of carbon dioxide into combined forms). The apparent cause of such rapid closure is inhibition of the uptake of potassium ions by the guard cells flanking the stomata; such uptake is necessary for stomatal opening.

Nitrogen. An approach to increasing crop production involves the use of fertilization to achieve plants with higher yields. An important nutrient for plant growth is nitrogen, which also is of concern to environmentalists because it is found to leach very quickly from farmland into streams and wells when applied heavily as nitrate fertilizer. Too much nitrate in drinking water is thought to cause problems in very young children because it is reduced to nitrite, which competes with oxygen for the hemoglobin necessary for circulating oxygen through the body. This is of special concern in regions of intensive agriculture, where excessive nitrogen fertilizers are employed.

Researchers such as Barry Commoner are attempting to find ways to reduce the necessity of excess fertilization and yet maintain high yields of corn. He has found that adding a carbohydrate source such as ground corncobs to the soil allows bacteria to utilize nitrogen from fertilizers more efficiently, thereby allowing the actual amounts of nitrogen to be decreased. Crop plants may then, in turn, utilize the nitrogen, which is released from decaying bacteria in the form of ammonium ions. Nitrogen in this form is released over a longer period of time than when applied in the form of nitrate fertilizer, thus diminishing the leaching problem.

Another approach to the nitrogen problem is to develop plants that can supply their own nitrogen by taking it from the atmosphere. This process, nitrogen fixation, is a well-known characteristic of legumes and some other plants, which maintain a symbiotic relationship with nitrogen-fixing bacteria, such as *Rhizobium*, that inhabit nodules on the plant roots. Normally, such bacteria are transferred through the soil from one plant to another, or seed is purposely inoculated before planting. However, new approaches to inducing nitrogen fixation were being contemplated. One was to find or develop strains of nitrogen-fixing bacteria that would live symbiotically in cereal plants previously unable to host them. Another was to transfer the genes necessary for the nitrogen-fixing ability from the bacteria to cells of the plant itself.

Pollen. Increased interest in plant pollen was rewarded during the year by the development of techniques for the removal of pollen surface coats. These coats, the exines, are composed of a rigid structural substance called sporopollenin. While

198

removing the coats is a remarkable feat in itself, it has also allowed experimental manipulation of the pollination processes themselves.

In nature, only certain pollen grains, usually from the same species, are allowed to grow pollen tubes through the styles and into the ovules of pistils after landing on the stigmas of those female organs. In this way fertilization is accomplished between pollen and ovule. It is believed that surface pollen substances are used by plants to distinguish pollen of their own species. If such substances are removed, identification is lost.

Two kinds of experimentation are suggested by the removal of the pollen surface coats. One is based on the lowered barriers to hybridization, offering the opportunity to create useful hybrids never before possible. The other would make use of adding inhibitory substances to naked pollen to cause its rejection by stigmas of its own species, thus enforcing hybridization.

Cotton. For several years, C. A. Beasley and his associates at the University of California (Riverside) have been reporting their ability to grow fertilized cotton ovules in an artificial environment (in vitro) by adding appropriate nutrients. Such ovules produce apparently normal fibers, which are the outgrowths of ovule epidermal cells. During the past year these workers reported the successful growth of unfertilized ovules in vitro. The new technique demanded the addition of phytohormones for the development of fibers and for ovule enlargement, thus demonstrating that these hormones had been produced by the previously cultured fertilized ovules.

Several hormones function together in ways that depend on their relative concentrations. Generally, indoleacetic acid and / or gibberellic acid promote fiber induction and growth; kinetin promotes ovule enlargement; and abscisic acid tends to be antagonistic to the first three. As the researchers pointed out, the in vitro substitution of phytohormones offers greater possibilities of developing an understanding of the embryonic development of plants and the formation of seeds.

Photosynthesis. Research on photosynthesis, the process by which solar energy is made available to both natural and agricultural ecosystems, may contribute to an increase in food production. Scientists were increasing their understanding of the ecology of photosynthesis with respect to plants growing in arid regions where, for several decades, the basic photosynthetic physiology and corresponding leaf-cell morphology have been known to be modified under certain environmental conditions.

The standard photosynthetic process involves two phases. One is characterized by the trapping of light energy (thus called the light phase), the evolution of oxygen, and the production of energy-containing molecules. The second phase, the Calvin-Benson cycle, fixes carbon dioxide on a five-carbon molecule (ribulose diphosphate) and ultimately yields the three-carbon (C_3) compounds used in cellular processes. This whole complicated process is called C_3 photosynthesis because of the resulting three-carbon compounds. It occurs in the chloroplasts of green plants.

Variations in carbon dioxide fixation occur in plants existing under specific environmentally induced stress conditions, including some combination of high temperature, low relative humidity, and high illumination. In one variation, the Hatch-Slack pathway, carbon dioxide is fixed by a three-carbon molecule (phosphoenolpyruvate) to form a four-carbon (C_4) dicarboxylic acid, which then donates the carbon dioxide to the Calvin-Benson cycle. Plants utilizing this variation are often called C_4 plants. Because of the C_4 variation they are able to function with a lower internal concentration of carbon dioxide, thus making it possible for their leaf stomata to remain closed in order to decrease water loss in arid environments.

In another variation, carbon dioxide is fixed on a three-carbon molecule to form malate (malic acid, a C_4 dicarboxylic acid), the process occurring at night when open stomata would not result in as great a water loss as during the day. The accumulation of malate during the night essentially stores the fixed carbon until it can be cycled during the day. Because accumulation of malate increases acidity and the process has been found in all species of the family Crassulaceae, which includes stonecrops and houseleeks, it is called crassulacean acid metabolism.

Both variations are employed by plants that must conserve water and thus operate with closed or only partially opened stomata during the day (xeric plants). Such closure not only reduces diffusion of water vapor but also that of carbon dioxide, resulting in a lowered concentration of carbon dioxide within leaves.

Many xeric grasses of economic or ecological importance employ C_4 photosynthesis, among them corn, sugarcane, and crabgrass. The connection between the physiological problem of lowered carbon dioxide concentration and structural adaptation is the subject of research on a number of kinds of plants. An example is the work of Olle Björkman and associates of the Carnegie Institution of Washington, first on *Atriplex*, a genus of desert saltbushes, and later *Tidestromia*, another desert shrub. Chloroplast distribution as well as photosynthesis is modified in these plants. Fortunately for the researchers, some species of *Atri-*

plex carry on typical photosynthesis while others employ the C_4 form, and these two groups may be used for comparative and hybridization studies. In normal C_3 plants, chloroplasts are of a single type and are distributed throughout the mesophyll (internal) tissue. In C_4 plants, the C_3 chloroplasts are confined to sheath cells surrounding the vascular bundles of the leaves. The same leaves contain C_4 chloroplasts in a ring of mesophyll cells surrounding the C_3 sheath cells, and the remainder of the mesophyll contains almost no chloroplasts. Thus, C_4 plants function with a very low concentration of carbon dioxide in the mesophyll, allowing for a more rapid diffusion of carbon dioxide into the leaf through partially closed stomata.

Björkman and his colleagues established the fact that structure and physiology in plants are independently and genetically induced. Crosses between C_4 and non-C_4 species of *Atriplex* produced an F_1 generation intermediate between the two species in structure and in the amount of C_4 enzymes produced. Among plants of the F_2 generation, however, some were similar to their C_4 grandparents in structure but were not as similar in function.

Crassulacean acid photosynthesis, while having some physiological features similar to C_4 photosynthesis, utilizes opened stomata at night to secure carbon dioxide and minimize water loss. S. R. Szarek and his colleagues found an interesting variant in a species of *Opuntia* (cactus), which functions as an ordinary crassulacean plant only for short periods after precipitation. At other times, which are marked by the need for water re-

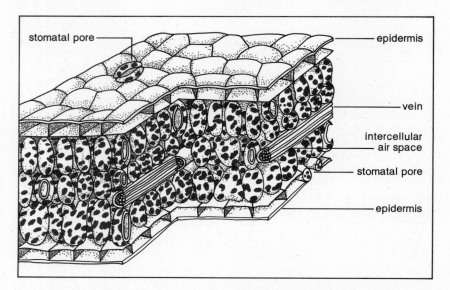

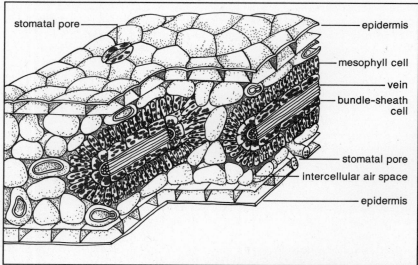

Leaf structure of the three-carbon plant Atriplex patula *(top) demonstrates usual pattern in which all cells are of a single type. In four-carbon* Atriplex rosea *(bottom) the leaf contains two types of cells, mesophyll and bundle-sheath.*

tention, the stomata of such plants are kept closed at all times and the carbon dioxide yielded in cellular respiration is recycled by being fixed in the cactus stem tissue rather than allowed to escape. Thus, the plant functions photosynthetically at a reduced rate during periods of drought.

Further evidence that many environmental factors influence photosynthesis comes from experiments on *Kalanchoe*, a crassulacean succulent, by J. C. Lerman and O. Queiroz. They were able to show that *Kalanchoe* switched from C_3-dominant activity to C_4-dominant activity when plants were grown first under long-day conditions and then changed to short-day conditions. This observation was made possible by the fact that the relatively rare carbon isotopes, ^{13}C and ^{14}C, are not incorporated to the same extent in both forms of photosynthesis; C_3 processes discriminate against the heavier isotopes, causing a lower proportion of them to be present than is found in air, while products of C_4 processes have only a slightly lowered proportion of these isotopes. Analysis of extracts containing photosynthetic products reveals whether C_3 or C_4 photosynthesis dominates at the time.

The practical aspects of understanding photosynthetic patterns lie in the future incorporation of stress-adapted physiology into agricultural plants. One way to accomplish this is by hybridization, in which wild species with adaptive photosynthetic systems would be crossed with domestic species having other favorable characteristics, such as yield. Another possibility is the selective development of already existing wild species that have the desired photosynthetic system. By either of these means productivity of arid land may be initiated or increased.

—Albert Smith

Chemistry

Among the outstanding developments in chemistry during the past year were successful tests of a new process for deriving low-sulfur fuel oil from coal, the development of a plastic that will degrade in landfill or compost, and the synthesis of compounds that show promise as anticancer agents.

Applied chemistry

Chemists and chemical engineers produced during the year more economical routes to clean energy, new sources of edible protein, better ways to renovate waste water for human consumption, and alternatives to the solid waste problem.

Clean fuels. By 1974 it seemed evident that a clean environment did not have to be sacrificed because of the energy crisis; the technology to clean up coal either before or after use was on hand at costs not significantly different from the current high price of petroleum, and the momentum of research was such that further improvements were likely. Alternative fuels derived from shale oil or sand tars were much further removed from reality.

Coal can be adapted to today's energy needs in several ways, but excessive sulfur in it must be removed at some point in the process. The coal can be changed into a fuel oil, a low-energy gas for electrical utilities, a pipeline-quality gas (synthetic natural gas) for home consumption, or a clean solid fuel. Most of the processes were in developmental stages, although a few were on the market.

In a highly successful pilot-plant stage in 1974, the U.S. Bureau of Mines' Synthoil process yielded a very-low-sulfur fuel oil from coal. This oil appeared to be less costly than petroleum-derived fuel oils. Unusually good data were obtained in trials with bituminous and cheaper coals: an inexpensive Kentucky coal with 5.5% sulfur and 17% ash was converted to oil with only 0.17% sulfur and 0.7% ash. Moreover, the conversion of carbon was 98%, and energy conversion efficiency was 78%.

The Synthoil process requires expensive hydrogen, but uses better reactors and catalysts than did earlier counterparts and thus can operate at much lower pressure with less consumption of hydrogen. Pulverized coal is slurried with some of its own product oil and pumped at high velocity (to create turbulence) into a reactor containing hydrogen and immobilized catalyst pellets. There the coal is liquefied and desulfurized with a high yield of oil. Hydrogen sulfide gas is converted to disposable sulfur, and the unused hydrogen is recycled.

The most economical coal gasification process under consideration in 1974 was the Hydrane process developed by the U.S. Bureau of Mines. The method is simple, but a large hydrogen generator is needed. Raw coal is reacted with hydrogen to form methane directly, rather than processing the coal by the indirect method of forming town gas and following that with extensive methanation. In addition to methane, the method may be able to produce other products, including complex aromatic liquids.

Several techniques to remove sulfur from coal prior to use were described early in 1974, and each method claimed to be commercially viable, leading to a clean fuel competitive in cost with the method of cleaning the stack gas after the com-

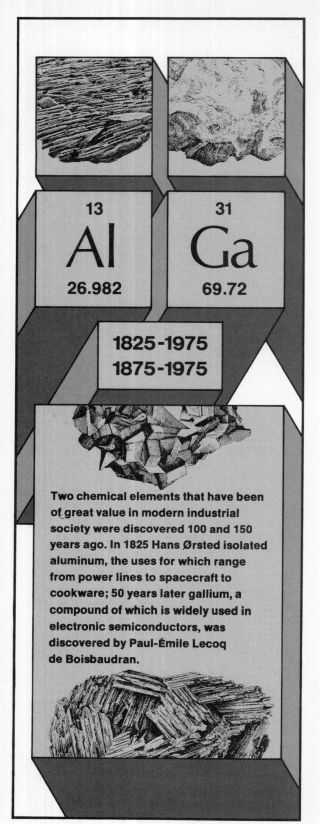

13
Al
26.982

31
Ga
69.72

1825-1975
1875-1975

Two chemical elements that have been of great value in modern industrial society were discovered 100 and 150 years ago. In 1825 Hans Ørsted isolated aluminum, the uses for which range from power lines to spacecraft to cookware; 50 years later gallium, a compound of which is widely used in electronic semiconductors, was discovered by Paul-Émile Lecoq de Boisbaudran.

bustion of the coal. High-gradient magnetic separation techniques developed at the Massachusetts Institute of Technology were applied to coal slurries. This magnetic process for cleaning coals can remove practically all the liberated inorganic (pyritic) sulfur and a portion of the other minerals. Without superconducting magnets, however, the cost of the operation is tied to the cost of electric power.

About 80% of the sulfur in coal can be removed by chemical leaching of pyritic sulfur with aqueous ferric sulfate solutions at moderate temperatures (90°–130° C). Called the TRW Meyer's Process, this relatively inexpensive treatment is particularly applicable to the Appalachian coal reserves, in which most of the sulfur is inorganic rather than organic. The pyritic sulfur (iron sulfide) is converted to iron sulfate and elemental sulfur, which are then removed from the coal by filtration and vaporization. Costs of the process are generally lower than those projected for stack-gas scrubbing, liquefaction, and other sulfur-oxide control methods. This process would result in the sulfur content of about one-third of the coal in the Appalachian region being lowered to the 0.6–0.9% levels required to meet the standards set by the Clean Air Act for new power plants.

A fuel system based on methanol was proposed in 1973. Methanol could be used as an automobile fuel to replace or blend with gasoline, as a substitute liquefied natural gas for gas turbines and power plants, as a component in fuel cells for the generation of electricity, or as a convenient energy storage medium. An alternative end product of coal gasification, methanol can be formed by catalytic conversion after the sulfur has been removed from the product gases. It can be made from coal using present-day technology at a cost per BTU that is competitive with gasoline prices. As an automobile fuel, methanol increases gas mileage and burns more cleanly than gasoline.

Renovated water. A major problem suggested for several years and established in early 1974 is the unreliable nature of standard chlorine disinfection treatment of sewage and waste water prior to disposal in waterways. Not only is chlorination ineffective against viruses but it also can generate chlorinated organic compounds of questionable toxicity. Current sewage treatment techniques are not equipped to remove such compounds, and the incoming water to communities downstream from the disposal point generally contains these chemicals to some degree.

An increasing number of reports show biorefractory (nonvolatile) organic chemicals present in tap water. Carcinogenic (cancer-producing) polynuclear aromatic hydrocarbons were discovered in

Rhine River water. The tap water of Evansville, Ind., contained 13 organic compounds, including bis (2-chlorisopropyl) ether from an industrial outfall 150 mi upstream. Only two-thirds of this chemical was removed by standard treatment of the incoming water supply.

The first definitive proof of the formation of chlorinated organic compounds was reported in early 1974 by a chemist at Oak Ridge (Tenn.) National Laboratory, who used a radioactive chlorine isotope to "tag" and identify the compounds that incorporated chlorine. Robert L. Jolley found more than 40 chlorine-containing organic constituents in the chlorinated effluent from the primary stage of a sanitary sewage treatment plant.

Several alternatives to standard chlorination of sewage were proposed in 1973. Mark McClanahan of the Georgia Institute of Technology urged the use of a dual disinfectant, with ozone as the active agent in raw recycled water. Because ozone has a short existence in water, another disinfectant would be required to provide the residual protection needed throughout the distribution system. For this purpose he recommended chlorine. The prior use of ozone should reduce the amount of organic compounds that could be readily chlorinated.

Bromine as the disinfecting agent was recommended by several chemists during the past year. Among its advantages are that it does not react readily with sewage to form brominated organic compounds and, in contrast to chlorine, it reacts with ammonia to form bromamines that are far superior to chloramines as virucides and are too short-lived to cause fish kills.

Bromine presently costs twice as much as chlorine, but only one-half as much bromine as chlorine is required to achieve the same effectiveness. In 1973 Dow Chemical Co. introduced bromine chloride as a source of free bromine in water at a lower cost than bromine itself. Bromine chloride, available as a gas at low pressure, promised to be easier to use than liquid bromine. Compared to chlorine, bromine chloride is more soluble in water, is more effective over a wider pH range, and shows higher biocidal activity without fish toxicity hazards.

Food supply. According to Aaron M. Altschul of Georgetown University, conventional agriculture is no longer capable of supporting larger populations or wealthier nations at the present quality of protein intake or with the present manner of achieving red meat. New technologies such as fortification with amino acids or direct utilization of vegetable proteins may alleviate the pressure on agricultural capability to produce protein.

Two new technologies supplying protein were announced in 1973. Chemists of the U.S. Department of Agriculture's Western Regional Research Center in Berkeley, Calif., devised a way that appears commercially feasible to extract white, bland, edible, high-quality, and potentially inexpensive protein from green alfalfa. The process, called PRO-XAN II, can be adapted to yield either an insoluble heat-precipitated protein that may be suitable for use in cookies, snack foods, gravies, soups, stews, pastas, milk substitutes, preprocessed foods, and meat extenders in ground meats and sausage; or a soluble protein with foam-stabilizing properties that may allow its use in whipped toppings, soups, sausages, and meat extenders. As of 1974 production of the insoluble form of white protein from alfalfa was in a successful pilot-plant stage, while preparation of the soluble form was still in the laboratory.

Easier production of agricultural products in less developed countries could be provided by a simple single-stage system that ensures uniform seed distribution while securing the soil for generation of the root system. Such a system was developed by Canada Wire & Cable Ltd. in the form of a water-soluble, flexible, inert polymeric sheet containing seed, fertilizer, herbicides, and other additives necessary for the rapid development of a food crop or of erosion-controlling ground cover. The system requires no particular skills for efficient arable farming and should be useful for transporting correctly proportioned seed-fertilizer mixtures. The seed sheets are made of polyvinyl alcohol with carbon black added to accelerate germination by increasing the absorption of surface solar radiation.

Solid wastes. The first biodegradable plastics, dependent upon "controlled photooxidation" in sunlight, were reported in late 1972 at a meeting of the American Chemical Society. In these plastics, specific chemical groups that could be activated by the energy of ultraviolet radiation were incorporated into the polymer backbone. When excited by sunlight, these groups ruptured the polymer chain, forming smaller segments that bacteria could attack effectively. A plastic cup made of such material can disintegrate in 15 days of exposure to sunlight.

In 1973 the first plastic designed to degrade in landfill or compost was announced by Gerry J. L. Griffin of Brunel University, Uxbridge, Eng. Thin films of polyolefin plastic with starch as a filler were used. The starch degraded rapidly, leaving a porous film readily attacked by microorganisms and saturated with oxygen. Sunlight is not required for disintegration of this plastic. A fourth biodegradable plastic was introduced in early 1974 by William Bailey of the University of Maryland. Incorporating the amino acid glycine, this polymer breaks down to give harmless natural substances.

In early 1974 objection to biodegradable plastics surfaced because of the shortage of petroleum and natural gas from which the plastics are made. This most recent reflection about the disposal of plastic wastes suggests that such wastes should be used as a resource and be recycled rather than dissipated by means of incineration or biodegradable landfill.

—Dorothy Plack Smith

Chemical dynamics

Progress in the area of chemical dynamics during the past year proceeded along predictable lines. Instrumental improvements permitted the resolution of individual rotational quantum transitions in a favorable case. The observation of optical emission from collisionally excited species permitted determination of the quantum states of the products in certain ion-molecule reactions for the first time. The reaction of atoms with oriented mole-

Time of flight is plotted against number of ions to produce graph depicting a collision of beam of Li$^+$ ions with H$_2$ molecules at 2.7 eV. Arrows show calculated locations of the elastic and inelastic maxima corresponding to the transitions between rotational energy states in the target molecule.

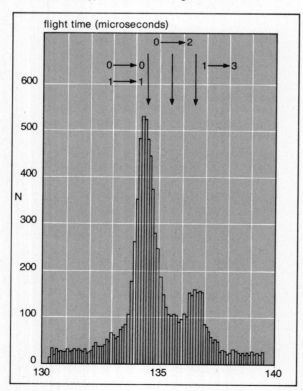

cules was studied, and advances in the theory of reactive collision processes promised to yield a theoretical technique to complement molecular beam experiments.

Rotational transitions. Dynamicists in the United States and other nations were intensively studying the conversion of translational energy of a projectile atom or ion into internal energy (electronic, vibrational, and rotational) of the molecule with which the projectile collides. (Translational energy is the kinetic energy associated with the translational motion of a molecule, in which every part of the molecule moves parallel to and in the same direction as every other part.) The process for an ion-molecule inelastic collision (one in which part of the kinetic energy of the colliding particles changes into another kind of energy) can be stated by the equation $A^+ + BC \rightarrow BC^* + A^+$. The ion A^+ is assigned an accurately known amount of kinetic energy and then permitted to collide with a molecule BC. During the collision a portion of the kinetic energy of A^+ is converted into internal energy of the molecule BC, and then the kinetic energy of A^+ is measured.

From an accurate measure of the energy loss of A^+, one can determine which internal mode of the molecule BC was excited. The results of work on Li$^+$-H$_2$ collisions were recently published by a group of scientists from Göttingen, W.Ger. The Li$^+$ ions were produced by surface ionization and energy selected with an electrostatic analyzer, resulting in an Li$^+$ ion beam of very narrow energy and angular spread. This ion beam then collided with a beam of H$_2$ molecules that were predominantly in the rotational energy states indicated by $j = 0$ and $j = 1$. The kinetic energy of the Li$^+$ ions was measured by accurately determining the time of flight over a known path. Ion detection was achieved with an electron multiplier. A typical result, shown in the figure, gives the measured time of flight spectrum for Li$^+$ colliding with H$_2$. The laboratory ion energy was 2.7 eV, and the flight path was 115 cm. The arrows show the calculated locations of the elastic and inelastic maxima corresponding to the rotational state transitions to higher energy levels, $j = 0 \rightarrow 2$ and $j = 1 \rightarrow 3$, in the target molecule. The measurement time was about 100 hours for the spectrum shown. Not only was it necessary to give much attention to stability and angular and energy resolution but also the earth's magnetic field had to be compensated for over the entire length of the ion path. The measurements represented a significant experimental triumph.

Optical emission. Optical emissions resulting from chemical reactions between neutral colliding species have been used to determine the quantum states of the reaction products. (Quantum states of

Type of process	Reaction observed
excitation	$N^+ + NO \rightarrow N^+ + NO^*$
reaction	$N^+ + NO \rightarrow N_2^{*+} + O$
charge exchange	$Ne^+ + CO \rightarrow CO^{*+} + Ne$
dissociative charge exchange	$Ne^+ + CO \rightarrow Ne + C^{*+} + O$
	*Indicates the light-emitting species.

a molecule are states characterized by discrete energy levels.) In 1974 for the first time such results for ion-molecule reactions involving ion beams were reported by scientists at Göttingen. In their experiment ions with a few eV of energy collided with a target gas in a reaction chamber. The collision region was viewed using a fast grating spectrometer with quartz optics. The spectral region covered ranged from 2000–4500 Å. The photons were detected with a photomultiplier tube. Signal-averaging techniques for times up to 15 hours were necessary to observe the signals. One of the reactions that was studied in some detail was

$$N^+ + NO \rightarrow N_2^{*+} + O.$$

The N_2^{*+} is the excited ion, and the emitted light results from the transition

$$N_2^{*+} (B\ ^2\Sigma_u^+) \rightarrow N_2^+ (X\ ^2\Sigma_g^+) + h\nu.$$

The spectroscopic nomenclature, in parentheses, identifies the quantum states involved in the transition, $X\ ^2\Sigma_g^+$ being the ground state of N_2^+.

Each peak in the spectrum is a superposition of bands belonging to a sequence with fixed changes between vibrational quantum numbers. Much higher optical resolution would be needed to resolve this structure further, and so clearly there is much more work to be done before one can completely identify all of the product ion states.

In addition, the same research group reported luminescence from the ion impact processes shown at the top of this page. Scientists hoped that eventually this type of experiment would enable investigators to identify the quantum states of products resulting not only from chemical reactions but also from collisional processes in general.

Oriented molecules. Some intriguing experiments involving the reaction of a beam of atoms with oriented molecules were reported during the past year. It is possible to orient molecules (such as CH_3I) that possess a dipole moment by passing them through an electric field of proper geometry (hexapole field). Earlier experiments with a beam of potassium atoms, K, and CH_3I had indicated that the reaction occurs preferentially when the I is oriented toward the oncoming K. More recent studies with $K + CF_3I$ indicated that KI is formed regardless of whether the I or the CF_3 faces the oncoming potassium atom. However, a study of the angular distribution shows that if the I is facing the

oncoming K the product KI is scattered in the backward direction in relation to the center-of-mass system of coordinates, whereas if the CF_3 group faces the oncoming K, the product is scattered in the forward direction. (A center-of-mass system is a system of polar coordinates used in describing processes involving moving swarms of particles, with the moving common center of mass as the origin and the path of that center as the polar axis.)

The explanation for this behavior is believed to be the following. An oriented molecule should not be considered as one lined up with its symmetry axis pointing in the direction of the incoming potassium atom. Instead, the molecule is actually precessing about the total angular momentum vector, which passes through the center of gravity of the system and is inclined with respect to the symmetry axis. This means that the line joining the center of gravity and the carbon of the methyl group or the line from the iodine atom to the center of gravity sweeps out a cone. In the case of CH_3I, the precessional cone has a wide base and effectively shields the iodine from the potassium atoms coming from the methyl group direction; therefore, preferential reaction does not take place from this direction.

The mass and angular momentum conditions in the case of CF_3I are such that the precessional cone generated by the CF_3 group is much narrower at the cone base, and the result is that the iodine is not completely shielded by the motion of the CF_3. Consequently, the potassium atom approaching from the CF_3 end and just grazing the group still can "see" the iodine and pick it off to form KI as it goes by the CF_3I molecule. These KI molecules will be scattered in the forward direction. When the CF_3I molecules are so oriented that the I atom faces the K beam, KI is again formed but scattering by the CF_3 group causes this KI to be scattered in the backward direction.

Future outlook. The immediate future outlook for chemical dynamics can be summarized by the phrase "more of the same." Many more reactions of the types described above will have to be studied in even greater detail before chemists can round out their understanding of the collisional processes. From the theoretical point of view, the ideal goal is a complete quantum mechanical calculation of the reaction cross sections, angular and energy distributions of the products, and other parameters of collisional processes. Many able scientists were working on this, but it is a difficult task and its accomplishment remains some years in the future.

—Walter S. Koski

Chemical synthesis

On the frontier of chemical synthesis, routes to fabricating molecules of ever increasing complexity are being developed. For example, syntheses of medically important substances from readily available starting materials are being rapidly accelerated by the application of new reagents and experimental techniques. Similarly, a broad variety of substances important for biological, commercial, and/or exploratory reasons is becoming more easily available. The rapid introduction of organometallic and enzymatic reagents combined with the most advanced methods for molecule protection, chromatographic separation, and instrumental analysis have greatly enhanced progress over the past several years. Such stimulating advances are not confined to the hydrides of carbon (organic chemistry) but extend to the chemistry of some 20 other elements.

Organometallic compounds. In the field of organometallics, certain reagents derived from copper (1), mercury (2), thallium (3), and iron (4) were being rapidly adopted for, respectively, alkylation reactions, cyclopropane syntheses, halogenation

and ring-contraction reactions, and the preparation of aldehydes. Various hydrides of boron continued to enjoy wide use, and John Hooz at the University of Alberta reported a new and potentially useful boron reagent (5) early in 1974. The new compound, lithium dimesitylborohydride, showed promise in effecting stereospecific reductions of the type shown.

Discoveries of other novel organometallic compounds proceeded at a fast pace during the past year, but their utility remained to be determined. The isolation of a crystalline salt with a sodium anion (6) provides an example. In January 1974, J. L. Dye and colleagues at Michigan State University described the preparation of this salt by reacting sodium metal with the bicyclic polyoxadiamine shown in the diagram. The cyclic polyether apparently traps a sodium cation in a cage complex, and the resulting salt can be isolated as crystals with a shiny metallic appearance.

Tetracarbonylnickel played a key role in a new synthesis of one member of the vitamin K_2 series described by K. Sato at Yokohama National University, Japan. Reaction between an allylic bromide (7) and the nickel carbonyl led to an alkylnickel bromide complex that readily condensed with bromide (8). After saponification (hydrolysis by alkali) and oxidation, vitamin $K_{2(45)}$ was obtained in good yield. This vitamin is used medically to stimulate blood clotting and is widely distributed in algae, some fungi, plants, and bacteria.

The most active testing ground for new reagents derived from copper, zinc, boron, aluminum, thal-

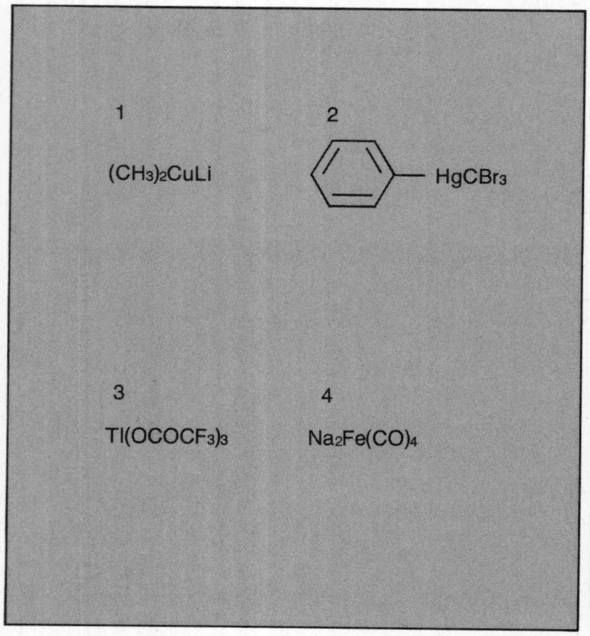

1

$(CH_3)_2CuLi$

2

HgCBr_3

3

$Tl(OCOCF_3)_3$

4

$Na_2Fe(CO)_4$

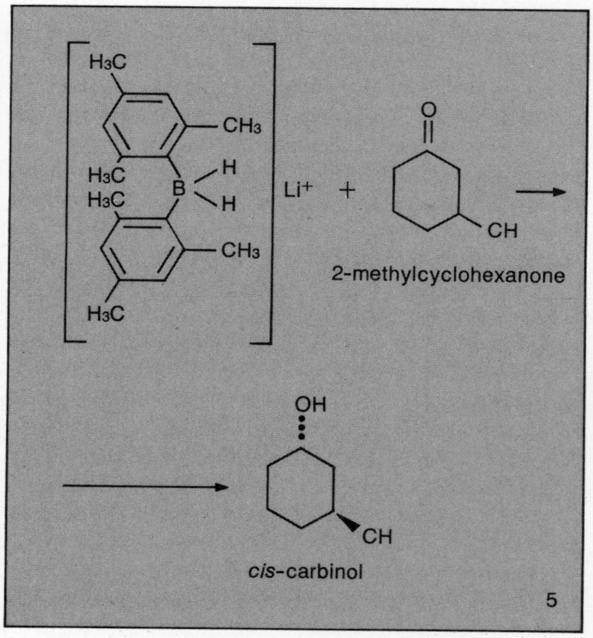

2-methylcyclohexanone

cis-carbinol

5

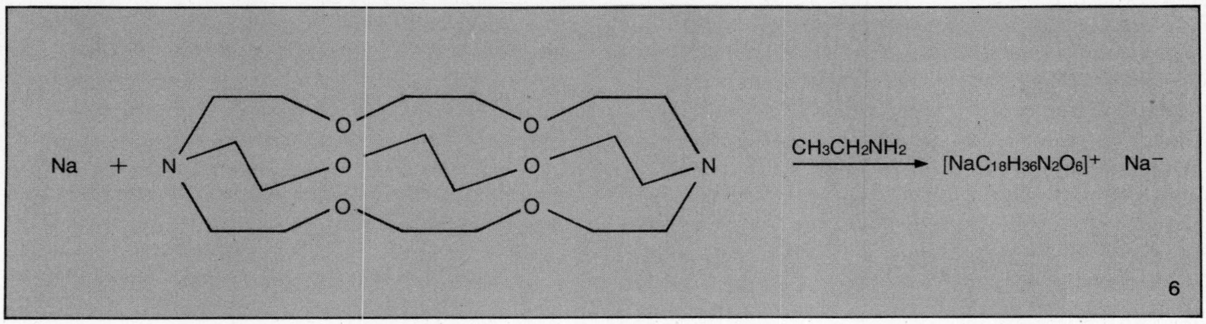

$$Na + N\text{(crypt)} \xrightarrow{CH_3CH_2NH_2} [NaC_{18}H_{36}N_2O_6]^+ \quad Na^-$$

6

lium, and phosphorus has been in research on prostaglandins. The prostate gland produces a series of substances such as prostaglandin E_2 and $F_{2\alpha}$ that exhibit an amazing diversity of biological activities. One of the series, prostaglandin E_1, is even a potent bronchial dilator in aerosol form, and tests were under way to determine its effectiveness in the treatment of bronchial asthma. Because of the exceptional academic and commercial interest in these natural products, practical methods for their synthesis were under study in a number of universities (principally E. J. Corey at Harvard University) and pharmaceutical company laboratories. The result over the past year was an almost continuous flow of ingenious syntheses. The Corey group completed new routes to the prostaglandins, and these were complemented by new approaches from other laboratories. For example, a route to prostaglandin $F_{2\alpha}$ was devised by R. B. Woodward of Harvard University (and the Woodward Research Institute in Switzerland) starting with 1,3,5-trihydroxy-cis-cyclohexane (9). At the University of Wisconsin, J. B. Heather had made use of the interesting copper reagent (10) in a new synthesis of prostaglandin E_2.

Antibiotics and chemotherapy. The tetracycline-type antibiotics are drugs used for the treatment of infections caused by various bacteria. Chlortetracycline (aureomycin), oxytetracycline,

and tetracycline are the best-known members of this group. The antibiotic activity of these compounds is believed to be caused by their ability to block protein synthesis in the offending microorganism. A group under the direction of Hans Muxfeldt at the University of Stuttgart in West Germany completed the chemical synthesis of chlortetracycline (11). By using a nitrile as a starting point Muxfeldt and his researchers developed a relatively short path to chlortetracycline, suggesting that this approach could be made economically feasible.

Developing effective chemotherapy (drug treatment) of virus infections required as of 1974 much more effort by synthetic chemists. However, the prospects of discovering drugs for the treatment and cure of many important virus diseases of some duration seemed quite good. With acute virus infections such as the common cold, influenza, and rubella (German measles), the symptomatic course is quite short, and by the time a diagnosis can be made death or recovery has already occurred. But with acute infections of a somewhat longer duration, such as infectious mononucleosis, measles, tick-borne encephalitis, and hepatitis, there is time to make the necessary diagnosis and begin treatment. The situation is even more favorable for chronic virus infections such as warts, herpes, and, possibly, multiple sclerosis and some forms of cancer (for example, Burkitt's

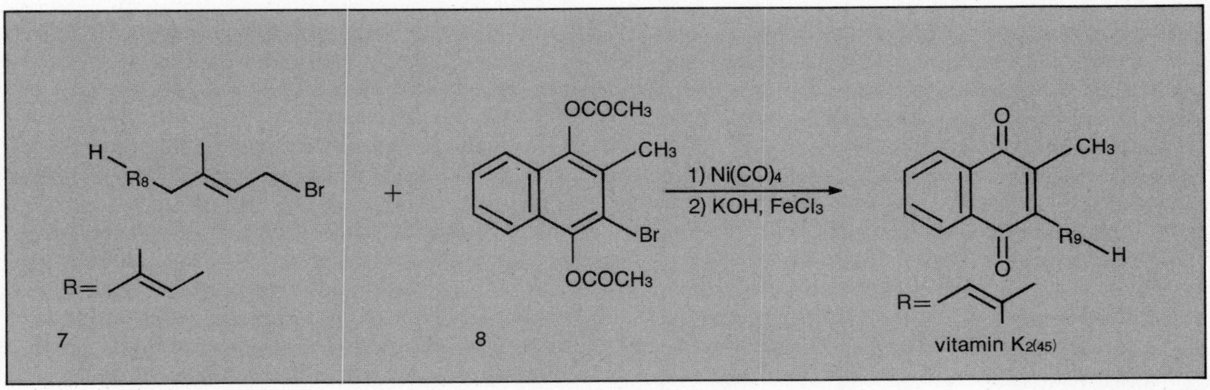

7 8 vitamin $K_{2(45)}$

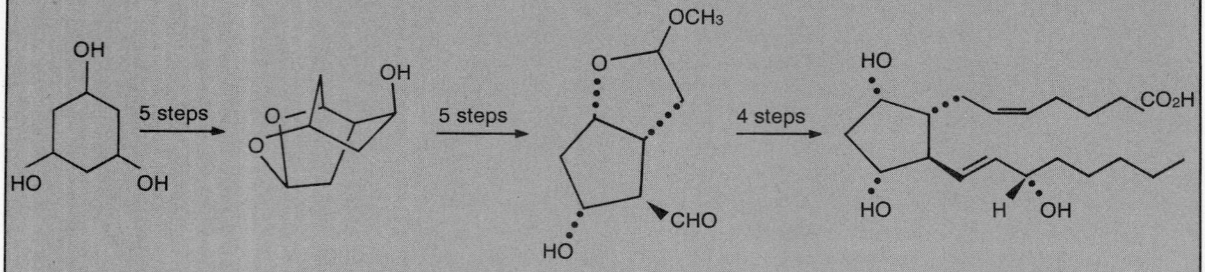

prostaglandin F$_{2\alpha}$

9

lymphoma). Attack on DNA (deoxyribonucleic acid) virus infections was well under way, but most virus diseases are caused by RNA (ribonucleic acid) viruses and much needs to be accomplished in the way of chemical synthesis. By mid-1974 only about six compounds with potential application to human viral infections had been found. Thus, the penicillins and tetracyclines of this problem area still await discovery.

In October 1973 the synthesis of the racemic methyl ester of elenolic acid, which has a broad range of antiviral activity, was reported by R. C. Kelly at the Upjohn Co. The synthesis began with cyclopentadienylsodium and was directed through a series of synthetic manipulations to the racemic methyl ester of elenolic acid.

Li$^+$Cu$^-$

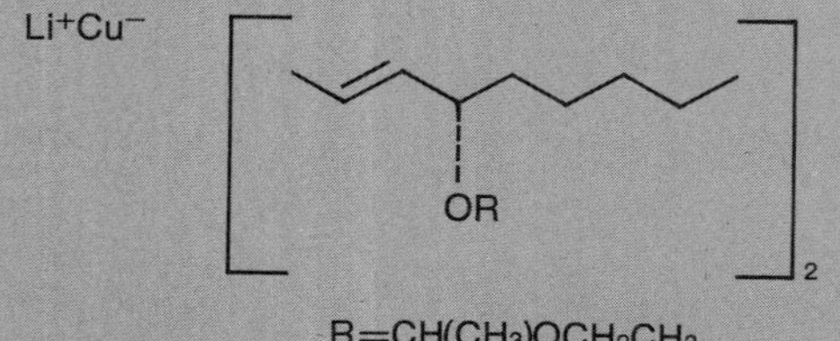

R=CH(CH$_3$)OCH$_2$CH$_3$ 10

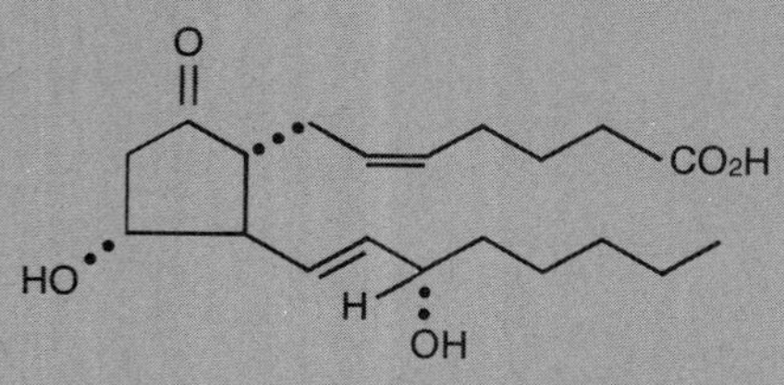

prostaglandin E$_2$

Anticancer agents. In the U.S. National Cancer Institute's cancer conquest program, anticancer agents were being isolated and studied from a variety of plant, microorganism, and animal products. Such living organisms have the capability of synthesizing potentially useful medicinal agents far beyond the present level of scientific design abilities. An interesting example is provided by the Chinese tree *Camptotheca acuminata*. Extracts of the tree have been employed in ancient Chinese medical practice for treatment of cancer, and when this plant was processed in the National Cancer Institute's program, camptothecin was isolated and found useful in the treatment of several animal cancers. Since the tree was native only to China, it became necessary in the West to prepare camptothecin by synthesis. Two dozen or more groups in various parts of the world undertook this problem.

The result over the past two years was a number of camptothecin syntheses. Late in 1973 syntheses completed at Colorado State University and at the University of Rochester showed promise of being adaptable to the preparation of camptothecin on a large scale. Acronycine, isolated from bark of the Australian tree *Acronychia baueri* Schott, represents another experimentally useful substance with broad antitumor activity, and a total synthesis was described by F. N. Lahey at the University of Queensland in Australia.

Some microorganisms are capable of producing compounds with anticancer activity. One of the most useful examples to date is that of adriamycin, which showed promise during the year as an effective agent in the treatment of human cancer, particularly of the solid tumor variety. In 1973, through the combined efforts of chemists at an Italian pharmaceutical company, at the University of Manitoba, and at the National Cancer Institute's

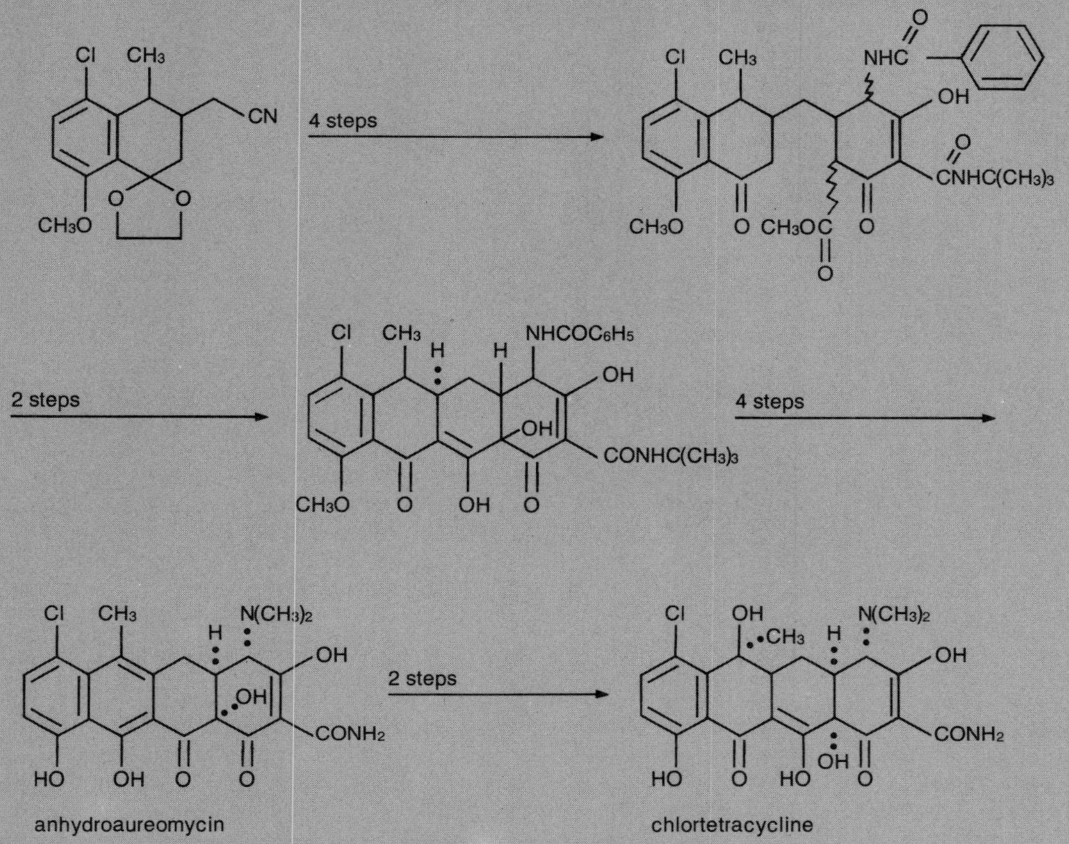

anhydroaureomycin

chlortetracycline

11

laboratory at the Stanford Research Institute, the total synthesis of adriamycin was completed. For commercial production, however, much shorter chemical routes to synthesis must be developed, and this was an objective of programs being conducted at the National Cancer Institute. In the Soviet Union, a substance structurally related to adriamycin was isolated by G. F. Gause and M. G. Brachnikova and their colleagues at the Institute of New Antibiotics in Moscow and was being evaluated in the Soviet Union's cancer conquest program. Again, prospects for human treatment appeared hopeful.

Hormones. Toad poison constituents represent a group of physiologically active steroid hormones in which progress in synthesis has been made recently. Toads of the genus *Bufo* (tailless amphibians of the family Bufonidae) have behind the eyes a pair of skin glands that contain venom. For at least several millennia, the Chinese have used dried venom from a common Chinese toad. The preparation is known as Ch'an Su and has been used in Asia for treating local inflammations, can-

ker sores, toothache, and sinusitis. Today we know that the principal constituents of such toad venoms are steroids of the type represented by bufalin, telocinobufagin, and marinobufotoxin. Closely related steroids such as scillarenin also have been found in plants.

This group of bufadienolides, derived from both toad venoms and plants, exhibits potent biological activity. The cardiac action of bufalin, for example, has been found about equal to digitoxigenin (from digitalis), and in terms of local anesthetic potency is about 90 times more active than cocaine. Another of these toad venom constituents, resibufogenin, is used in modern medicine as a respiratory stimulant. At Arizona State University, George Pettit and his colleagues studied the ability of bufadienolides such as telocinobufagin to inhibit growth of cells derived from a human epidermoid carcinoma (cancer) of the mouth. They wanted to make these substances more available by synthesis and completed an extension of their earlier syntheses of bufalin and scillarenin to the first syntheses of telocinobufagin and marinobufotoxin. Except for scillarenin, all are constituents of Ch'an Su.

In related studies, Pettit and his group also completed the synthesis of the cardiac active steroid, periplogenin, starting with digitoxigenin. Digitoxigenin occurs in the plant *Digitalis purpurea* as a sugar derivative known as digitoxin, which is the modern drug of choice for treating heart failure. The drug is particularly useful for treating those suffering from a combination of heart failure and atrial fibrillation.

At the University of California, a group led by Choh Hao Li began work directed at a complete synthesis of the human growth hormone (somatotropin). The hormone consists of approximately 190 amino acids, and its synthesis presented a formidable task. However, the Li group completed synthesis of several fragments, such as residues 95–136, by employing the polymer support method. As of 1974, researchers hoped that a fragment of the hormone containing less than 40 amino acids may still have all the growth-promoting activity. If such fragments could be synthesized easily, up to 20,000 children in the United States alone would benefit greatly. Presently, children affected by an improperly functioning pituitary gland, or lack of the pituitary due to surgical removal, generally are not treated with the natural hormone because of the very limited supply available from humans at autopsy.

—George R. Pettit

Structural chemistry

Modern structural chemistry depends almost entirely on spectroscopic methods that report to the observer the ways in which different types of radiation, usually electromagnetic, interact with the bonds between atoms. Chemists were also increasingly concerned with the exact geometrical arrangement of the atoms in space. The advances in structural chemistry that occurred during the past year continued to emphasize the use of analytical instrumentation.

Small molecules. The precise determination of the structure and internal energy of a molecule is a lengthy and expensive labor. Judging from past experience, the future of chemistry will be closely related to how much detailed information can be obtained about the exact shapes of molecules, the lengths of the bonds between their atoms, and the exact angles between those bonds. Such calculations have only become practical in recent years by using large, fast, digital computers; these have mostly been limited to small molecules whose structures are already known, so that they serve as a check on the computational methods.

An example of what is presently possible for a very large system if enough computer time is available was demonstrated by scientists at IBM laboratories when they reproduced the molecular geometry of the hydrogen-bonded guanine-cytosine base pair. This complex is of considerable importance in view of its occurrence in DNA, the nucleic acid that plays a major role in transmitting heredity characteristics. It was suggested that mismatching of hydrogen-bonding interactions during protein synthesis may provide a mechanism for altering the genetic code and thus lead to mutations. The entire calculation required eight days on an IBM 360/195 computer. Because 72,000,-000,000 two-electron integrals were required, 100,000 integrals per second were computed, sorted, processed, and retrieved. In spite of this enormous computing power it was still not possible to study the effect of simultaneously moving two of the bonding hydrogens. The binding energy was computed within 15% of the experimental value.

Theoretical chemists are frequently accused of calculating the answers to problems that are already solved. No such criticism could be leveled at the prediction by John Pople (Carnegie-Mellon University) of the structure of the hydrogen-bonded dimer of hydrofluoric acid. Eight months after the prediction, based on pure mathematical physics, a research group at Harvard University published an experimental microwave spectroscopy study that agreed in every detail with the predicted geometry.

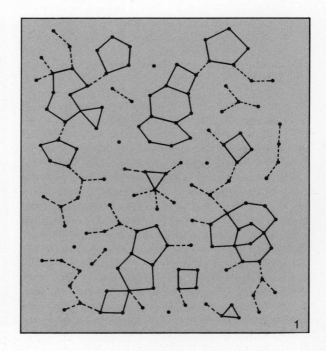

A different approach to calculating the geometries of molecules and also amounts of energy stored in them avoids quantum mechanics entirely. Instead, the method of "molecular mechanics" uses the information that has already been collected about the bond angles, lengths, and energies of well-studied molecules to predict the corresponding parameters for molecules that have not yet been examined or even for hypothetical molecules. The computer treats the atoms and bonds of a molecule like a collection of balls and springs, each of which has an appropriate force field. Calculations by scientists at Princeton, Yale, and the University of Georgia yielded results as accurate and precise as experimental values: bond lengths to within $\pm$ 0.01 Å and bond angles to $\pm$ 1°. This method is much faster and cheaper than quantum mechanics and can be applied to large molecules. It is especially useful for comparing the energies of different conformations of complex ring systems.

Another application of computer calculations to structural problems was reported by scientists at the Bell Laboratories and Argonne National Laboratory. This was an attack on the important question of how water molecules interact with each other in ordinary liquid water. The physical properties of water are so unusual and its importance to terrestrial life so great that it has been a perennial target for investigation by theoreticians. All theories of water structure recognize that it is an extended three-dimensional network of molecules held together by hydrogen bonds. Some theories consider that small clusters of five or six molecules are being made and broken continuously with a flickering, cooperative interaction. The normal lifetime of any cluster is very short, measured in millionths of a second, but the overall effect as molecules align themselves with one cluster after another would be to provide structure to the entire ensemble. In a recent computer study relating 216 water molecules through coupled equations of classical motion, different polygons were generated (1), and it was found that the distribution of polygon sizes depended closely on the value assigned to the hydrogen bond strength. No support was found for earlier theories that had considered liquid water to contain localized regions of icelike structure mixed with unbonded individual molecules.

One of the most important entities in chemistry is the proton (the bare hydrogen nucleus). So great is the charge density on the proton that it is never found free in solution but is transferred from one acid to another. For some time it has been realized that the hydroxonium ion (H_3O^+) formed by the addition of a proton to water is the fundamental unit in aqueous acids. Because hydrogen atoms are small and light they do not scatter X rays well, and their positions in hydroxonium salts were only inferred indirectly from X-ray crystallographic studies. The use of neutron diffraction corroborated these inferences by direct demonstration that H_3O^+ in the monohydrate of p-toluenesulfuric acid is non-planar and that all three O-H bond lengths and angles are equivalent.

Unstable intermediates. A perennial problem of organic chemistry is whether cyclobutadiene (2a) is too unstable to exist. The question is of great importance in view of the extraordinary stability of benzene (2b) and many other cyclic conju-

gated polyolefins. Until cyclobutadiene could be made and isolated the question of its stability could not be answered definitely. In the last decade a number of outstanding scientists presented strong but indirect evidence that the molecule can be made, though it is very unstable at room temperature. New experiments in the last year provided near certainty that this elusive compound was made by the photolysis reaction (2c) using matrix isolation methods (freezing the precursor molecule in a frozen, glassy matrix of solid nitrogen or argon) at 20° K (−253° C).

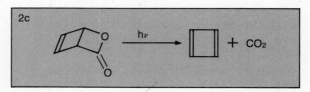

Infrared spectra of the material were consistent with this structure, and labeling experiments using isotopic tracers also supported it. Warming above this very low temperature causes the material to

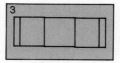

yield a dimer (3), while further photolysis leads to acetylene, H-C≡C-H, the expected product of cracking (2a) in half. The evidence is therefore strong that cyclobutadiene can be made and that it is very unstable.

Another closely related unstable molecule that was finally prepared by matrix isolation and subjected to infrared spectrometry is "benzyne" (4). This was prepared in a solid argon matrix at 8° K

211

and upon warming produced products that had previously been postulated to arise from it.

An unstable species whose possible transient existence challenged organic chemists for a generation is the 2-norbornyl cation (5). Originally

proposed to explain the peculiar rate differences between various norbornyl derivatives, this species became the center of a furious controversy as to whether it could exist as a true molecular ion and, if it did, whether its stability would affect the rates of reactions in which it might be an intermediate. Powerful new evidence for the existence of the ion was presented in recent months by George Olah of Case Western Reserve University using ^{13}C nuclear magnetic resonance (NMR)

spectroscopy at low temperatures and X-ray photoelectron spectroscopy. Since the time scale of the latter method is less than that of a molecular vibration, it can be used to eliminate the possibility that several classical ions (those in which all bonds to carbon atoms are of the normal sigma type and in which the positive charge is localized on a single carbon atom) are interconverting at a rate too fast for resolution on the NMR time scale. Olah claimed that the question of the existence and structure of the norbornyl cation was settled. This did not however completely resolve the issue of its role as a reaction intermediate.

Large molecules. Progress toward establishing the detailed structures and reactive sites of very large biomolecules continued steadily. By 1974 nearly 50 enzymes and large protein systems had been completely worked out by X-ray diffraction. In some cases the binding sites in the presence of the substrate molecule were established. New applications of NMR spectroscopy to the study of enzyme-substrate interactions were beginning to show exactly which functional groups are in very close contact in the complex and thus could be responsible for the enzyme catalysis.

One of the most interesting characteristics of

Blue spirals of chemical activity appear in a shallow solution of bromate, bromide, malonate, and an indicator dye when the surface of the solution is touched with a hot filament. Pictures were taken after 1 minute (upper left), 4½ minutes (lower left), and 8 minutes (right).

living systems is their ability to make extremely complex molecules at rapid rates and with almost unerring stereochemical accuracy. The production of proteins by DNA is sometimes likened to fitting a glove to a hand. Yet the comparison between the performance of the simplest cell and the most advanced synthetic chemistry laboratory is really more like that between high-speed precision casting of intricate parts and knitting mittens by hand. Some progress was made in applying biochemical principles by chemists at UCLA, who created complex "host" molecules with disymmetric cavities into which "guest" molecules of closely specified size, shape, and chirality (handedness) could fit. This was being developed in the direction of constructing an amino acid resolving machine that could separate right- and left-handed isomers.

New methods and applications. Closely related to methods used for structure determination are those used to detect molecules of known structure. A recent report appears to have advanced the limits of such detection enormously. It describes a bioanalytical system that is claimed to have unlimited sensitivity through the use of enzymatic cycling. The activity of a single molecule of glucose-6-phosphate dehydrogenase was detected in support of the author's claim.

Going from the very small to the very large, scientists made rapid advances in radio astronomy by using emission in the millimeter-wave range. This type of spectroscopy had by 1974 enabled the discovery of nearly 30 types of simple organic and inorganic molecules distributed in enormous clouds in interstellar space. Further evidence that the chemical requirements for life may exist in extraterrestrial systems is found in reports of traces of amino acids in samples of moon rocks.

Not all progress in structural chemistry occurs at the research frontier of science. Often it is determined in the marketplace at the time that well-known methods are reduced to convenient practice and packaged as commercial instruments. The rapid advancement of computer-controlled X-ray diffraction is an example. X-ray crystallography has long been considered the ultimate weapon in attacking molecular structure problems, but until recently it has been a costly, time-consuming technique restricted to a few experts. Noteworthy, therefore, was a report that F. A. Cotton had settled the structure of a complicated new molecule within 60 hours using only 50 micrograms of material at a total computing cost of $157. He did this with a packaged commercial instrument. A conventional study by other means would have taken longer, cost more, and not given an unequivocal answer.

—Edward M. Arnett

Communications

Communications took on increased importance during the year as the energy shortage threatened to make people less mobile. There was continued growth in the many roles played by communication in modern industrialized society. Science and technology made further improvements in the transmission of speech, writing, pictures, and documents and of data vital to the world of business.

Technical advances. Rapid progress continued in the emerging art of fashioning hair-thin glass fibers for transmitting digitized information by light impulses and in the development of associated electronic devices and techniques. Such optical communication offered attractive advantages over conventional electrical media (such as cables), especially with regard to size and interference from other signals. (*See* Feature Article: FIBER OPTICS: COMMUNICATIONS SYSTEM OF THE FUTURE?)

A new kind of solid-state descendant of the transistor, the charge-coupled-device (CCD), reached the stage of refinement required for use as the light-sensitive target in a video camera. Recently constructed CCD chips had a rectangular array of 256 by 220 cells, each capable of converting the incident light level to an electric charge that could be stored and, on command, shifted to the neighboring cell. This shifting capability made it possible for the device to carry out the image-scanning function normally performed by the sweeping electron beam of a cathode-ray tube.

The reduction in size and weight was impressive, a complete camera weighing less than one pound. Demonstrations were held of a color TV camera that used three of the tiny rectangular array CCD chips and of a high-resolution facsimile scanner that used a chip with 1,600 cells in a linear array. Before the CCD or similar solid-state technology could gain commercial acceptance in the video camera market, however, it would have to prove its superiority over highly refined vacuum tube technology, such as the versatile, inexpensive Vidicon. (*See* Year in Review: ELECTRONICS.)

A new kind of visual communications system, the remote blackboard, was introduced into experimental service for educational purposes. The instructor writes with a stylus on what looks like an ordinary blackboard. Electrical signals are generated denoting the position of the stylus, and these signals are communicated to distant locations, where they control an extremely well-focused laser beam. The laser beam, in turn, "writes" on a heat-sensitive film, which is simultaneously projected on a screen. Thus, as the instructor writes and draws, every detail appears, like magic, on the distant screen. The information signals transmitted are digital data sequences, representing the locations of the stylus as successive horizontal and vertical coordinates. For smaller audiences, the laser and screen at the receiving end can be replaced by a TV set. The instructor's voice is transmitted separately by telephone line.

Telephony. The total number of telephones in service throughout the world increased by 7.4% during the most recently tabulated year. This brought the total to just under 313 million, nearly double the figure ten years earlier. The U.S. continued to lead with 132 million telephones, and Americans completed more than 180,000,000,000 calls. Japan and the United Kingdom followed with 34 million and 18 million phones, respectively. China made its first appearance on the list of countries having more than half a million telephones, bringing the total to 39. Bell System telephones connected with 98% of the world's telephones, and overseas calls from the U.S. grew by 28% in just one year—to more than 50 million calls annually. Of the world's 21 metropolitan areas having more than one million telephones, 8 were in the U.S.: Baltimore, Md.; Chicago; Detroit; Houston, Tex.; Los Angeles; Minneapolis-St. Paul, Minn.; New York City; and Philadelphia. The leader in number of phones per person was Washington, D.C.

In the U.S., new technical developments in transmission systems helped interconnect the growing number of telephones and the increasingly talkative users. At the current growth rate, the number of voice-band channel-miles would double approximately every seven years. L5, the jumbo jet of transmission systems, began commercial service early in 1974, linking Pittsburgh, Pa., and St. Louis, Mo., a distance of more than 800 mi. This carrier system could accommodate 108,000 two-way voice channels on ten pairs of coaxial cables. Each cable carried 10,800 individual one-way voice channels multiplexed in three "jumbo groups" of 3,600 each. A jumbo group, in turn, was made up of six master groups; each master group comprised ten super groups (60 channels), and each super group was made up of five groups (12 channels each). Along its route, each cable had amplifiers or "repeaters" spaced at one-mile intervals. Like the jumbo jet airliner, the new communications system about tripled the capacity of its predecessor.

Perhaps even more significant, at least in the long run, was the introduction of a digital transmission system, capable of handling 96 voice channels, that used a high-speed stream of more than six million binary pulses per second (megabits/sec). Despite this very high rate, the transmission medium was a relatively inexpensive twisted-wire cable, with signal regenerators spaced up to 4.5 km apart. Although voice signals are inherently analog in nature—that is, continuously variable—they can readily be "digitized" by so-called analog-to-digital converters, which have been greatly improved over the years. Similarly, a Picturephone or other video signal can be digitized and sent within the six megabit/sec stream. Such digital coding, using only two levels of recognized pulse voltage, proved advantageous in preserving the information in the face of interference and similar impediments. Moreover, the increasingly voluminous computer-generated data signals are intrinsically digital in form, and so the

Top ten in world telephones					
Countries		**Cities**		**Density by country**	
Total number of telephones		(* means greater metropolitan area)		Telephones per 100 people	
United States	131,606,000	New York	5,922,128	Monaco	75.30
Japan	34,021,155	*Tokyo	5,155,069	United States	62.75
United Kingdom	17,570,904	*Los Angeles	5,067,189	Bermuda	60.39
West Germany	16,521,149	*London	3,970,938	Sweden	59.29
Soviet Union	13,198,700	*Paris	2,937,409	Liechtenstein	55.30
Italy	11,345,497	*Osaka	2,479,726	Switzerland	53.95
Canada	10,987,141	Chicago	2,389,073	Canada	49.98
France	10,338,000	Philadelphia	1,574,692	New Zealand	44.61
Spain	5,712,549	*Detroit	1,414,424	Denmark	37.93
Sweden	4,829,047	*Minneapolis-St. Paul	1,315,800	Iceland	37.10

new transmission system was well matched to them.

Lower-speed (1.5 megabits/sec), shorter-distance, digital transmission systems had been experiencing a phenomenal rate of growth in the U.S. over the past dozen years. The new system, with four times the speed and ten times the distance capability, was the second step in a planned hierarchy of digital transmission systems that would culminate in an almost 300 megabit/sec system on a coaxial cable carrying more than 4,000 voice channels. Although such cables were capable of spanning coast to coast, it was expected that they would be used primarily in large metropolitan areas. Meanwhile, digital radio systems also appeared, and an extensive field test was begun of the millimeter waveguide, with 60 times the capacity of the 4,000-channel coaxial cable. These small hollow tubes were expected to be the basis of transcontinental communications in the 1980s.

Other telephone-related developments, more visible to the user, included a portable conference telephone that allowed classroom-size groups of people to listen and talk with their hands free, new compact speaker-phones for individual hands-free use, and telephones with built-in memories for automatic dialing of frequently called numbers. One of these was a solid-state memory-equipped instrument called the Touch-a-matic. At the touch of one of 31 buttons, the set immediately called the party whose number had been recorded previously by the user (by dialing it manually after pressing the record button). In addition, one memory slot remembered the last manually dialed call, so that, in case of a busy signal or no answer, the caller could try again by touching button no. 32. Improvements were introduced even in such prosaic things as wire and cable insulation. Central office cable, for example, was reduced by 25% in volume and 9% in weight as a result of a new type of polyvinyl chloride plastic insulation, one layer of which replaced multiple layers of plastic, cotton, and lacquer.

Data communication usage. Business and industry continued to increase their use of centralized electronic data processing. The result was vigorous growth in the transmission of digital data —the kind of numerical and alphanumeric language spoken and understood by computers and other business machines. Data communication made its appearance, for example, in some large department stores, where "point-of-sale terminals" that could handle a credit card purchase, including updating the customer's balance stored in a centralized computer, began to replace conventional cash registers (*see* Year in Review: COMPUT-

ERS). In other cases, data communication was used primarily to determine whether the customer's credit standing was good. An impressive installation that entered service during 1973 was the elaborate, nationwide Wizard of Avis system, which enabled a salesclerk to handle an entire car-rental transaction by merely "typing" the essential information into a distant central computer. The latter made all the appropriate choices on the basis of information stored in its memory and responded with data messages that activated the clerk's printer and filled out the entire contract.

Another growing use of data communication concerned the handling of payroll, inventory, or parts ordering within far-flung corporations. Such applications could involve large batches of data, making it desirable to maximize the speed of data transfer. A rate of 9,600 bits per second was the highest offered commercially for use over voiceband telephone channels, by far the most prevalent data-communication medium. For such purposes the voiceband channels were fitted at each end with a so-called modem (modulator-demodulator) or data set, which converted the digital messages into voicelike signals suitable for transmission over the analog channels and back into digital form at the receiving end. Improvements in modems included the introduction of large-scale integrated, solid-state technology, as well as diagnostic features that allowed users to pinpoint failures in their data systems.

Data transmission involving mobile radio also began to appear. For example, in a system put on demonstration trial in St. Louis, Mo., police cars automatically sent digital messages describing their compass heading and odometer reading. These messages were received at a central minicomputer installation, where the location of each reporting vehicle was continually computed and displayed on a color TV receiver. The display was designed to be accurate within 50 ft. In addition, up to 99 different coded messages could be sent manually in ten milliseconds by punching two of ten pushbuttons on a keyboard. Such messages could be treated at various levels of priority, with voice communication reserved for emergencies.

Digital data communication networks. While data communication of all kinds experienced great growth, particular emphasis continued to be placed on the development of networks for the transmission of digital data. Construction of the initial phase of the American Telephone and Telegraph Company's Dataphone Digital Service, linking Boston, Chicago, New York City, Philadelphia, and Washington, D.C., was completed in mid-1974, and it was expected that 96 cities would be included in the network by the end of 1976. For

Television camera utilizing solid-state charge-coupled devices (CCDs) measures only about 15 cm (6 in.) in length and weighs less than 1 lb.

shorter distances, already existing digital channels (operating at 1.5 megabits/sec) were used. To bridge long distances, a new technological development called Data Under Voice, or DUV, was employed. DUV permitted a 1.5-megabit pulse stream to hitchhike on an already existing microwave radio beam, typically carrying more than 1,000 voice conversations. This could be achieved because the frequency spectrum of the signal modulating the radio carrier was mostly empty below about 560 kHz and because sufficient compression could be applied to the 1.5-megabit pulse signal to make its spectrum fit into the available frequency band. The compression was accomplished by encoding the two-level pulses into a multilevel signal—seven levels in this particular case. The high-speed pulse streams were subdivided so that four lower speeds were available to users: 56, 9.6, 4.8, and 2.4 kilobits per second.

Meanwhile, other communications carriers were either constructing or planning new networks to offer similar types of service. Western Union International proposed a new international digital data service based on the same speeds. The Canadian Dataroute, entering its second year of service, also offered lower speeds for use with keyboard machines such as teletypewriters. Another independent digital network, under construction by Datran, had its nucleus in the southwestern U.S. The most recent trend was to provide so-called packet communication service, in which a customer's message was treated as one or more pack-

ets, about 1,000 or so bits in length, with its destination address and other relevant information included among the bits. The packet could route itself through a special communications network, which generally utilized minicomputers at the junction or branch points.

Other developments. The first of the packet communication applications was acted on favorably by the Federal Communications Commission, and additional applicants entered the arena. This appeared to be in line with the FCC's policy of achieving a diversity of attractive services by encouraging competition. Controversy arose, however, over the FCC's attempt to foster competition within the tightly regulated telecommunications business. Competition among carriers with differing geographic coverage showed signs of leading to the demise of nationwide price averaging, whereby the heavy-traffic communication routes subsidized the more costly rural ones and also paid for a portion of residential telephone service. Thus an important debate was under way, possibly involving far-reaching economic and policy issues.

As transoceanic television and telephone transmission by satellite became more commonplace, progress continued in the planning of domestic satellites to compete with the long-distance land-based communications channels. The Canadian Telesat system was already operational, and Western Union and other major U.S. carriers were planning similar systems. Western Union launched the first U.S. domestic satellite in April 1974 (*see* Year in Review: ASTRONAUTICS AND SPACE EXPLORATION: *Earth Satellites*).

The beginning of 1974 marked the start of another four-year round of work by the Comité Consultatif International Télégraphique et Téléphonique (CCITT), an international body meeting periodically in Geneva. Without the standardization that this organization sought to achieve, such instant global communication as direct dialing from one country to another would be difficult if not impossible. Work on setting up standards was progressing in the areas of transmission, signaling, switching, data signal formats, data transmission networks, and many related details.

—Ernest R. Kretzmer

Computers

Events in 1974 demonstrated once again the rapid evolution, if not revolution, in the development and application of computers. The retail store point-of-sale computer system was an example of an application that will become a part of everyday life within a few years.

Point-of-sale systems. The traditional cash registers of retail stores are about to be replaced by console terminals that are connected to minicomputers. These systems can do all the arithmetic that cash registers can do and, in addition, record the pertinent information for the store's records. The records are maintained inside the computer and are continuously updated by the point-of-sale information. Consequently, management can have up-to-date information on such matters as sales, inventories, special items, and profits. Furthermore, the expense of keeping the store's books can be substantially reduced.

A more novel aspect of the minicomputer systems is their ability to read directly the name and price of the items as they are sold to the customer. The name and price information is put on the items in a special computer-readable form (similar to the magnetic ink numbers on checks). This information is read by passing a magnetic wand over the item or by sliding the item across an optical laser reading device. The wand approach is more flexible for bulky items (clothes, hardware, appliances) with a tag attached. Supermarkets primarily sell smaller items, and in such situations it is practical simply to slide the hamburger, cookies, and cabbage over a reading slot at the checkout counter as they are being placed in a bag. The main advantage to the customer is faster and more accurate service, and a fringe benefit is that the point-of-sale tape receipt can contain the name as well as the price of each item purchased.

Sidney Harris

"7×8=52, 9—3=5 . . . things like that."

The two technologies used to operate these systems are optical character reading and magnetic sensing. In the optical systems the information is encoded in a set of wide and narrow stripes on the item, and the laser reading device decodes it. The widespread use of point-of-sale systems will require a standard approach to marking items.

The first point-of-sale systems consisted of one minicomputer connected to just a few sales positions. However, the newer systems can be connected together in storewide or citywide networks. Indeed, one could foresee the time when within a few hours after the last of his hundreds of stores close the president of Sears, Roebuck and Co. can find out just how successful was the Labor Day clearance of lawn mowers.

Personal computers. While the minicomputers are exciting, the potential of the even smaller hand-held calculators is enormous. These devices were the subjects of two significant developments in 1974. First was a dramatic reduction in price of the hand-held calculators and second was the introduction of a hand-held computer. The difference between a calculator and a computer is that the user can write programs for a computer but can only push the buttons on a calculator.

The simple calculators do all the arithmetic operations (addition, subtraction, multiplication, and division), and millions of these have been sold. The economies of automated production and large sales allowed the price of many models to fall below $50. The proliferation of these calculators may have a direct impact on our educational system.

217

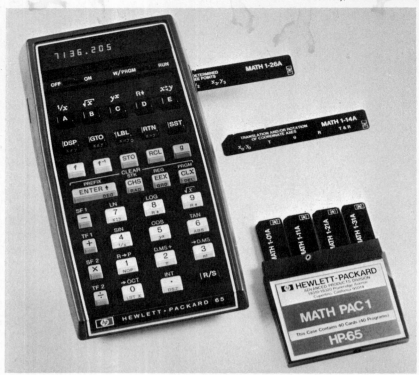

HP-65, the first hand-held computer, was marketed in 1974 by the Hewlett-Packard Co. Users of the 6-in. by 3-in. device can expand its capability with six prerecorded programs, designed for mathematical, statistical, engineering, medical, and surveying applications.

Several universities banned them from examinations because of the advantage they give to students who use them as compared to those who have only a slide rule or pencil and paper. Some people remarked, only half in jest, that because of the hand-held calculators it is no longer necessary for their children to learn to add well. Of course, it can be argued that it would be unwise to depend completely on a device that might suddenly break down at an inopportune time, but the proposal does indicate how our future lives can be fundamentally changed by advances in computer and electronic technology.

The first hand-held computer was marketed in 1974 by Hewlett-Packard Co. (the HP-65) at a price of $795. This 6-in. by 3-in. device contained the equivalent of 75,000 transistors and was in many ways comparable (even superior) to the huge computer installations of the middle 1940s. While the HP-65 has limited programming capabilities, so did those giant forerunners of modern computing. The HP-65 allows only 100 instructions in a program, and there are only 14 words of memory available. Even so, the solutions to a great number of problems in science, engineering, and finance can be programmed and saved on a plastic magnetized card. Then, at some later time, the plastic card can be slipped into the computer, data relating to a new problem entered, and a solution to

the new problem obtained in a few seconds. About 300 programs may be obtained with the HP-65 for the handling of such problems as solution of triangles, solution of polynomial equations (quadratic through fifth degree), analysis of variance, survey engineering calculations, medical science computations, and electrical circuit analysis. The user can combine these standard programs with his own customized ones to obtain a powerful and sophisticated computing capability.

It seems reasonable to expect significantly more powerful hand-held computers to become available within a few years at prices below $500. More powerful in this context means that longer programs will be allowed, the size of the memory will be expanded, and the program control operations will be improved.

Parallel computers. At the opposite end of the spectrum from the hand-held devices are the large, extremely fast computers. By 1970 it was easy to see that the absolute limit on the speed of computations was being approached. The speed of the flow of electricity in wires (which is about the speed of light) limits the speed at which a computer can operate. By 1970 one million instructions per second (MIPS) were common, and these computers had critical components that could not be separated by more than about 25 cm. By 1974 the improved technology could produce compo-

218

nents that operated at 10 MIPS, and 40 to 50 MIPS appeared to be the limit in speed, assuming that the entire computer processor is about 1 cu cm in size (equivalent to a small grape). However, it was very difficult to incorporate a substantial (100,000 words) memory within 1 cu cm, thereby supporting the conclusion that the upper limit on the speed of the traditional computer is about 10 to 30 MIPS for general-purpose applications.

By 1974 four computers had been designed to overcome this limitation on speed: the ILLIAC IV (designed at University of Illinois and built by Burroughs Corp.), the STAR-100 (Control Data Corp.), the ASC-Advanced Scientific Computer (Texas Instruments Inc.), and the STARAN (Goodyear Aerospace Corp.). They were similar in that each had entered serious field testing, each was expensive ($15 million–$40 million), each hoped to achieve operational speeds of 150 MIPS (at least in certain situations), and each was based on the idea of having many of the computations for a problem executed simultaneously by arrays of processors. This method, called parallel computation, achieved a gain in speed. There are significant variations among these computers in the way parallel computation is achieved, but each can have at least 64 essentially identical operations taking place at once. In particular, the access to memory is also made in parallel so that it is not necessary to put a large memory into a very small space in order to achieve continuous high-speed operation.

The advantages to be gained by parallel computation are still largely unknown. In some important applications parallelism can increase the computation speed almost indefinitely. It is important to note that in some situations there is no substitute for speed. That is, one hour of computation at 100 MIPS cannot be replaced by 100 hours at 1 MIPS. Examples of this include weather prediction (it does no good to take 48 hours to compute a 24-hour weather forecast), air traffic control (it does no good to learn that 10 minutes ago two planes were going to collide within 2 minutes), credit checking (it does no good to learn that the person who walked out of the store 20 minutes ago with a new TV set is using a stolen credit card). On the other hand, there is some conjecture that the average, general-purpose computation can make very little use of parallelism because these computations must be performed sequentially.

Computer software. The problem of developing computer programs, or software, has long been a painful one. By 1974 software accounted for 75% of the total cost of a typical application. Thus, when one buys or rents a computer, a large portion of the price goes for the basic software to make it operate. The magnitude of this problem

becomes apparent from the fact that $11 billion (over 1% of the gross national product) was spent in 1974 on software.

The software problem came to a head in 1973–74 and began to receive much more widespread and intensive study. The two main questions were, of course: Why does software require so much time, effort, and expense to develop? and What can be done about it? The basic source of the difficulty is that computer programming is a complex, high-level intellectual activity.

Several systematic studies of software development projects showed that the majority of the effort was not spent on analyzing the problem and writing the programs but rather on checking and testing the programs after they were first written. The reason for this is that the testing revealed a continuous stream of errors made in the problem analysis and programming phase. A logical conclusion is that much more time should be devoted to problem analysis and program design. Another cause of software difficulties is the common practice of buying and/or designing a computer (hardware) and delaying software development until the hardware is delivered. Many financial disasters could have been avoided by starting on the soft-

Point-of-sale minicomputer and laser reader automatically read the name and price of groceries as they slide over the reading slot (under the can).

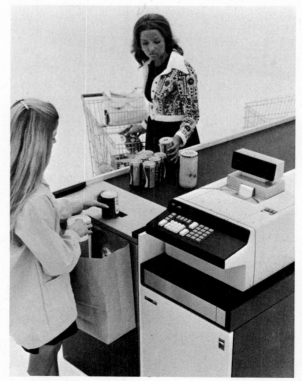

Courtesy, IBM Corp.

ware well in advance of considering the hardware. After all, the software cost is three or four times that of the hardware, and so it is reasonable to consider it first.

Structured programming. One of the keys to efficient software development is proper organization. The need for this becomes obvious when one realizes that many systems involve millions of instructions written by thousands of programmers. Although the need to be organized has long been recognized, only recently have the concepts involved been crystallized into what has become known as structured programming. The general idea in this approach is to force all the problem analysis and written programs to have a structure that satisfies a number of rules. Examples of such rules include never allowing a "GO TO" statement and never permitting any subprogram to have more than about 200 statements. The structured programming rules force the logical structure of a program to be branched like a tree, and are intended to force people to think more clearly.

Intertwined with structured programming was an effort to achieve higher level languages. The higher the level of the language, the less instruction is required for a particular task and, therefore, the fewer the errors made. Higher level languages, however, are not cheap. Elaborate translation programs are needed to produce the machine language instructions of the computer, and writing and operating these programs are expensive. Nevertheless, overall software development costs may be significantly reduced by the use of higher level languages.

In the future the computing profession will experience an increasing variety of interacting tools and concepts to overcome the software bottleneck. As of 1974 the eventual combination to be preferred was not clear, but the outcome is certain to be much more efficient and cheaper software.

—John R. Rice

See also Year in Review: ELECTRONICS.

Earth sciences

The search for new sources of power, an effort intensified by the Arab cutbacks in oil production, occupied the attention of many earth scientists during the year. Other research was concerned with such subjects as lunar rocks, early fossil remains, and earthquake prediction.

Geology and geochemistry

The energy crisis that began late in 1973 served as a warning about the consequences of exploit-

ing a finite energy supply and stimulated vastly increased expenditures for research, planning, and development. The increase in research in geology was directed toward both short-range needs and long-range plans.

The search for energy sources. Alternatives to petroleum as a source of liquid hydrocarbons were receiving new attention. The U.S. Department of the Interior arranged the first oil shale lease near Rangely in northwestern Colorado. Two environmental problems were causing concern there. The processing of oil shale takes huge volumes of water in a part of the country where water is already in short supply. Also, one process leaves a very fine powdery ash that is greater in volume than the shale before it is mined and is difficult to dispose of. However, one consortium successfully tested a new process that permits the shale to be kept underground, in this way eliminating the disposal problem.

Geothermal energy showed promise as an important source of power in the future. Both dry steam and wet steam sources were being used. The three best-known geothermal plants—Larderello in Italy, Matsukawa in Japan, and the Geysers in California—are all dry steam installations. It is difficult to estimate how important geothermal energy can be. Present estimates are that it can furnish in the range of 0.5–20% of the world's power needs by 1985. (*See* Feature Article: POWER FROM THE EARTH.)

In a new method for producing natural gas, three nuclear bombs were detonated about one mile below the earth's surface in western Colorado in an experiment called Project Rio Blanco. Three atomic devices stacked one above the other about 450 ft apart were detonated to produce a cylindrical chamber in sandstone known to contain gas-bearing pockets. It was estimated that the pocket would produce more than 20 billion cu ft of gas over the next 20 years. The initial production, however, was disappointing.

Ore deposits. By 1974 the production of some metals had already reached the critical stage. Adequate supplies of iron, aluminum, chromium, and nickel are assured until well into the 21st century, but some authorities believed that mercury, lead, silver, and zinc may become critically scarce in the next 20 years.

Some hope for increasing the earth's supply of metals lies in the billions of tons of them contained in manganese nodules that occur over many areas of the ocean floor. Besides manganese these nodules contain nickel and copper in higher concentrations than most ores being mined today, and also some cobalt and iron. Only a few areas have enough nodules to make mining feasible, the

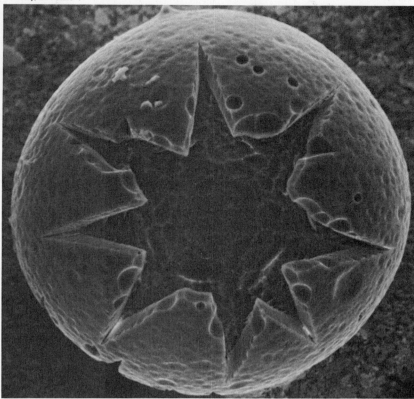

Microtektite with a star-shaped pit was recovered from sediments in the Caribbean Sea. These glass particles are thought to have fallen onto the earth in a single event about 35 million years ago.

richest deposits occurring in the Pacific Ocean in a belt running roughly east-west at about latitude 9° N southeast of Hawaii in waters about 5,000 m deep. Whether or not they can be mined from deep waters at a profit depends on whether new marine technology can be developed. Also, areas of crustal plate boundaries are of particular interest in the search for metallic ores on the seafloor. Analyses of manganese-rich rocks dredged from the rift valley in the middle of the Mid-Atlantic Ridge indicate that the concentrating of metals on the ocean floor might be more efficient than previously believed.

Extraterrestrial geology. The last exploration by men on the moon, Apollo 17, took place on Dec. 11–14, 1972. However, analyses of the rocks and the geophysical data obtained by these and earlier astronauts continued. Among the major findings was that all lunar rocks are differentiates. That is, they have a chemical composition that could only have been produced by fractionating processes in magma (molten rock). No primitive undifferentiated lunar material has been found. All the rocks that had been sampled as of mid-1974 had been completely melted at one time or another.

In general, the rocks of the moon are breccias and anorthosites in the highlands and basalt in the maria. The age of the highland rocks falls with re-markable consistency at 3,900,000,000 to 4,000,-000,000 years, indicating that a major cataclysm affected the moon at that time. The basalts of the lunar maria, generated by radioactive heating in an iron-rich region of the interior of the moon, are rich in iron and titanium. Their ages range from 3,150,000,000 to 3,850,000,000 years.

The unmanned U.S. scientific space probe Mariner 10 passed by the planet Mercury between March 23 and April 3, 1974. It found that Mercury greatly resembles the moon in having areas of lava plains, rugged landscapes with overlapping craters, scarps, and ridges. However, unlike the moon, there are irregular scarps up to one kilometer high extending for hundreds of miles.

When Mariner 10 passed by Venus on Feb. 5, 1974, it did not get photographs of the planet's surface because of the heavy cloud cover. However, scientists of the Jet Propulsion Laboratory at Pasadena, Calif., made a ground-based radar map of Venus and showed that it, too, is pocked with craters.

Marine geology. The cruises of the research ship "Glomar Challenger" continued to add an annual store of information about the geology of the oceans. The bottom of the Philippine Sea was found to be a complex of fractured trenches and

221

volcanic activity. The oldest rocks there are only 60 million years in age, which is surprisingly young and contrasts with the 100-million-year age of the oldest rocks on the sea floor that lies to the east of the Mariana Trench. This finding confirms that the trench is the junction of two major crustal plates.

Drilling in the Pacific Ocean by the "Glomar Challenger" helped elucidate the evolution of the Atlantic Ocean. This paradox came about when researchers discovered in the northwest Pacific an unusual magnetic band that corresponds to a similar band on both sides of the floor of the Atlantic Ocean. From the width of the band and the age of the rocks in it, scientists could confirm an episode about 115 million to 150 million years ago when seafloor spreading was about twice as fast as normal. This correlates well with the equivalent magnetic bands in the South Atlantic which occurred about 125 million to 130 million years ago when the Atlantic Ocean was a narrow strip of water. Both of the bands in the Atlantic and the one in the Pacific are assumed to be due to the same episode of accelerated seafloor spreading.

Man has now dived into the trench in the middle of the Mid-Atlantic Ridge. Xavier Le Pichon and other scientists went down in the French bathyscaphe "Archimède" and reported many fresh lava flows and the unexpected fact that parts of the wall of the trench are vertical.

Paleontology. Pterosaur tracks of ten different kinds were described by William L. Stokes as occurring in two localities in the Navajo Sandstone from the Lower Jurassic period in Utah. These are probably the oldest known evidence of pterosaurs (flying reptiles) in the world. The tracks are quite numerous at both sites.

The diving bird *Hesperornis* has long been known from the Cretaceous rocks of Kansas. P. D. Gingerich studied a skull of *Hesperornis* at Yale University and further substantiated the evolution of birds from theropod dinosaurs. The theropods were bipedal dinosaurs with typically reptilian pelvic bones. One rather unexpected finding was that birds are not as closely related to the Ornithischia, the order of dinosaurs with a pelvic structure that did resemble that of birds.

Trilobites, the abundant marine arthropods of

Tubular lava photographed at a depth of about 6,300 ft lies in the axial valley of the Mid-Atlantic Ridge. The tube, approximately 1 ft in diameter, is probably less than 10,000 years old.

Courtesy, Woods Hole Oceanographic Institution

the Lower Paleozoic Era, were shown to have remarkable eyes with lenses made of calcite, probably capable of producing clear images over a wide depth of field. According to Kenneth M. Towe of the Smithsonian Institution, the crystal orientation is consistent from specimen to specimen, showing that the calcite was a product of biomineralization and not due to replacement by calcite after burial in the rock.

A pair of trilobite specimens from the Cambrian Period were found in the Carolina Slate Group in Stanly County, N.C., by Joseph St. Jean, Jr., of the University of North Carolina. They probably belong to the genus *Paradoxides.* No identifiable invertebrate fossils are otherwise known from the Piedmont of the Carolinas. These specimens provided much-needed proof that at least part of the extensive, but inadequately known, Carolina Slate Group is of Cambrian age.

The sudden appearance of the fossil remains of invertebrates in the earliest part of the Cambrian has long been one of the great puzzles of the history of life on earth. According to Steven M. Stanley of Johns Hopkins University, these invertebrates were part of a burst of evolution that was triggered when new forms of life replaced the availability of natural resources as the limiting factor in the environment. As envisioned by Stevens available environments were clogged with algae. Very slowly evolving invertebrates finally reached a point where they could consume significant amounts of algae. This almost literally left the algae with room to diversify and they did. This led, in turn, to diversification of the herbivorous invertebrates that grazed on the algae and of the newly evolved carnivorous invertebrates that fed on the herbivorous ones.

The oldest known occurrence of cell diversification was found in Precambrian rocks that are about 2,200,000,000 years old, located in Transvaal Sequence in South Africa. Lois Anne Nagy of the University of Arizona reported enlarged cells in the filaments of the blue-green algae making up a stromatolite. Older stromatolites, presumably algal structures, had been found in the Bulawayan Group of Rhodesia (2,700,000,000–3,100,000,000 years), but the cellular structures of the organism creating them are unknown.

Other developments. Many microtektites have been found in sediments in the Caribbean. These are particles of glass that were shaped by passage in the atmosphere while molten and resemble the larger tektites found in Texas and Georgia. According to Billy P. Glass of the University of Delaware, all of these tektites, as much as a 100 million tons, fell in a single mysterious event about 35 million years ago.

Lonar Crater in India was established as the only known terrestrial impact crater in basalt. Kurt Fredriksson of the Smithsonian Institution and geologists of the Geological Survey of India found on the crater rim impact breccias as well as shock fragments that show signs of sudden heating. The fact that the bedrock is basalt offers opportunity for comparisons with lunar craters that craters in sedimentary or granite terrains do not provide.

—Walter H. Wheeler

Geophysics

As the world's population grows, man becomes increasingly aware of the diverse natural hazards and limited material resources of his planet. It is not surprising, then, that geophysics—the science charged with studying the physics of the earth—received active attention during the year.

Geophysics and energy. Nowhere was the limited quantity of the earth's resources brought into sharper focus than in the critical shortage of energy, particularly of petroleum, that became increasingly apparent during the year. Virtually all the shallow, easy-to-find, reserves of petroleum in the contiguous U.S. had been found, and these reserves were being exhausted. Consequently, the search for petroleum had become much more sophisticated. The exploration targets, at greater depths in the earth's crust or offshore on the continental shelf, required more intensive application of geophysical prospecting techniques than ever before.

Of the approximately $1 billion per year being spent worldwide on geophysical exploration techniques, about 90% was allocated for the seismic reflection method used primarily in petroleum exploration. In this method, elastic waves are generated by explosions or other means in shallow holes or in water, and transducers at the surface record the reflections and diffractions of these waves from structures within the earth's crust. The interpretation of these reflections involves an immense technical job of information processing and advanced numerical calculation. During the year a new field system was introduced capable of recording 1,000 channels of data and processing these data immediately in the field. This system represented a tenfold increase in the volume of information that could be collected at one time. In addition, the field-processing capability allowed field operations to be modified on the basis of the data evaluations as the work progressed.

Perhaps the most significant of the new data-processing techniques was one that reportedly indicated the presence or absence of hydrocarbons within a given structure. This technique was based on a correlation between variations in the intensity of the waves reflected from a particular rock layer and the amount of hydrocarbons present. Previously, geologic and geophysical techniques could only determine whether or not a structure was favorable for the accumulation of hydrocarbons. If this new technique proved practical, it could significantly reduce the cost of petroleum exploration by increasing the ratio of successful to dry wells. Currently, only about one out of every eight exploration wells drilled at sites chosen on the basis of favorable geologic and geophysical evidence is successful.

Geophysicists were also deeply involved in finding other forms of energy. During the year a great deal of effort was expended in developing techniques for discovering and proving sources of geothermal energy. A variety of techniques were tested in the Long Valley area of eastern California, a volcanic caldera of Pleistocene age. While more intense, widespread geothermal activity had been associated with this caldera in the past, by the 1970s geothermal activity occurred along recently active faults. Gravity, magnetic, seismic refraction, seismic noise, heat flow, microearthquake, and a variety of electrical techniques, as well as geologic and geochemical studies, were applied to this area by the U.S. Geological Survey in an attempt to understand the structure and processes of the caldera and to determine which techniques held the most promise. (*See* Feature Article: POWER FROM THE EARTH.)

Studying volcanoes with satellites. Although there are more than 500 historically active—and therefore potentially hazardous—volcanoes throughout the world, not more than a few may be erupting at any one time. Extensive studies at a few volcanoes have suggested that instrumental indications of an impending eruption precede even the visual signs, but the establishment of observatories to make routine measurements at a large number of potentially active volcanoes would be prohibitively expensive. A group from the U.S. Geological Survey in Menlo Park, Calif., took a long step toward the solution of this problem by using the Earth Resources Technology Satellite (ERTS) to monitor relatively simple remote instrument packages on 15 volcanoes in the Western Hemisphere.

Instrumentation included earthquake counters and tiltmeters, since an increase in the number of local earthquakes and a tilting of the earth's surface associated with the swelling of the volcano as it fills with magma appeared to be two of the most useful premonitors of an eruption. Counts of earthquakes and measured changes in tilt at the selected sites were relayed via the satellite back to Menlo Park for analysis. Preliminary results were extremely encouraging. Six days prior to the Feb. 22, 1973, eruption of Volcano Fuego in Guatemala, the rate of earthquakes counted at sites 5 and 15 km from the volcano increased by a factor of 15.

The Geodynamics Project. The International Geodynamics Project was designed to increase understanding of the physics and history of the dynamic processes that shape the earth. It was motivated by the need to consolidate and extend the essentially interdisciplinary efforts that had led to the discovery and substantiation of the concept of plate tectonics.

During the year the U.S. Geodynamics Committee of the National Academy of Sciences published a report detailing priorities for the Geodynamics Program in the U.S. The report gave high priority to studies of the fine structure of the crust and upper mantle, particularly by seismologic techniques. A detailed study of the Mid-Atlantic Ridge was ranked high because this actively spreading plate boundary is relatively accessible. High-temperature and pressure measurements of the elastic and anelastic properties of lower crustal and upper mantle materials were noted as being very important for the interpretation of seismologic data. Finally, since the question "What makes the plates go?" was fundamental to the Geodynamics Project, emphasis was placed on theoretical, numerical, and experimental models of the dynamic processes of plate motion.

The report also recommended research in geomagnetism, the rotation of the earth, the thermal regime of the earth, and the low-velocity zone in the earth's mantle. Extending the geomagnetic time scale backward from its present limit of about 70 million years to as much as 225 million years would make it possible to decipher the history of the older parts of the ocean floor. The relationship between plate tectonics, polar wandering, and the rotation of the earth might suggest driving mechanisms for the plates. Heat flow and thermal distribution at depths within the earth are almost certainly related to the motions of the plates, while broad-scale mapping of the low-velocity zone in the earth's mantle would help to determine the shape of the region below the plates.

Earthquake prediction. Optimism about the possibilities of earthquake prediction was sustained and even increased in 1974, as the U.S. began the first year of an integrated program of earthquake hazard reduction research. During the year virtually all federal earthquake research activities were consolidated in the U.S. Geological Survey. New observations of the variation in the ratio of compressional (P) to shear (S) wave velocity preceding earthquakes were obtained, as well as observations of the variation of the velocity of P waves alone. The observations tended to support the hypothesis of dilatancy, which argues that, before the actual occurrence of an earthquake, numerous small cracks open in crustal rocks as they are stressed to near the point of failure. These cracks reduce the P velocity but leave the S velocity relatively unchanged.

The highlight of the year was the successful prediction of the approximate time, location, and magnitude of a small earthquake in the Blue Mountain Lake region of the Adirondack Mountains of New York state by a team from Columbia University. While observing a small swarm of earthquakes, they noticed a decrease in the ratio of the shear velocity to the compressional velocity. Based on the spatial extent of this anomaly and their experience with previous earthquake swarms in the area, they predicted that a magnitude 2.5–3.0 earthquake should be expected within the next few days. Two days after the prediction was made, a magnitude 2.6 earthquake occurred at the predicted location.

Substantial progress toward the prediction of earthquakes along the San Andreas Fault in central California was made during the year. Despite several earlier unsuccessful attempts to observe temporal variations either in the ratio of wave velocities or in the P-wave velocity alone, workers at the Geological Survey's National Center for Earthquake Research obtained solid evidence for the decrease in P velocity in regions surrounding the foci of two moderate earthquakes (magnitude 4.7 and 5) by observing the travel times for local earthquakes. This new evidence, taken together with the previously unsuccessful efforts, placed strong constraints on the volume of material in which the change takes place. It appeared that for moderate earthquakes, during which slip occurs over surfaces with dimensions of a few to perhaps ten kilometers, the volume within which the velocity changes cannot be more than two to three times this dimension.

An array of ten tiltmeters was established along the central San Andreas Fault. Results from tiltmeters had long been subject to skepticism, but recordings from this array showed substantial evidence for coherent deformation across the array, both at the time of and preceding relatively small local earthquakes. These results were puzzling in that the magnitude of the observed tilts greatly exceeded that expected from elastic models of the earthquake source. Explanation of this discrepancy awaited further observations and interpretation.

Also of interest to earthquake prediction were studies of the seismological environment of earthquakes. It had been reported earlier that the rupture—or slip—surfaces of great earthquakes fill so-called "gaps," areas that had been relatively quiescent during the preceding years. Detailed studies of moderate earthquakes within California and of larger earthquakes around the world showed that while the earthquake rupture surface fills an area of relative quiescence, the hypocenter, or point at which the rupture begins, commonly occurs in a region where there has been some activity.

While international exchange and cooperation in seismology was fairly common, the seismologi-

Tiltmeter installation monitors signs of a possible eruption at Pacaya volcano in Guatemala. Measurements are transmitted from the site via satellite.

Courtesy, U.S. Geological Survey

cal community in the U.S. was particularly excited about joint fieldwork with Soviet geophysicists and geologists. Pursuant to an agreement signed during the visit of Pres. Richard Nixon to Moscow in 1972, a joint U.S.S.R.-U.S. Working Group on Earthquake Prediction was established. Beginning in the summer of 1974, teams of U.S. seismologists and Soviet scientists planned to begin a joint field study of earthquakes and the variation of seismic velocities in the Pamir and Tien Shan mountains in Soviet Central Asia. Joint fieldwork in California along the San Andreas Fault was also planned.

—Robert L. Wesson

Hydrology

Attitudes toward water problems changed during 1973–74, as the environmental crisis was followed by the land-use crisis, only to be overtaken by the energy crisis. All three crises, however, were really elements of the same problem. The basic relationships between water availability and human demands had not changed. The question remained: How do we maintain an optimum living environment for an expanding population within the constrictions imposed by a virtually finite amount of available fresh water.

All water resources systems consume energy and some produce it. Energy from the sun drives the hydrologic cycle; gravity moves water through irrigation systems; and man-made energy from fossil fuels is used to drive the pumps that withdraw water from surface and subsurface sources. In turn, the production of energy uses water: the fall of water produces hydroelectricity, and the mining and processing of coal, oil, gas, and other power-producing natural resources require tremendous quantities of water. Wise water management and development of water resources as means for reducing the amounts of energy expended to put water to man's use (and the amounts of water used to produce energy) were just beginning to receive the attention they deserved.

The concern for energy sources led to a worldwide search for geothermal centers capable of using the natural heat generated within the earth to produce energy. Steam, heated by internal heat sources, was already producing power commercially in a few places, and heat from the earth's crust had been used much longer for such purposes as heating buildings, growing crops, and air conditioning. It appeared, however, that geothermal heat would be useful primarily for supplying comparatively small amounts of energy locally. (See Feature Article: POWER FROM THE EARTH.)

Advances in research. Although no major breakthrough occurred in 1973–74, there was a growing sophistication in the use of modeling techniques to approximate complex water-resources conditions. Modeling of water-quality factors, particularly groundwater quality, was still in the early stages. Nonetheless, computer simulation techniques were being used increasingly, and the need for three-dimensional models and models incorporating political, economic, and social factors was widely accepted.

One area of accelerating development was the hydrology of snow and ice. Snow and ice are important sources of water supply; they form the largest reserve of fresh water in the world, and while they hinder transportation, they form the basis of a relatively new recreational industry. They are also important factors in the ecosystems of the polar regions. Snow and ice together store approximately 80% of the earth's freshwater supply, a quantity that represents about 1,000 times the average annual discharge of the flow of the world's rivers.

Glacier inventories for the North American continent were rapidly being completed, and combined heat, ice, and water-balance studies were being conducted on glaciers in three international sample areas in the Western Hemisphere. One spin-off from this increasing interest in snow-and-ice hydrology was the idea of an Arctic Circumpolar Geophysical Program (ACGP). The proposal for ACGP was scheduled to be submitted to the International Union of Geodesy and Geophysics at its quadrennial meeting in 1975.

Environmental factors. Some of the largest rivers in the U.S. are also the most polluted. A recent study of 22 rivers by the Environmental Protection Agency reported that, among those studied, the dirtiest were the Arkansas, the Hudson, the lower Mississippi, the middle and lower Missouri, the middle and lower Ohio, and the lower Red. Those considered the cleanest included the lower Colorado, the Columbia, the upper Mississippi, the upper Missouri, the Snake, the Susquehanna, the Tennessee, the Willamette, and the Yukon. Pollution by sewage and bacteria had shown an overall decrease in the past five years, but pollution by such materials as phosphorus and nitrogen was rising.

Sediment is one of the common pollutants, and the large-scale shifting of sediment loads can cause major modifications in a stream's flood regime. A new flood insurance law, passed in 1973, would force property owners in areas subject to periodic flooding to buy federal flood insurance as a condition of continued or subsequent federal assistance. In addition, communities would be

Courtesy, U.S. Environmental Protection Agency, photo, D. L. Lewis

Cutaway view of the simulator stream channel, part of the Aquatic Ecosystem Simulator facility at the U.S. Environmental Protection Agency's Southeast Water Laboratory in Athens, Ga. Designed to permit studies of the chemical and microbial constituents of aquatic ecosystems under controlled conditions, the facility is part of a program to predict water quality in major U.S. river basins. The paddles are used to induce turbulence in water flowing through the channel.

forced to adopt zoning and land-use practices designed to minimize flood damage and to flood-proof existing subsurface utility lines. However expensive implementation of the law might be, it was hoped that the cost would still be less than the $4.5 billion paid to flood victims in the past five years.

Droughts and floods. The world was gradually becoming aware of the magnitude of the nearly six-year-old drought in Africa. The drought had affected a belt of countries from Senegal on the west coast to Ethiopia in the east. It was estimated that about 6 million people were either starving or in danger of starvation and about 25 million were strongly affected. Millions of cattle had died and thousands of hectares of land had been abandoned. Furthermore, reports of less than normal rainfall in Zaire suggested a general movement of drought conditions southward. During the year, droughts were also reported in Japan, Cyprus (the worst in 100 years), India, China, the Middle East, and Rhodesia. In the Western Hemisphere, droughts in Honduras and Costa Rica were said to be the most severe in 50 years.

The scale, severity, and gradual expansion of the drought in the afflicted area of Africa sug-gested to some that the drought represented a southward shift of climates. (*See* Feature Article: THE ADVANCING SAHARA.) However, a preliminary analysis of a limited amount of drought data suggested that this drought was not a unique phenomenon. On the average, a drought of this severity can be expected about once in 50 years, though in Senegal and Mauritania the drought was of a severity that can be expected to return only once in 100 years. All available information suggested that a six- or seven-year sequence of dry years was probable.

The difficulty is that the semiarid environment is extremely delicate, and the damage done during an extended drought probably cannot be recouped in a similar period of greater than average rainfall, even if one directly follows the other. During and after a period of dry years, erosion effects virtually ineradicable changes, and the soil and vegetation need a period of rest and recuperation. Human demands, however, press for an early return to grazing and farming. The slowness of recovery and the pressures imposed by men and their animals combine to make it appear that the desert is slowly encroaching on its adjoining environments.

Although droughts dominated the hydrologic events of the past year, particularly devastating floods occurred in Vietnam, Tunisia, Brazil, and the northern part of the Indian subcontinent.

International. The World Meteorological Organization (WMO) celebrated its 100th anniversary in 1973. Originally, WMO was composed of representatives of national meteorologic organizations, but beginning in 1959 it also showed a strong interest in hydrology. WMO's current Operational Hydrology Program was designed to emphasize research, development, and training in those operations that are concerned with collection, processing, and transmission of hydrometeorologic information.

The International Hydrological Decade was in its last year. UNESCO was initiating an open-ended follow-up program to begin Jan. 1, 1975. Principal elements of the new International Hydrological Program would be water-balance studies, influence of man on the hydrologic cycle, transfer of new technologies and methodologies, education and training in hydrology, and technical assistance to less developed countries.

—L. A. Heindl

Electronics

Progress in electronics during 1974 was characterized by steady advancement on several fronts rather than by singular major breakthroughs that might portend future revolutionary developments. Nevertheless, development continued so rapidly that the achievements were not only significant in themselves but implied continued progress. In particular, the advancing technology of integrated circuits, the ever increasing intrusion of optical techniques into electronics, and the still promising area of cryogenics could be singled out as typical arenas of action.

Integrated circuits. Probably the one single achievement of the year that could be identified as a milestone was the practical availability for the first time of n-channel MOS (metal oxide semiconductor) integrated circuit technology. Available in the laboratory for several years, the fabrication technology to produce n-channel field-effect transistors in the integrated circuit configuration became sufficiently developed to make commercial applications possible. (A field-effect transistor consists of a conducting channel formed from a strip of n-type or p-type semiconductor and a gate electrode—an electronic device having an output that is controlled by voltages applied to the gate. An n-type semiconductor is one in which the electron conduction—negative—exceeds the hole

conduction—absence of electrons.) Integrated circuit technology, about a decade old in 1974, combines active and passive circuit components on a microscopic scale into a single solid silicon wafer, or chip. The capability to produce complicated circuitry involving thousands of transistors and other components in a single unit led to many electronic circuit applications not previously feasible. As explained below, n-channel MOS significantly increases versatility in design, application, and fabrication of such circuitry.

Two general types of integrated circuit technologies are currently in use. The older is based on the conventional bipolar junction transistor, while the most recent approach uses field-effect transistor design. Although the two are overlapping and complementary in their characteristics, the latter approach has a number of cost fabrication advantages. For example, the field-effect transistor can be fabricated by relatively simple vapor deposition of a metal pattern on the semiconductor chip, commonly labeled MOS technology. By contrast, more elaborate impurity-diffusion steps are required in the bipolar transistor case.

For detailed chemical reasons, it is much easier to fabricate satisfactory p-channel (in which the hole conduction exceeds the electron conduction) MOS configurations than n-channel devices, although by using laboratory processes both have been available for some time. The electrical

Amoeba is shown outlined against background of part of an 8,192-bit experimental p-channel semiconductor memory unit.

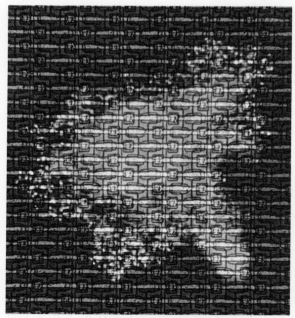

properties of n-channel field-effect transistors are somewhat better than p-channel units, primarily because of the greater mobility of electrons over holes in the semiconductor. More important, however, is that the availability of n-channel MOS makes possible complementary MOS (or CMOS) circuitry, an advantage analogous to the npn-pnp complementary symmetry in bipolar transistor circuitry that has been available since the discovery of the junction transistor.

The advantages of complementary symmetry stem from the symmetrically inverted bias voltages and signal voltage polarities inherent in npn compared to pnp transistors. Circuit design using the two types is simpler and more flexible than with either type alone. Now CMOS integrated circuits, with all the flexibility and fabrication cost advantages of bipolar circuitry, have become commercially practical.

Of particular significance is that CMOS integrated circuitry makes possible semiconductor memory arrays having very desirable properties for digital computer applications. The electrical characteristics of field-effect transistors are such that the combination of an n-channel unit and a p-channel unit results in semiconductor memories with very low power requirements. This means that significant amounts of information can be stored economically and that relatively inexpensive semiconductor memories can be more widely used in computer applications.

Microcomputers. The developments in semiconductor memories and MOS integrated circuits described above were directly responsible for progress during the year in microcomputer development and application. By 1974 it had become appropriate to think of electronic digital computers in three size categories. The first of these was the familiar major computer installation involving a room full of equipment capable of the most extensive data manipulation and computational tasks. The second category, minicomputers of the size of a refrigerator or home television set, perform less extensive but no less sophisticated calculations and are often designed for one or a few specific applications. Only recently has the third size, microcomputers of pocketbook dimensions, become a reality. The familiar hand calculator is only the beginning of microcomputer technology.

For example, an important future application of microcomputers is in automotive electronics. Electronics had already invaded the automobile, yielding improved safety, performance, and convenience features. Electronic ignition, solid-state voltage regulators, and reliable silicon rectifiers that enable efficient alternators to replace direct-current generators were commonplace. But in re-

Two inventions by Italian physicist Count Alessandro Volta have helped provide the basis for today's electrical and electronic industries: the electrophorus, in 1775, generated static electricity, and the voltaic pile, in 1800, was the first electric battery.

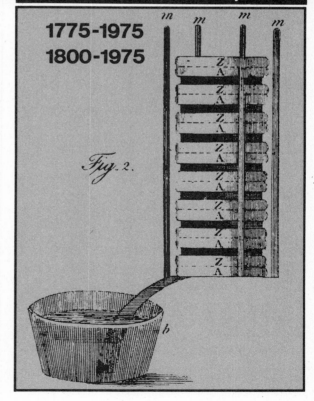

1775-1975
1800-1975

Fig. 2.

Fairchild Camera & Instrument Corp., Space & Defense Systems Division

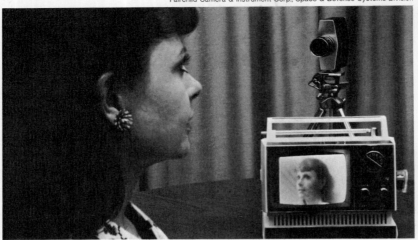

MV-100 television camera, designed by Fairchild Camera & Instrument Corp., is based on charge-coupled-device technology. Its major uses will probably be in the security, surveillance, and instrumentation markets.

cent months microcomputer antiskid controls, seat belt safety features, improved radios, and thermostatic controls were demonstrated. As the uses for electronics in automobiles further expanded, industry leaders could foresee that ultimately $100 of the purchase price would be directly attributable to electronics. A nagging problem still to be overcome completely was to reach the exacting reliability standards required in automotive applications, both with regard to long-life operation and also, perhaps even more importantly, with regard to manufacturing very large numbers of units.

Solid-state television cameras. Integrated circuit techniques also produce the image-dissecting element in completely solid-state television cameras. In such cameras the televised scene is imaged on a semiconductor wafer, and the image is scanned electronically by processes that occur within the chip. The way this comes about is as follows. The photoconductivity of silicon results in the formation of an electrical charge pattern corresponding to the optical image, and this charge pattern is sensed by a suitable array of electrodes disposed over the area of the chip. Integrated circuitry juxtaposed in the silicon chip is used to switch from element to element of the image in order to scan the scene. This is analogous to the scanning of the photoconductive target by the electron beam in a conventional vidicon camera tube.

Actually, however, it proved to be a considerable challenge to fabricate the necessarily complicated switching circuitry on a small enough scale and with sufficient reliability to approach the resolution of a typical television image, which encompasses approximately 250,000 picture elements. Among the more promising approaches was the so-called charge-coupled-device (CCD), in which

an electrical charge is caused to step successively between specific regions in the semiconductor chip in response to electrical pulses applied to a single pair of external terminals. A suitable arrangement of horizontal and vertical electrodes provides two-dimensional scanning of the electrical charge pattern in order to dissect the image.

Both the electrode pattern in the chip and the external scanning signals of CCD units are much less complicated than in the case of conventional transistor-design integrated circuitry. Even so, the difficulty in the past has been to fabricate sensitive elements with sufficient resolution. Several devices developed early in the year achieved resolutions corresponding to 10,000 or more picture elements. In particular, a practical TV camera the size of a package of cigarettes corresponding to nearly 60,000 elements was demonstrated. This is more than sufficient for many applications other than home entertainment television. It is, for example, sufficient for Picturephone use and for viewing industrial areas for security purposes.

This achievement was followed almost immediately by a similar unit capable of achieving a resolution of acceptable commercial television standards, 200,000 elements. Not to be outdone, proponents of the switching circuitry approach fabricated a "dense" MOS unit that demonstrated resolutions equivalent to 500,000 elements.

The advantages of small size, reliability, and low power requirements characteristic of solid-state electronics are realized in these demonstration cameras. Clearly, many new TV applications will arise with this capability and, equally clearly, practical solid-state television cameras with satisfactory resolution can be confidently predicted.

Optical electronics. Although optical electronics had its roots in the discovery of photoconductivity in selenium, which led to light-sensitive electric devices as early as 1900, the real intrusion of optical techniques into electronics can be reckoned from the discovery of the laser in the early 1960s. Lasers generate coherent light, that is, electromagnetic waves at optical frequencies that are not different, in principle, from radio waves. In the effort to transmit information in ever increasing quantities, the frequencies characteristic of optical waves are more attractive than those of radio waves. This is so because the maximum information rate in communications signals is limited directly by the frequency of the carrier wave, and optical frequencies are many orders of magnitude higher than conventional radio signals.

When considered in connection with the above discussion, the techniques of optical electronics are a natural extrapolation of conventional electronic circuitry. Most particularly, the transmission of coherent light waves along thin glass fibers is analogous to the use of metallic waveguides at microwave frequencies and to cross-country transmission lines at power frequencies. Optical fiber transmission is the most attractive way to guide light signals with their tremendous information-handling capability over appreciable distances. (*See* Feature Article: FIBER OPTICS: COMMUNICATIONS SYSTEM OF THE FUTURE?)

Cryogenics. The first prototypes of superconducting cryogenic cables for electrical power transmission were tested during the year. The practical promise of superconductivity, which eliminates power losses in electrical conductors because the electrical resistance vanishes, had been already achieved in electromagnets used in research applications and in prototype rotating electrical machinery, such as motors and generators. Applications to long-distance power transmission and to levitated high-speed ground transportation were complicated by the difficulty of maintaining the very low temperatures necessary for superconductivity.

During the year search continued for materials that exhibit superconductivity at the highest possible temperature. A particular niobium-germanium compound that remains superconducting at temperatures as high as 23° K (−250° C) was discovered. The significance of this achievement is that liquid hydrogen boils at 20.4° K so that, in principle, the new compound could operate with liquid hydrogen as a coolant. All other materials discovered to date must employ liquid helium to reach sufficiently low temperatures, and this is not only expensive but introduces difficult thermal design problems. (*See* Year in Review: PHYSICS, *Solid-State Physics*.)

Future possibilities. Perhaps indicative of things to come is a theoretical investigation of practical rectifier action within a single organic molecule. It has been known for some time that effective n-type conduction or p-type conduction is possible in organic molecules, depending upon their molecular structure. Combining the two types in a single molecule but separating them by a saturated-bond neutral region produces, in effect, an organic p-n junction. Though such an achievement was still a gleam in the eyes of theoretical solid-state physicists in 1974, their calculations demonstrated the expected rectification properties. It remained a long way to the development of practical semiconductor devices of molecular dimensions, but perhaps the first step was taken in 1974.

—James J. Brophy

See also Year in Review: COMPUTERS.

Energy

In late 1973, energy moved to the center of the world stage and commanded the anxious attention of the leaders of the world's producing and consuming countries alike. For the third time in two decades, the outbreak of hostilities between Arabs and Israelis in the volatile Middle East interrupted the flow of oil to the consuming countries. The October war and subsequent actions by the Arab oil producers and exporters to reduce production levels; embargo exports to the United States, The Netherlands, and several other nations; and increased oil prices to more than three times their prewar levels combined to create a world energy crisis and demonstrated the excessive dependence of Japan, Western Europe, and other highly industrialized nations on Arab oil. Despite a partial rollback of the Arab production cuts and on March 18 a lifting of the embargo of shipments to the U.S., energy remained a critical problem.

From the mid-1960s through the early 1970s gross energy consumption in the U.S. grew at annual rates of 4–5%. At the same time, energy demand in the remainder of the world increased at nearly twice those annual rates. This demand was largely met by increased supplies of petroleum, with most of the new supplies of oil coming from the Middle East and North Africa. These regions contain over two-thirds of the world's proven oil reserves, and the petroleum there can be produced at costs that are very low relative to other energy sources. In 1973, Japan and Western Europe depended on foreign oil imports, mostly from the Arab states, for the bulk of their energy. The U.S. received directly and indirectly, about one-third of its oil imports and about 7% of its total energy from these sources.

Energy

The Arab oil production cutbacks and embargo, coupled with the sudden and great increase in world oil prices, had and would continue to have serious impacts on the world economy. Although the total world oil supply was only reduced by about 7%, or about 4 million bbl per day, at the peak of the cutbacks, it greatly slowed economic growth in the U.S., Japan, Western Europe, and other highly industrialized countries. Moreover, it was expected that oil would continue to be in short supply, even if all producer countries restored production to the pre-embargo levels.

The enormous oil price increases posed serious international monetary, trade, and balance of payments problems for the oil-importing nations. The World Bank estimated that consuming nations would have to pay three to four times more for oil in 1974 than in 1973. Revenues of the 11 major oil-exporting countries were expected to increase from about $23 billion in 1973 to $85 billion in 1974, $100 billion in 1975, and more in succeeding years.

In an effort to cope better with the world energy supply and economic crisis, U.S. Pres. Richard Nixon on Jan. 9, 1974, called for an extraordinary international conference of oil-consuming countries to be followed by similar talks with less developed nations and then with the oil-producing states. Consequently, foreign ministers from the 13 largest oil-consuming nations met in Washington, D.C., on Feb. 11–13 to formulate a cooperative energy action program. The conference produced no dramatic results to alleviate the world energy crisis and its threat to economic order, nor did it set the stage for a "confrontation" with the oil producers that many feared would develop. Instead, agreement was reached among all nations except France to cooperate in seeking solutions through joint actions on oil-pricing agreements, the conservation of supplies and lowering of consumption levels, the exchange of technologies to develop alternate energy sources, and the exploration of other areas of cooperative effort. France split with its eight Common Market partners and Japan, Canada, Norway, and the U.S., and determined to follow a "go it alone" policy of seeking bilateral oil deals with the producer countries. Despite an extension of the conference from two days to three days in an effort to produce unanimous agreement on a cooperative plan, France maintained its dissenting stance. Nevertheless, the conference did produce an agreement among most major industrial nations to explore together ways to deal with the present economic crisis and to meet long-term energy needs.

In the U.S., the Arab oil embargo brought to a head very quickly the energy crisis that had been developing for many years. For the first time since World War II, the American public experienced fuel allocations and continuing shortages of gasoline, fuel oils, and other petroleum products. The government instituted a number of voluntary and mandatory measures designed to increase fuel supplies, reduce energy demand, allocate fuels to essential uses, and share the remaining supplies on an equitable basis.

In November 1973, President Nixon announced Project Independence and set as an essential na-

Drivers form miles-long line for gasoline at a service station in New Haven, Conn., a scene repeated many times throughout the U.S. during the severe gas shortage of early 1974.

tional goal the achievement of energy self-sufficiency by 1980. He established the Federal Energy Administration and empowered it to plan, manage, and direct all energy policy and programs. William E. Simon, deputy secretary of the treasury, was designated administrator of the agency and promptly became popularly known as the "energy czar." The government subsequently announced a five-year, $10 billion energy research and development program to accelerate the development of new energy technologies and sources.

Despite a warmer than normal winter in both Europe and the U.S., the moderation of Arab oil production cutbacks, "leakage" in the embargo, and a substantial reduction in demand through conservation efforts, the energy crisis had a serious and growing impact on most of the world. In the United Kingdom it was combined with a work slowdown and eventual strike of coal miners, bringing the nation's economy to desperate straits. The workweek was reduced to three days and new elections were held. After the elections and a change in government leadership, the coal strike was settled and the economy began to recover. In the U.S., the impact appeared greatest in February, when long lines developed at gasoline service stations as supplies became increasingly scarce. The lifting of the embargo in mid-March permitted increased allocations, and the gasoline supply situation improved after that time.

Other significant developments affecting energy during 1973 and early 1974 included: (1) world energy demands and consumption declined significantly as consuming nations conserved supplies in response to the oil supply cutbacks and price increases; (2) U.S. energy use in 1973 increased by 4.8% despite the energy crisis, although the rate of increase was less than that of the previous year and oil conservation efforts sharply reduced demand after the Arab oil embargo; (3) world and U.S. oil prices rose rapidly, and oil companies reported record profits and increased earnings; (4) legislation to permit construction of the long delayed trans-Alaska pipeline was passed, and permits to allow work to begin were issued; (5) U.S. production of natural gas expanded only slightly and may have peaked in 1973; (6) petroleum and gas shortages shifted attention to coal, and U.S. coal demand increased by 8.4% during 1973; (7) nuclear power plants supplied about 4% of U.S. electricity in 1973, a share that was expected to increase sharply in future years; and (8) U.S. federal lands were leased for the purpose of determining the economic and environmental feasibility of oil shale and geothermal steam as new energy sources.

Petroleum and natural gas

The critical dependence of the highly industrialized oil-consuming countries on producing nations that have exportable surpluses and the resulting economic power of those "seller" nations were revealed in late 1973. As a political action to influence a settlement of the Arab-Israeli war, the Arab oil-producing nations announced production cutbacks and embargoed oil shipments to the U.S. and The Netherlands. Their action was effective because the Arab nations controlled more than 60% of the world's proven oil reserves and were the only countries capable of meeting soaring world demands, especially those of Western Europe and Japan. At the same time, the Organization of Petroleum-Exporting Countries (OPEC) unilaterally abrogated existing agreements and took control of world oil prices. Posted oil prices were increased from about $3 per bbl in September 1973 to nearly $12 per bbl by the year's end. Spot crude oil prices ranged as high as $22 per bbl. Non-OPEC exporting nations raised their prices to the high new world levels, and Canada imposed an export tax in steps to $6.40 per bbl for crude oil shipments to the U.S.

In the U.S. controlled domestic oil price levels were increased in response to the world oil price increases and as a spur to increase output. The price of old oil was permitted to rise to $5.25 per bbl, while all new oil developed plus a matching volume of old oil was exempted from price controls. Also, oil from stripper wells, those producing less than 10 bbl per day, was decontrolled. Thus, nearly a fifth of all U.S. oil was decontrolled and the price rose to about $10 per bbl by early 1974.

Despite the supply disruptions, worldwide oil demand rose by nearly 9% to average nearly 55 million bbl daily during 1973. In the U.S., 1973 demand averaged 17.3 million bbl per day, an increase of 4.8% over that of 1972. Of this total, 36%, or 6.3 million bbl per day, was imported; however, only an estimated 2.5 to 2.7 million bbl per day were imported directly and indirectly from the Middle East. At the start of the embargo, U.S. demand was averaging about 18.5 million bbl per day. Therefore, the Arab denial affected about 15% of total oil supply and only 7% of total energy. (In 1973 in the U.S., petroleum accounted for 46% of total gross energy consumption; coal, 18%; natural gas, 31%; hydroelectric power, 4%; and nuclear energy, 1%.) Nevertheless, the reduced supply had increasingly severe impacts on the economy, and Americans suffered the personal inconveniences and higher prices caused by fuel scarcity. The government called for a number of voluntary conservation measures and imposed mandatory allocations of propane, middle petroleum distillates, gasolines, and other fuels to deal with shortages. The federal government called for Sunday closings of gasoline stations and imposed a 55 mph speed limit, while many states adopted some form of gasoline rationing. These measures produced substantial savings and, coupled with much warmer than normal weather and some leakage in the embargo, allowed the essential energy needs of the nation largely to be met. However, the adjustments were painful to many. As the fuel shortages deepened, independent truckers called a nationwide strike and enforced it with violence and roadblocks on the highways until they obtained rate increases and assurances of more fuels.

On March 18, 1974, the Arab oil-producing states, with the exception of Libya and Syria, agreed to lift their five-month embargo against the U.S. for a trial period ending June 1. This announcement and the resumption of oil shipments alleviated the shortages, and the economic dislocations began to ease as the energy squeeze lessened in the U.S. It was a welcome relief to motorists and other petroleum consumers, but short supplies and the threat of further oil supply interruptions remained a disquieting future possibility.

Natural gas marketed and consumed throughout the world reached an estimated 45 trillion cu ft (Tcf) in 1973. As in the past, the U.S. accounted for slightly more than half of this total, as supply shortages limited consumption to an estimated 23.3 Tcf. The Soviet Union continued in second place with consumption of approximately 8 Tcf.

U.S. production of natural gas had about peaked by 1974 and was expected to decline rapidly unless exploration and development of new reserves were greatly accelerated. Regulation of wellhead gas prices at artificially low levels by the U.S. Federal Power Commission (FPC) caused a decided decline in the rate of new reserve additions in recent years. At the same time, production increased to the point that existing reserves were inadequate to sustain supplies. Many industry observers believed that partial or complete deregulation is necessary to develop the undiscovered potential gas resources, which have been estimated at 50 times current annual production. Legislation to accomplish this was being considered by Congress. Besides stimulating exploration and development, higher gas prices would curtail demand by reducing the amounts burned under boilers and in other uneconomic uses.

In view of the deteriorating domestic gas supply, much effort was being exerted to develop alternate and supplementary sources. Plans to import liquefied natural gas (LNG) from Algeria, Nigeria, the U.S.S.R., and Venezuela continued to be developed. These sources could provide 2 to 3 Tcf per year by 1980. Also, plans were advanced for the development of gas pipelines to tap the large known gas reserves of the Alaskan North Slope, estimated to be in excess of 55 Tcf.

Coal

The world supply shortages of crude oil and petroleum products shifted attention to coal as an alternative means of meeting energy needs. Coal, once the world's principal source of energy, supplied an estimated 3,000,000,000 tons, only 25–30% of world commercial energy, in 1973 even though it is the most abundant fossil fuel.

Known coal reserves of the U.S. could supply the nation's energy needs for several hundred years at current levels of consumption. Yet coal supplied only 18% of the country's energy needs in 1973, largely as a boiler fuel for electricity generation. During 1973 U.S. coal consumption was an estimated 563 million tons, an increase of 8.4% over that of 1972. Coal production decreased by 1% to 596 million tons during 1973.

To resolve the nation's increasingly critical energy situation, a greatly expanded utilization of its enormous coal resources seemed essential. Coal's share of the U.S. energy supply has declined steadily since the 1940s, when it comprised more than half the consumption. Price competition from natural gas and imported fuel oil caused it to lose Midwest and East Coast markets during the 1950s and 1960s. More recently, environmental constraints imposed under the 1970 Clean Air Act limited the combustion of coal, most of which is high in sulfur content. This act requires that primary (health-related) air quality standards for sulfur oxide emissions be met by 1975 and that more stringent secondary (general welfare) standards be achieved within a reasonable period of time. In compliance with regulations issued by various states and approved by the U.S. Environmental Protection Agency, many utilities switched from coal to oil to comply with the limitations on sulfur content.

In order to help relieve the fuel oil shortages, the federal energy office in December ordered those utilities capable of switching back to coal to do so. Consequently, by the end of February an estimated saving of 80,000 bbl per day of fuel oil was being realized by East Coast utilities.

Over the longer term, the urgent and important role of coal to domestic energy self-sufficiency was recognized in President Nixon's energy messages throughout 1973 and early 1974. He called for major programs to expand coal use through: (1) a massive expansion of research and development programs to develop technologies that can produce clean synthetic fuels from coal, and processes that can remove contaminants before, during, or after combustion; (2) the relaxation of existing environmental constraints on coal utilization where primary health standards are not breached; (3) the accelerated leasing of federal coal lands for development; and (4) the enactment of mined-area reclamation and restoration legislation to promote the best mining practices.

Coal research and development programs were expected to command nearly $3 billion of the proposed $11.4 billion federal energy research and development program through 1979. Researchers, however, did not expect advanced coal conversion technologies, such as coal gasification, to be in significant operation before 1980. Thereafter, such plants were expected to provide increasing amounts of clean energy to the U.S. economy.

Electric power

Electricity is by far the most convenient and cleanest form of energy at the point of consumption. The five major sources of electrical generation are: (1) fossil fuels, such as oil, gas, and coal; (2) falling water, or hydroelectric power; (3) nuclear energy; (4) geothermal energy; and (5) solar energy. The fossil fuels are by far the dominant source, accounting for nearly 75% of world electricity generation. Hydroelectric power accounts for about 25% and nuclear energy about 1%.

World electric energy production is estimated to have been about 5.5 trillion kw-hr in 1973. Power generation by utilities in the U.S. was an estimated 1.75 trillion kw-hr, an increase of 5.1% over 1972. In the major countries of the world, electric power growth rates have ranged from 5 to 15% during recent years, averaging 8%. Increased rates of growth are anticipated in the future. Electricity use in the U.S. increased from 15% of gross energy consumption in 1960 to 26% in 1973, and it was expected to rise to 35% by 1985 and 42% by the year 2000.

In the future, most new electrical generating capacity is expected to come from nuclear power, although coal and other fossil-fuel plants will continue to provide the majority of electricity to the end of the century. Possible large-scale development of geothermal and solar electricity generation systems offers hope for almost infinitely large new electrical energy sources, but major technological breakthroughs are required for such achievements.

Nuclear power plants supplied about 4% of U.S electricity in 1973, these installations having a total capacity of about 25,000 Mw. Additional plants under construction are expected to raise this capacity to 120,000 Mw by 1980. The European Economic Community had 11,000 Mw of nuclear capacity in operation, and reactors under construction in EEC countries totaled 33,000 Mw.

Most nuclear reactors in use or being built are converter types that yield less fissionable material than they consume. These light-water and pressurized-water reactors are fueled by uranium and convert only about 1.5% of the potential uranium energy into power. The second generation high-temperature gas-cooled reactors (HTGR) are more efficient in fuel use because they start on uranium but use thorium as a fertile material to produce fuel. The first large-scale HTGR went into operation in 1973, and five others were ordered.

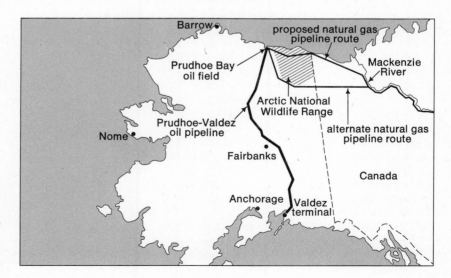

Oil pipeline route extends from the oil and natural gas fields of Alaska's North Slope 789 mi south to Valdez. A proposed natural gas pipeline across the Arctic National Wildlife Range has raised protests from conservationists.

Energy

Breeder reactors, which produce more fissionable fuel than they consume, were being developed in many nations. The world's first commercial breeder reactor was placed in operation by the Soviet Union in 1972 at Shevchenko on the Caspian Sea. It was designed to produce 150,000 kw of electricity and desalt 30 million gallons per day of seawater. A second Soviet breeder plant, with a capacity of 600,000 kw, was under construction. Breeders were also being built in France and Scotland.

The U.S. actively pursued the development of its first fast-breeder demonstration plant but experienced legal and technical delays. This plant was designed to demonstrate the commercial feasibility of the liquid-metal fast breeder. Technical problems in pumping the liquid-sodium coolant and in the swelling of fuel elements caused concern. Also, the U.S. Court of Appeals ruled that the Atomic Energy Commission (AEC) must prepare an environmental impact statement on the consequences of widespread deployment of breeder reactors. The order did not affect the demonstration plant but did require the AEC to examine fully and disclose possible dangers of the entire breeder program.

Hydroelectric power generation continued as a major source of world energy in 1973; however, the best sites for generating such power had been largely developed. For example, by 1974 almost one-third of the estimated total potential hydro-electric power capacity of the U.S., 176,000 Mw, had been developed. Most of the remaining undeveloped capacity was in Alaska or remote Western areas, far from population centers. World hydroelectric capacity is approximately 270,000 Mw. Of this total, the U.S. accounts for about 21% followed by Canada with 13%, the Soviet Union with 10%, and Japan with 7%.

Other energy sources

The energy crisis focused attention on the accelerated development of nonconventional energy sources. Most significant among these are oil shale, geothermal steam, and the sun.

Oil shale deposits in the western U.S. contain a potentially recoverable resource of 600,000,000,-000 bbl of oil, more than 15 times the proven U.S. oil reserves and equal to the total world proven oil reserves. During 1973, the U.S. Department of the Interior instituted a prototype leasing program on six selected tracts to permit the development of demonstration plants by private industry. These operations were expected to establish the economic and environmental feasibility of large-scale oil shale mining, processing, and recovery. Six 5,000-ac tracts of public land (two each in Colorado, Utah, and Wyoming) were selected for competitive leasing by mid-1974.

The first competitive leasing of U.S. federal geothermal steam resource areas took place in Jan-

Uranium pellets to be used in the fuel rods on nuclear power reactors move through a factory in Wilmington, N.C. A new computerized manufacturing system simultaneously tracks and records the production history of each tray of pellets as it progresses through the factory.

Courtesy, General Electric Research and Development Center

Pilot project to extract oil from shale rock got under way early in 1974 when Gulf Oil Corp. and Standard Oil Co. of Indiana won the bid to develop a 5,000-ac shale region in northwestern Colorado.

uary 1974. Private bidders offered the government $6.8 million for 20 tracts in the known geothermal resource areas of California. Two tracts in the Geysers area, where geothermal steam was already utilized in electrical generation plants with a capacity of 400 Mw, received top bids of $4.5 million. Derived from the molten rock of the earth's core, geothermal energy offers the potential of essentially inexhaustible power if commercial technologies can be developed to tap this source. (*See* Feature Article: POWER FROM THE EARTH.)

Solar energy offers an ultimate and almost infinite source of potential energy, but by 1974 it had undergone only very limited commercial development. However, greatly increased research programs to develop solar heating and thermoelectric conversion technologies were announced during the year. The announced federal research budget for solar energy research in the U.S. will raise funding from $13.8 million in 1973 to $50 million in 1974, and even higher funding levels were con-

templated for future years. Expanded efforts by many European countries were also announced. (*See* Feature Articles: POWER FROM THE SUN; POWER FROM THE WIND.)

Outlook

The outlook for the coming year is that the world energy crisis will continue as petroleum supplies remain uncertain. The severity of shortages and the price of energy will be largely determined by the amount of new supply that the Arab producer nations make available for export. In the U.S., especially, energy problems are expected to continue. Energy conservation efforts and allocation programs will determine the quantities and distribution of individual oil products. It is hoped that Project Independence and other government initiatives will accelerate domestic energy resource development and set the stage for the attainment of national self-sufficiency by the mid- to late-1980s.

Other developments that are likely to occur are as follows: (1) coal production is expected to increase more rapidly than in previous years as it is substituted for oil and gas wherever possible; (2) existing environmental standards will be relaxed to permit greater use of coal and other fuels in situations where primary health standards are not expected to be jeopardized; (3) exploration and development of oil and gas areas will increase rapidly in response to higher prices and greater access to resource areas; (4) construction of the long-delayed trans-Alaska pipeline will begin; and (5) natural gas will remain in short supply as production begins to decline.

—James A. West

Environmental sciences

Perhaps the most noteworthy feature of the year was that many problems previously of concern only to environmental scientists were brought to the attention of the public in a most immediate way. The cutbacks and embargoes instituted by the Arab oil-producing nations in late 1973 and soaring prices for all commodities suggested that mankind was, in fact, approaching limits to growth. Less widely noticed but equally important were the widespread crop failures due to weather phenomena—perhaps caused by man's activities —and the continuing declines in birthrates in the developed countries. In short, the year may have marked the beginning of an abrupt conversion of civilization, worldwide, from a growing to a stable state.

Also noteworthy was the great concern with strategic social issues that surfaced at national and international scientific meetings, which previously had been more concerned with specialized, tactical problems. The national meetings of the American Association for the Advancement of Science held in Washington, D.C., in late December 1972 and in San Francisco in March 1974 had many symposia dealing with large-scale environmental issues: Should economic growth be limited? Is the growth in technology limited? What social changes will occur under zero growth? What will the future be like? Similarly, the 14th International Congress of Entomology (Canberra, Austr., August 1972) was characterized by a degree of concern with strategic issues for mankind that once would have been surprising.

Indeed, the amount of such discussion and the intensity that accompanied many controversial exchanges combined to give a feeling of impending crisis. Certain facts suggest that this feeling was an accurate index of the speed with which the world was changing. For example, completely plausible predictions often appeared retrospectively to have been quite wrong, even a few months after they were made. The U.S. Department of Commerce in late 1973 issued the document "U.S. Industrial Outlook 1974," which predicted that, for the automobile manufacturing business, "In 1974, a small decline from 1973 is anticipated with production down 2 percent to 9.6 million units." However, the Arab oil embargo brought the seven-decade-long love affair of Americans with the automobile to an abrupt halt, and during the first three months of 1974 automobile assembly rates were off about 34%.

Limits to growth. The Arab oil embargo and soaring food prices conveyed very important messages. Clearly, no embargo by foreign powers would have been relevant unless demand for cheap energy in the U.S. was outstripping domestic supply. Further, if cheap supplies of energy were approaching exhaustion in a country as large as the U.S., it was reasonable to project that worldwide exhaustion would follow within a few decades. The only permissible conclusion was that cheap energy would increasingly be replaced by expensive energy. The implications of this for the U.S. were far-reaching because, to a degree that would surprise most people, the entire character of American life was dependent on very cheap energy.

For example, the low cost of food had been made possible by the availability of cheap energy to produce and drive farm machinery and manufacture and distribute fertilizer. As energy prices rose, food prices would increase. Farmland would

rise in value, which meant that urban sprawl would become prohibitively expensive. Furthermore, urban sprawl would be inhibited not only by the high prices of farmland at the perimeter but also by the rising cost of energy needed for long commutes to and from work. As a result, U.S. cities might gradually become more compact, in the fashion of European or Middle Eastern cities.

Even in the short run, projections of sharply rising energy costs had several consequences. There was more motivation for regional mass rail transportation, and such systems were already being installed in the San Francisco Bay region, Washington, D.C., and Atlanta, Ga. Indeed, such regional rail systems were perceived as more than means of conserving energy per passenger mile; they were being planned as regional land use-shaping forces and as tools to aid in making whole regions more competitive economically. Indicating the broad significance being attached to regional transportation systems, the National Academy of Engineering set up a committee to advise on the design of comprehensive research programs to assess the complete spectrum of environmental impacts of such systems.

Because of crop failures in several important producing regions, and because of the great need of the U.S. to export massive quantities of food to pay for its huge and rapidly rising fuel imports, the U.S. exported more than a billion bushels of wheat in 1972–73, representing about 70% of all the wheat grown, and the result was a massive increase in domestic food prices. Since food was being exported to pay for imported fuel, part of which would be used to help farmers raise more food, attention was focused on the inherent inefficiency of this cycle.

David Pimentel at Cornell University and John Steinhart at the University of Wisconsin, among others, made careful examinations of the energy flow through U.S. agriculture. The startling result that emerged from this research was that U.S. agriculture had become very energy-inefficient: it took up to seven kilocalories of energy input into agriculture to obtain one kilocalorie of output in the form of food. The explanation for this inefficiency (which was increasing from year to year) was that the progressively higher proportion of the population leaving farms and rural towns to migrate to the cities had to be compensated for by increased substitution of energy for labor in farm work and by increased transportation overhead from city to farm (farm supplies) and from farm to city (farm produce).

Climate and food production. During the year there was a great increase of interest in the effect of climate on crop production and in the effect that

Supporters of the computerized Bay Area Rapid Transit (BART) system hope it will lure motorists away from the overcrowded freeways of the San Francisco Bay region.

man, in turn, might have on climatic changes. There were several climate anomalies in the 1972–74 period. The great Soviet wheat shortage of 1972 was caused by widespread drought south and east of Moscow unlike anything that had occurred since record keeping began. Excessive rainfall and cold weather delayed the 1972 harvest in much of the U.S. Most seriously, there was partial failure of the agriculturally important (rainbringing) monsoons of sub-Saharan Africa, Pakistan, India, and Indonesia. The U.S. meteorologist Reid Bryson pointed out that, historically, cooler periods in the earth's history have tended to coincide with dry times in this monsoon belt. Further, it appeared that cooler periods can be caused by atmospheric loading by fine particulate material, which may originate from volcanic eruptions or from the activities of man.

At present the earth is in a cooling trend, possibly because the worldwide increase in industrialization is building up the particulate load in the atmosphere. The effect of this particulate matter is to backscatter incoming solar radiation so that only a portion of it is able to penetrate to and warm the surface of the earth. While the incoming radiation is scarcely ever decreased by more than 20%, this is sufficient to chill the surface of the earth by up to 2° F — enough to have a spectacular effect on crop production. The most significant point about this situation is that chilling is most pronounced in those plains areas of North America that the world is looking to for wheat supplies.

Birthrate decline. The decline in the U.S. birthrate that began about 1960 continued, and this phenomenon would have some rather startling effects in the future. The rate declined from 25 per 1,000 population in 1955 to 15 in 1973 — even lower than the 18.7 recorded in the depths of the great depression (1935). As a result, the natural rate of population increase fell from 1.57% per year in 1955 to about 0.56%.

More interesting implications are revealed when one examines the number of people born each year, rather than the crude birthrate. This number declined from 4,268,000 in 1961 to 3,256,000 in 1972, and the trend appeared to be continuing. This meant, for example, that when the people born in 1972 are 55, in the year 2027, they will be attempting to support on retirement the 66-year-olds born in 1961. Clearly, this scenario is incompatible with much economic growth, because the available capital would be severely drained by the cost of a relatively small labor force attempting to support a relatively large retirement-age population. It is noteworthy that the decline in the number of births since 1959 was the steepest in U.S. history; the decline from 1920 to 1935 was only about 60% as steep.

Among the more immediate effects of this birthrate decline were economic difficulties for those industries that service or supply products to the very young, from teaching and maternity services to toy, baby food, and diaper manufacturers. And this decline in birthrates might well continue. In-

239

flation and high unemployment rates among teenagers and persons in their early 20s were further dampening interest in reproduction in these critical age groups.

The possibility that falling birthrates would lead to population stabilization or decline in the near future was true only for developed countries, however. In most less developed countries, birthrates had been so far in excess of death rates in recent years that the populations had a grossly distorted age structure, with enormous numbers of very young people. The age-specific birthrates of these groups were unlikely to drop sufficiently to enable the populations to stabilize at less than about three times their current size. Famine and plagues would probably become the major controls on population trends in these countries long before such population sizes were reached.

Stability, resiliency, and predictability. In recent years environmental scientists had become very concerned about features of systems that describe their variation from a stable state. At a time when highly complex social and economic systems can be subjected to sudden and surprising environmentally induced perturbations, there was ample reason for such concern. Gradually, it had

U.S. Census Bureau projection for the year 2000 indicates a rise in the median age as the country heads toward zero population growth around the year 2040.

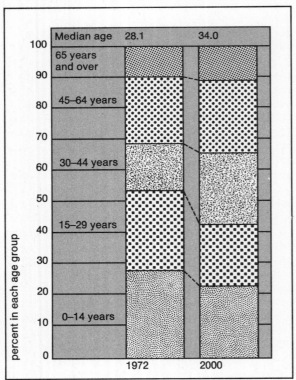

come to be recognized that the variation of complex systems can be described in terms of a number of different characteristics.

In 1973 C. S. Holling introduced the idea of separating the concepts of stability and resiliency. He proposed that resiliency be used to indicate the ability of systems to persist in the face of major changes in the forces that drive them, whereas stability is the ability of a system to return to a state of equilibrium after a temporary disturbance. The practical significance of these concepts is that man has now gone beyond the point of merely affecting the stability of social and planetary systems and is affecting their resiliency. Questions were being raised as to whether man could unwittingly affect society or the planet in such a way as to diminish this resiliency catastrophically. The fact that there are many systems that man understands only imperfectly—from the functioning of the economy to the determination of climate—suggests that such unwitting destruction is quite probable.

In any case, by 1973 most environmental scientists were prepared to accept two propositions about systems stability that had important implications for the future. First, it was widely believed that diversity within systems promotes stability of the entire system. Second, it was widely believed that too high a rate of energy or capital flow through systems tends to be inimical to the maintenance of diversity. One well-known example indicates why there is cause for alarm about decreasing systems diversity. Early in this century, the U.S. automobile industry had a very large number of small, competing companies (*i.e.*, great diversity). Gradually, however, as the volume of energy and capital flow through the industry increased, economies of scale and efficiency became so important to survival that only four companies persisted into the 1970s, and one of these, General Motors, had cornered approximately half of domestic automobile sales.

As theory predicted, the industry was vulnerable to wide-amplitude instability. In the face of repeated criticism that its cars were too large, inefficient, unsafe, and produced too much smog, GM showed itself to be remarkably resistant to change. When the Arab oil embargo occurred, the principal effect was a stunning decline in sales of GM cars—down 37% in the period between Jan. 1 and March 10, 1974. If there had been a large number of automobile companies trying a wide variety of corporate strategies (such as steam and electric cars), the percentage impact of the oil embargo on the entire industry would have been far smaller. This illustrates as simply as possible why environmental scientists argue that diversity within all types of systems (social, economic, biological, agricultural) is a positive asset.

Systems modeling. By about 1968 several groups had begun to use computer simulation with mathematical models of society to explore the consequences of various policies at regional, national, and global levels. The public received its first major introduction to this activity with the publication in 1972 of *The Limits to Growth,* the widely publicized and highly controversial first public document of the Club of Rome, an international group of concerned citizens.

By 1974 it appeared that much of the criticism directed against the book at the time of publication was ill-informed. For example, it was argued that mankind could not be anywhere near limits to growth because there had as yet been no widespread increases in commodity prices. After those statements appeared in print, the price rises occurred.

Further, the great volume of rather emotional criticism directed against the book tended to obscure a highly significant phenomenon: publication of *The Limits to Growth* stimulated an enormous increase in the number of groups doing such simulation modeling. Indeed, the book largely created a new field, and on a grand, international scale. Systems modeling groups were set up in a large number of universities, in many countries, in the International Institute for Applied Systems Analysis in Vienna, in the World Health Organization in Geneva, and in many corporate and government groups.

What had been a major debate within systems modeling groups seemed to have been resolved by 1974. Suppose one wishes to model phenomena at a particular level of aggregation (say a metropolitan area); does the computer simulation model need to mimic events in other levels of aggregation (*e.g.,* the world, the nation) or does it suffice to model only the level of concern? Increasingly, it appeared to be wise to make a multilevel (hierarchical) model. The reason is that many of the effects produced at any level of aggregation have their primary causes at some other level. For example, in the years ahead, farmland at the perimeter of U.S. metropolitan areas will almost certainly become more valuable and, hence, less likely to be converted to urban use. This will happen, not because of any phenomena occurring at the regional level but, rather, because the nation must sell more grain to pay for imported fuel. Thus the phenomenon occurring at a regional level can be understood (and simulated on a computer) only if the model incorporates certain causal pathways originating, in this case, at national and global levels.

Another new development in systems modeling concerned the use of normative models. For example, suppose it is clear that the behavior of a region or a country in the future can be predicted only if the model-builder can mimic the way a policy-maker would make policies on the basis of the data available to him about events occurring within his jurisdiction. There are two approaches to this. One is to sit such a person at the console of a computer and show him, on a television screen connected to the computer, data simulating the data that would become available, year after year, in the real world in each scenario. The policy-maker simulates his responses by typing instructions to the computer, indicating how policies would be used to cope with the unfolding situation. Then the computer simulates the effects of such policies on the system, and the policy-maker types new instructions indicating how to cope with the effects of his previous policies. A second approach is to build the behavior of the policy-maker into the whole computer simulation model. This was the method being used by Hartmut Bossel and Barry Hughes of the Mesarovic-Pestel World Model Project.

Perhaps the major finding of all simulation studies on world, national, and regional systems modeling was that, if anything, the rather gloomy prognosis of *The Limits to Growth* was too optimistic. The greater the detail built into these models, the more types of difficulties for the future were revealed. However, certain feedback control mechanisms were beginning to operate that could work against the occurrence of major catastrophes in the developed countries. The two most important were probably the decline in the age-specific birthrate and the increase in inflation, which would tend to inhibit profligately wasteful use of resources and the resulting pollution.

Energy utilization and city design. Because of the great concern about availability of energy, new research was done to indicate how to make the most efficient use of energy in urban transportation systems. The object of such research was to discover possible types of policies that would minimize gasoline consumption per person per year in cities. Two findings were not surprising, but two others were. As expected, increasing the price of gasoline or the availability of public transportation would cut down on transportation fuel wastage. (Public transportation delivers roughly ten times more passenger-miles of transportation per gallon of fuel than the private automobile.) The surprising results were that it could not be shown that the average distance between homes and places of work had a significant effect on total annual energy use in transportation, but it could be

Lead trap for use in vehicle exhaust systems is produced in Manchester, Eng. It uses a stainless steel filter coated with alumina to trap lead emissions.

shown that the availability of freeway miles per person in metropolitan areas did have an effect on gasoline use per person.

The interpretation placed on these results was as follows: It is not the "need to travel" (home–work commuting distance) that determines how much traveling people do by car in cities, but rather the "ease of travel" (the availability of freeways that make it easy to travel long distances at high speed with minimum congestion in urban areas). Thus, with energy running out, an entirely new light was shed on the advisability of building freeways. By making it very easy for people to travel long distances at high speed, we make it very easy for them to use gasoline in a frivolous, wasteful way; by making it easy to do something, one increases the probability that it will be done, whether it is necessary or not. If we generalize this principle, it raises questions about the desirability of doing a number of things designed to increase the convenience of performing various energy-wasting activities.

Communication of research results. It has been known for some time that the magnitude of disasters is strongly affected by the extent to which populations warned about them take steps to protect themselves. For example, loss of life due to floods, tidal waves, hurricanes, or earthquakes can be minimized if people can be made to believe that the threatening event may in fact occur. In many instances, much loss of life occurred because the potentially affected people could not or would not heed warnings given to them.

As mankind faces the possibility of certain types of unprecedented global or national disasters, the scientific study of appropriate techniques for communicating warnings of impending threatening events assumes great importance. The significance of such research is highlighted when we note that one response by newspapers to the Arab oil embargo was to raise the question of why no one pointed out to the public in advance the possibility that such an event might occur. The fact is that many people pointed out such a possibility and, more generally, large numbers of people had been expressing public warnings about depletion of fossil fuel energy supplies for decades. It is of great concern to determine why people and institutions did not respond to such warnings and to discover how such warnings can be delivered in the future so as to maximize the probability of a useful response.

Related to such research is the problem of the way in which the information flow in organizations is affected by the structure and dynamics of organizations. This problem is important because organizations must have time-to-respond characteristics that are socially meaningful, relative to the lead times needed to make a useful response to a crisis. For example, Lester Lees and others pointed out that it takes a long time for new sources of energy generation to become operational on a large scale within an economy. Specifically, an examination of the *Historical Statistics of the United States* reveals that from the time when a new source of energy was first used to the time when it was supplying 10% of national energy needs was 40–60 years.

But when the U.S. runs out of currently important sources of energy, it will be necessary to get substitutes in place very quickly. It seems certain that oil and gas will be unavailable within a few decades. However, if substitutes are to be ready when needed, the process of getting the technology for providing them in place must begin decades in advance. This raises important questions in a society in which all large private and public institutions are characterized by a conspicuous lack of long-range planning capability, or even a concern with such capability. Clearly, examining how institutions are designed and function with the aim of increasing their ability to plan ahead and respond quickly and vigorously to surprising situations is revealed as an important new problem of the environmental sciences.

—Kenneth E. F. Watt

Foods and nutrition

Food scarcities and the dramatic climb in food prices overshadowed all other aspects of food and nutrition during the year. In 1974 it was estimated that 800 million persons, or more than one-fifth of the world's population, suffered severe sickness or disabling diseases brought on by malnutrition or food shortages. Outright hunger and starvation were common in many areas. Severe famine, the immediate cause of which was adverse weather, existed in large parts of Africa, as well as in India and Latin America. More than 100,000 persons (actual numbers unknown) were reported to have died in sub-Saharan Africa in the famine of 1973–74. In rural India approximately 80% of preschool children suffered from malnutrition severe enough to reduce their growth rate.

Costs of basic grains and soybeans reached all-time highs in 1973, and this was reflected in increased food prices across the board. The situation in the U.S. was typical. Food prices as a whole rose 20% in 1973. In March 1974 a loaf of bread cost 8.6 cents more than it had a year earlier —an increase of one-third, the largest 12-month rise on record and equal to the total increase in bread prices over the preceding 19 years. Meat prices also reached record levels, although these later declined somewhat, and the prices of other basic foods rose considerably. The total cost of food to the consumer, including meals eaten away from home, increased from about $140 billion in 1973 to an estimated $160 billion in 1974, representing an average of approximately $760 per capita, also a record high. Some 15.3 million persons, or about 7% of the U.S. population, were receiving some form of food assistance from the government. Reserves of grain in the U.S. were at a 20-year low.

The international food problem had become increasingly severe, and the outlook for 1975 was not optimistic. It was becoming clear that world grain and legume supplies would need to be diverted more and more to direct human consumption rather than being cycled through animals. The importance of this in making the most efficient use of limited protein supplies was strikingly illustrated when the number of kilograms of grain con-sumed per person per year in meat-eating countries was compared with the corresponding figures for countries where meat formed a relatively small part of the diet. Thus, according to statistics published by the U.S. Agency for International Development, the average Canadian consumed 1,076 kg of grain (either directly or as meat) annually while the average person in central Africa consumed 121. Other representative figures were: the U.S. 850 kg; the Soviet Union 724; EEC countries 458; Argentina 454; Japan 275; Southeast Asia 172; and South America (excluding Argentina and Brazil) 149.

Undoubtedly, greater nutrient enrichment of foods would be necessary in the future. Theoretically, at least, the cost of such a program was not a problem, since all of a person's vitamin and mineral needs can be taken care of for only a few cents per month. However, the real problem was obtaining sufficient amounts of basic grains to provide calories and protein for expanding populations.

Advances in food science. In the field of food science, the outstanding developments continued to be in "fabricated foods"—foods made in factories, usually as cheaper replacements for more traditional varieties. In the winter of 1973–74, when meat prices reached an all-time high in the U.S., most of the hamburger used in school lunch programs was actually a mixture of ground meat and up to 30% "texturized" soybean protein. Retail grocery stores throughout the country were selling similar soybean-meat products, and the use of

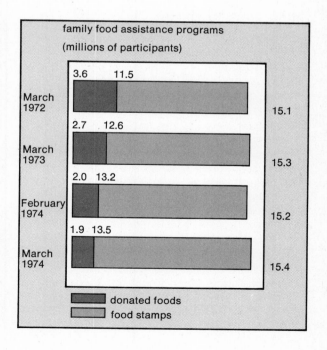

family food assistance programs
(millions of participants)

fabricated foods containing relatively lower-cost vegetable products was expected to increase at a steady rate. Widespread use of such products presented no real nutritional hazard as long as whole defatted grains or legumes were used rather than protein isolated from such sources.

A problem that remained to be solved before fabricated soybean foods could be fully accepted was flatulence, or intestinal gas production, in persons consuming them. Research was being conducted on means of eliminating or reducing this drawback. The high cost of processing, packaging, and marketing the products also tended to limit their use.

Alternate food sources were constantly being sought by food scientists. The production of single-cell protein, made by growing cells on waste products, continued to be a highly researched topic. The British Petroleum Co., Ltd., was reported to have several factories in Europe producing annually 20,000 tons of yeast raised on petroleum products. The yeast cost about $300 per ton to produce and thus was not yet competitive with soybeans and grains. Standard Oil Co. (Indiana) announced in 1974 that it planned to begin large-scale production tests on yeast grown for human consumption on ethyl alcohol derived from petroleum. The testing plant, scheduled for completion in 1975, would be capable of producing ten million pounds per year.

There was continued interest in the safety of food additives in 1974. The U.S. Food and Drug Administration (FDA) proposed new regulations for their classification and testing. Over 3,900 different compounds were now added to foods, consisting mainly of about 1,300 different flavors. The President's Science Advisory Committee on Chemicals and Health stated in 1973 that "there is no present evidence from epidemiological or other sources that points particularly to food additives as possible sources of hazards meriting special investigation." However, a study in 1974 indicated that artificial coloring and flavoring result in hyperactivity in some children.

As in previous years, hundreds of new types of manufactured foods reached the grocery shelf in 1974, though not all survived. Breakfast foods made from "natural ingredients" were very popular. Egg substitutes with no cholesterol were widely advertised and were used to a limited extent by persons who were advised by their physicians to consume low-cholesterol diets. An important new process that reduced the fat on the surface of potato chips, french fries, corn chips, and other snacks fried in fat was developed in 1974 by U.S. Department of Agriculture (USDA) scientists. The method reduced the fat content of

potato chips by about a third, or from 38% to 26%. Other new food products included tomato extenders (imitation dried tomatoes for manufactured foods), imitation honey, imitation meat products, synthetic caviar, chocolate-flavored peanut butter, imitation cheese, low-lactose milk products for persons intolerant to milk, dried molasses products, and commercial ready-to-eat salads.

Nutrition. There was a marked upsurge of interest in nutrition throughout the world in 1974, sparked by food shortages and inflation, growing concern over hunger and starvation, the availability of many types of processed foods, new labeling programs, and increased food advertising.

New NAS-recommended dietary allowances. A major event was the release, in May 1974, of the revised Recommended Daily Dietary Allowances of the U.S. National Academy of Sciences-Food and Nutrition Board. Revised every five years, the RDAs are the basic standards for nutritional intakes of all age groups in the U.S.

Some important changes were made in the 1974 report. Thus, for the adult reference man, the recommended daily intake of protein was reduced from 65 to 56 g, of vitamin C from 60 to 45 mg, of vitamin E from 30 to 15 international units, and of vitamin B_{12} from 5 to 3 mcg. Zinc was added as a recommended nutrient at a level of 15 mg per day. It was expected to be some time before these new standards were reflected in the "U.S. RDAs" used by the FDA for labeling standards, even though they differed considerably in some cases.

1974 nutrition survey. The National Center for Health Statistics of the U.S. Department of Health, Education, and Welfare released the first of what were announced as semiannual reports on the nutritional status of the U.S. Based on a population sample of 30,000 persons, the HANES (for Health and Nutrition Examination Survey) report showed that rather large percentages of the U.S. population had diets that were deficient in one or more important nutrients. In general, deficiencies were higher for blacks, low-income persons, women of childbearing age, and the aged.

Thirty percent of persons aged 60 to 74 with low incomes reported an inadequate intake of less than 1,000 calories a day. Among those in the same age group but with higher incomes, 16% ate at this level. Inadequate intakes were found for calcium (72% of black women of childbearing age), vitamin A, vitamin C, protein, and iron in one or more of the classes studied. Iron deficiency with anemia was found to be most marked in persons aged 1 to 17, with the problem greatest among young children.

National nutrition policy. A National Nutrition Consortium representing several food and nutrition groups was formed in 1974 in the U.S. with the aim of assisting in the development of a much needed national nutrition policy. The following proposals for such a policy, prepared by the Consortium, were discussed at hearings held in June 1974 by the U.S. Senate Select Committee on Nutrition and Human Needs, chaired by George McGovern (Dem., S.D.).

National goals and the program to achieve such goals should:

1. Assure an adequate wholesome food supply at reasonable cost to meet the needs of all segments of the population, this supply being available at a level consistent with the affordable life style of the era.

2. Maintain food resources sufficient to meet emergency needs and to fulfill a responsible role as a nation in meeting world food needs.

3. Develop a level of sound public knowledge and responsible understanding of nutrition and foods that will promote maximal nutritional health.

4. Maintain a system of quality and safety control that justifies public confidence in its food supply.

5. Support research and education in foods and nutrition with adequate resources and reasoned priorities to solve important current problems and to permit exploratory basic research.

Trace element research. The role of trace elements in the nutrition of animals and humans received more attention in 1974 than ever before. There was a growing recognition that, in spite of claims to the contrary, the total environment, including the soil, climate, and fertilization, as well as toxins in the environment, can affect the composition of plant foods and hence the health of animals and people that eat them.

Vanadium, silicon, and nickel were shown to be essential for growth and metabolism of animals. Vanadium deficiency, produced in chickens by USDA researchers, resulted in impaired reproduction, lower survival rates among the young, altered red blood cell levels, changes in blood lipid levels, and reduced body growth. Zinc deficiency was im-

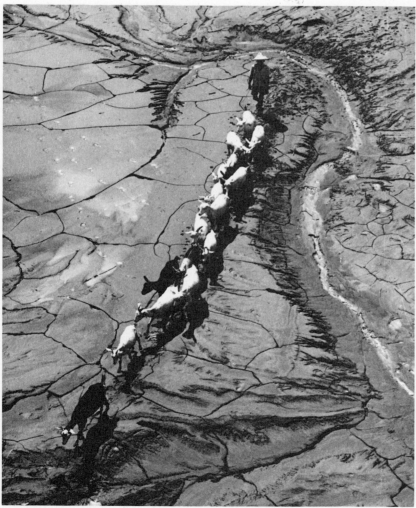

Keystone

Herdsman drives his cattle across the cracked and brittle ground of a dried-up riverbed in Taiwan. The river was a victim of an extended drought, one of the major causes of famine throughout the world.

"Wanna feel good? How about a massive dose of vitamin B$_{12}$?"

plicated as a public health problem in several countries. The addition of 40 mg of zinc per day (as zinc carbonate) to the diet of a group of schoolboys in Iran resulted in significant increases in growth, demonstrating that lack of this metallic element can be a principal limiting factor in the nutrition of children with high intakes of unleavened wholemeal bread.

Because many persons live in areas that lack central public water systems (in the U.S. about one-fourth of the population resides in such areas), alternate methods of water fluoridation were being sought. Workers at the National Institute of Dental Research showed that fluoridation of an individual school's water supply at four and a half times the recommended level for a community water supply reduced the incidence of decayed, missing, and filled teeth by 39% over a 12-year period. The children in the study had 65% fewer extractions.

Kenneth M. Hambidge of the University of Colorado presented evidence on the importance of chromium in glucose metabolism. He indicated that suboptimal chromium nutrition may be present in sections of the U.S.

Other nutrition advances. Increased recognition was being given to relationships between drug intake and nutrition. For example, a number of studies showed that women taking oral con-

traceptives usually have higher requirements for folic acid (a B vitamin also called folacin), vitamin B$_6$, and possibly copper and zinc. Whether these nutrients should be given as supplements or in food was a matter of debate.

Research continued on the claimed positive effects of very high doses of vitamins (megavitamins) on mental health and on colds and certain other illnesses. There was growing recognition that many of the vitamins, which serve as vital nutrients at the levels normally occurring in a well-balanced diet, do have certain pharmaceutical actions, including toxic actions, when taken in abnormally large quantities. A panel was established by the FDA to make recommendations in 1975 on effective, safe levels for vitamins and minerals sold over the counter in retail stores. In October 1973 the FDA ordered that pills and capsules containing vitamins A and D in quantities of more than 10,000 and 400 international units, respectively, could be sold only by prescription.

Recognition of widespread deficiencies of folic acid (which prevents one type of anemia and is especially important during pregnancy) gained much attention. Such deficiencies were found in the U.S. and many other countries. Food fortification with folic acid was proposed by scientists in South Africa and elsewhere.

The role of dietary roughage (fiber) in human nutrition was being studied following reports of higher incidences of heart disease, diseases of the digestive tract (including cancer of the colon and diverticulosis), and diabetes in populations eating low-fiber diets. These claims were not yet proven, but since the possibility existed, considerable research was being concentrated in this area.

There was a marked upsurge in the number of obesity treatment centers in urban areas. Many were being operated on a franchise basis. The health disadvantages of overweight, especially with regard to hypertension, were becoming increasingly clear, and obesity was considered to be a major public health problem in the U.S. However, the value of human chorionic gonadotrophin, used in the treatment of obese patients by some physicians at these centers, was still not a fully accepted practice.

Important advances on a number of other nutritional fronts included the discovery of practical ways to pre-digest lactose in milk, using natural enzymes, for persons intolerant to lactose; new methods of improving the nutritional quality of processed foods; methods of counteracting alcohol toxicity in man; research on the effect of diet in increasing longevity; more specific dietary requirements for certain amino acids, vitamins, and minerals; and better methods of assessing the nutritional status of populations. An important world-wide conference on nutrition, sponsored by the International Union of Nutritional Sciences and the Science Council of Japan, was scheduled to be held in Kyoto, Jap., Aug. 3–9, 1975.

Many unanswered questions in the field of foods and nutrition remained. The most vital one of all was: How can the world's growing population be fed adequately in the years to come when energy, tillable land, and other resources are becoming increasingly scarce?

Research was also needed to solve many current controversies, including the relationship of excess saturated fat, cholesterol, sugar, and calories to heart and vascular disease; the true role of fiber in the human diet; lactose intolerance and its relationship to the use of milk; the effects of very high intakes of vitamins; the relationship of nutrition to intelligence, infectious diseases, and cancer; and whether or not a fast growth rate in a child indicates optimal health and longevity.

—George M. Briggs

Honors

The following major scientific honors were awarded during the period from July 1, 1973, through June 30, 1974.

Aeronautics and astronautics

Daniel and Florence Guggenheim International Astronautics Award. The International Academy of Astronautics each year officially acknowledges an outstanding contribution to space research and exploration made during the preceding five years. The Academy named Maxime A. Faget recipient of the 1973 Daniel and Florence Guggenheim International Astronautics Award, which includes a $1,000 honorarium, for his accomplishments as director of engineering and development at the Lyndon B. Johnson Space Center in Houston, Tex. Faget's research was vital in developing the basic ideas and original design concepts used by the U.S. for its blunt-faced, manned spacecraft that reenter the earth's atmosphere at extremely high speeds. He was chief designer of the Mercury spacecraft and was responsible for the basic configuration of the Apollo command module and its pressure-fed hypergolic engines. His work also involved the development and integration of virtually all onboard systems used in the Mercury, Gemini, and Apollo vehicles: communications, guidance and control, propulsion, electrical power, docking, thermal protection, and life support.

Goddard Award. The American Institute of Aeronautics and Astronautics, which annually presents the Goddard Award for outstanding contributions in the engineering science of propulsion for energy conversion, named three men as co-recipients of the 1973 honor. Sharing the $10,000 honorarium donated by United Aircraft Corporation were Paul D. Castenholz of Rockwell International, Richard C. Mulready of Pratt & Whitney Aircraft, and John L. Sloop of the National Aeronautics and Space Administration (retired). The three were cited for "their significant contributions to the development of practical lox-hydrogen rocket engines which have played an essential role in the nation's space program and in the advancement of space technology."

Astronomy

Helen B. Warner Prize. The American Astronomical Society each year singles out a significant contribution to astronomy made during the preceding five years. The recipient of the $1,000 Warner Prize must be under 35 years of age and a resident of North America. The 1973 award was presented to Dimitri Mihalas, a specialist in the structure and composition of stellar atmospheres, whose most recent investigations at the High Altitude Observatory in Boulder, Col., centered on nonlocal thermodynamic equilibrium. He also gave a new dimension to his research by working out methods involving the use of large computers.

Henry Draper Medal. The National Academy of Sciences awards the Henry Draper Medal approximately every second year for notable investigations in astronomical physics. In 1974 the gold medal and a $1,000 honorarium were given to Lyman Spitzer, Jr., of the Princeton University observatory. In 1946 Spitzer accurately defined many scientific problems later researched by fellow scientists and suggested an extraterrestrial observatory, which was eventually launched with his help in 1968. It was Spitzer's studies of the chemical composition and interactions of interstellar matter that established the basis for subsequent study on processes leading to star formation. Having also been among the first to suggest magnetic containment as a means of achieving controlled nuclear fusion, he was for 13 years vitally concerned with plasma physics research at the Princeton Plasma Physics Laboratory.

Newton Lacy Pierce Prize. The American Astronomical Society announced a new award in 1973, the Newton Lacy Pierce Prize, and chose as first recipient Edwin M. Kellogg, one of several men who jointly designed the "Uhuru" (Explorer 42) satellite. With its launching off the coast of

Kenya on Dec. 12, 1970, "Uhuru" became the first X-ray astronomical observatory to orbit the earth. Its highly successful probing of the sky led to the discovery and growing understanding of X-ray emissions associated with clusters of galaxies. Because X rays generated in space do not penetrate the earth's atmosphere, "Uhuru" marked a new beginning in X-ray astronomy.

Biology

Franklin Medal. The Franklin Institute of Philadelphia has since 1914 presented its highest award, the Franklin Medal, to those workers in physical science or technology, without regard to country, whose efforts have done most to advance a knowledge of physical science or its applications. The 1973 medal was presented to Theodosius G. Dobzhansky, adjunct professor of genetics at the University of California at Davis, "for his intellectual leadership in advancing our experimental and theoretical knowledge of genetics and of evolution and of their interactions and for his recognition of the interactions of the biological and cultural factors in the unique human biocultural evolution."

Louisa Gross Horwitz Prize. Each year Columbia University administers the Horwitz Prize to acknowledge outstanding work in biology or biochemistry. In 1973 the $25,000 prize was shared by three biologists who achieved breakthroughs in tissue culture research.

Renato Dulbecco, assistant director of research at the Imperial Cancer Research Fund Laboratories in London, was chosen for developing the first plaque assay for an animal virus, poliovirus. The assay demonstrated that a single encounter between the virus and a cultured cell could lead to the production of a localized focus of infection. Later research provided convincing evidence that the genes of animal tumor viruses can insinuate themselves into the chromosomes of animal cells and that the viral genes then become part of the genetic complement of the host cell.

Harry Eagle, university professor at the Albert Einstein College of Medicine in New York City, was cited for developing a medium that permits the rapid growth of nearly all lines of mammalian cells, yet is simple enough in composition to allow nutritional experiments.

Theodore Puck, professor of biophysics and genetics at the University of Colorado Medical Center in Denver, received the award for being the first to demonstrate that individual animal cells in culture can give rise to colonies. As a result of his discovery, quantitative experiments on the genetics of mammalian cells became possible. With his associates, Puck also developed microbiological techniques for the study of inherited human disease.

Nobel Prize for Physiology or Medicine. A committee of the Karolinska Institutet in Stockholm awarded the 1973 Nobel Prize for Physiology or Medicine jointly to three scientists—Karl von Frisch, Konrad Lorenz, and Nikolaas Tinbergen—for "discoveries concerning organization and elicitation of individual and social behavior patterns." Although their studies in ethology directly involved animals and insects, the results have been extrapolated to human behavior.

Frisch, an Austrian zoologist, was the first to prove experimentally that fish can hear and that bees can distinguish odors and communicate with each other through the language of dance. His research showed that a scout bee entering its hive moves in a figure eight to announce a distant source of food and in a circle to declare a near source. The directional orientation and intensity of the dance further indicate the direction of the food and its amount. Frisch concluded that such dance patterns are hereditary inasmuch as bees from one colony cannot dance directions to those of another colony.

Lorenz, likewise an Austrian, proved that natural selection imposes certain innate behavior patterns on some animal species. Greylag goslings, for example, follow the first moving object they see after hatching; they will follow even balloons, boxes, or humans if the mother goose is taken away. Lorenz' work inspired research with primates that showed adult animals (and by extension, humans also) de-

Keystone

velop psychotic behavior in such unnatural environments as isolation and overcrowding. In a widely read and highly controversial book (*Das sogenannte Böse*, 1963; Eng. trans. *On Aggression*, 1966), Lorenz argued that aggression per se is a genetically programmed element in human behavior.

Tinbergen, a native of The Netherlands who established a department of animal behavior at Oxford University, was Lorenz' closest collaborator for more than 30 years. Tinbergen studied seagulls especially, devised comprehensive and ingenious experiments to verify his own and others' hypotheses, and applied his ethologic theories to human behavior. After noting that wild animals are often deterred from killing others of their own species when the intended victim uses certain gestures or facial expressions, he suggested that mankind's proclivity for war has been strengthened by the use of long-range weapons that all but eliminate face-to-face confrontations.

U.S. Steel Foundation Award in Molecular Biology. Administered by the National Academy of Sciences, the U.S. Steel Foundation Award may be given annually to a young scientist who has distinguished himself by some recent notable discovery in molecular biology. The 1974 award with its $5,000 honorarium was given to David Baltimore, a 36-year-old professor at the Massachusetts Institute of Technology. The citation described him as "a distinguished leader in virus research, who by his discoveries on the reproduction and enzymology of RNA viruses has greatly advanced the science of molecular biology."

Chemistry

Arthur C. Cope Award. Administered by the American Chemical Society since 1972, the Cope Award recognizes "outstanding achievement in the field of organic chemistry . . . that is different from any previously honored through a widely recognized scientific award." Donald J. Cram of the University of California at Los Angeles was given the 1974 gold medal, a $10,000 honorarium, and an unrestricted grant-in-aid of at least $20,000 for research at any university or nonprofit institution of his choice. Cram was chosen for several recent and significant innovations, including the genesis of "host-guest" synthetic chemistry, which imitates the way enzymes associate with substances in the body to catalyze reactions.

Garvan Medal. The American Chemical Society each year presents the Garvan Medal, established in 1936 and consisting of a gold medal and a $2,000 honorarium, to a U.S. woman chemist selected for her distinguished service to chemistry. In 1974 the prize was awarded to Joyce J. Kaufman of the Johns Hopkins University. She was cited for her theoretical work in quantum chemistry, including her application of computer calculations to predict the behavior of large drug molecules, such as those in morphine, that affect the central nervous system.

Nobel Prize for Chemistry. The Royal Swedish Academy of Sciences named Ernst Otto Fischer and Geoffrey Wilkinson co-winners of the 1973 Nobel Prize for Chemistry. Each worked independently to analyze ferrocene after reading the same

UPI Compix

Ernst Otto Fischer (opposite page), Geoffrey Wilkinson (left).

article in *Nature* that described the puzzling synthetic compound. They concluded that the material comprises a single atom sandwiched between two five-sided carbon rings to form an organometallic molecule. In effect, they explained how metals and organic substances can merge. Though practical applications of this knowledge are still rare, cleaner urban air could result if less toxic metals replace lead as a gasoline additive.

Fischer received his doctorate from Munich's Technical University, where he later taught; from 1959 he headed the university's Institute for Inorganic Chemistry. Wilkinson began his research on sandwich compounds in the early 1950s while teaching at Harvard University. In 1955 he returned to the Imperial College of Science and Technology at the University of London, where he had studied.

Priestley Medal. The American Chemical Society in 1922 established the Priestley Medal as the highest U.S. honor in the field of chemistry. The 1974 gold medal, presented annually for distinguished service, was given to Paul J. Flory, a physical chemist at Stanford University, for his work on polymers.

Earth sciences

Carl-Gustaf Rossby Research Medal. The American Meteorological Society annually presents its highest honor, the Rossby Research Medal, to a person who has made outstanding contributions to man's understanding of the structure or behavior of the atmosphere. The 1974 gold medal was given to Heinz H. Lettau of the University of Wisconsin for research "leading to a fuller understanding of the atmosphere's first mile. From his original concept of the stability length-scale to his pioneering contributions in boundary layer dynamics, turbulent transfer, climatonomy, and microscale surface modification, his work has been characterized by remarkable ingenuity and extraordinary dedication to purpose."

Cullum Geographical Medal. Since 1896, at irregular intervals, the American Geographical Society has presented its Cullum Geographical Medal to those who distinguish themselves by geographical discoveries or in the advancement of geographical science. The 1973 award was given to Bruce C. Heezen of the Lamont-Doherty Geographical Observatory of Columbia University for "distinguished contributions to the knowledge of the earth beneath the oceans." Heezen was prominent among those who proposed and substantiated the idea that previously known individual mid-ocean ridges are all part of the same worldwide system. He has also written extensively on sedimentary processes in the deep sea, on tecton-

ics, on seismicity, and on ocean-bottom currents as revealed in submarine photographs. His recent maps of the Antarctic and subantarctic regions have been hailed as major accomplishments.

Massey Medal. The Royal Canadian Geographical Society annually presents the Massey Medal, established in 1959, for outstanding personal achievement in the exploration, development, or description of the geography of Canada. The 1974 award was given to Frederick Kenneth Hare, head of the Institute of Environmental Studies at the University of Toronto, for the excellence of his scientific writings, in particular for his contributions to the understanding of climatology in Canada and for the role he has played in the development of geographical research in Canada.

Penrose Medal. The Geological Society of America has, since 1927, annually presented its Penrose Medal to someone whose original contributions have significantly advanced the geological sciences. Winner of the 1973 award was M. King Hubbert, a research geophysicist with the United

Photoreporters

States Geological Survey. His work on the physics of underground fluids, including the motion of groundwater, entrapment of petroleum under hydrodynamic conditions, and fluid behavior in petroleum reservoir engineering, was cited as the basis for significant additions to the world's store of available energy.

Medical sciences

Albert Lasker Medical Research Awards. The Albert and Mary Lasker Foundation each year presents a number of cash prizes in recognition of medical research in diseases causing death or disability. In 1973 two men shared the Albert Lasker Clinical Cardiovascular Research Award, which carries with it a $10,000 honorarium.

Paul M. Zoll, clinical professor of medicine at Harvard Medical School and associate in medical research at Beth Israel Hospital in Boston, was cited for inventing the pacemaker, for developing the theory and technique of continuous cardiac monitoring of heart rhythm, for introducing the use of externally applied electric stimulation to resume heartbeats, and for the application of alternating-current countershocks to stop ventricular fibrillation.

William B. Kouwenhoven, professor emeritus of electrical engineering and lecturer in surgery at the Johns Hopkins University in Baltimore, was honored for developing devices for both open and closed chest defibrillation, for originating techniques of external cardiac massage, and for confirming the fact that an electric shock can reverse ventricular fibrillation of the heart.

Distinguished Service Award. Since 1938 the American Medical Association has annually presented its highest honor, the Distinguished Service Award, to one of its most respected members. William F. House, a Los Angeles otologist (physician treating diseases of the ear), was named winner of the 1974 medallion for effectively combining the private practice of medicine with basic research, clinical research, and higher education. He directs what is believed to be the largest graduate and postgraduate program in otology in the world.

Kittay International Award. The Kittay Scientific Foundation named two psychiatrists as recipients of the $25,000 Kittay International Award for 1974. John Cade of Australia and Mogens Schou of Denmark were cited for their psychopharmacological use of lithium, which proved to be highly effective in controlling and preventing manic psy-

Konrad Lorenz (opposite page), Nikolaas Tinbergen (left), Karl von Frisch (above).

Photos, UPI Compix

chosis. Cade was the first to discover the efficacy of lithium in treating human manic excitement; Schou's further research established beyond doubt that lithium could be used safely to treat and prevent its recurrence. Their combined experiments, called the most important development in psychiatry during the past 20 years, revolutionized this aspect of psychiatry by shifting the focus from environment to biochemistry.

Physics

Edward Longstreth Medal. The Franklin Institute of Philadelphia established the Longstreth Medal in 1890; it is awarded for inventions of high order and for particularly meritorious improvements and developments in machines and mechanical processes. In naming Gerhard W. Goetze of the Westinghouse Electric Corp. as recipient of the 1973 silver medal, the Institute cited his "conception and development of the Secondary Electron Conduction Tube, which plays an important role in television, night surveillance and ultraviolet astronomical observations."

Howard N. Potts Medal. The Franklin Institute annually awards the Howard N. Potts Medal for distinguished work in science or the arts. The 1973 recipient was Howard Vollum, chairman of the board of Tektronix, Inc., who was cited "for his personal contributions to the design and manufacture of the Tektronix Oscilloscope, an instrument of unusual versatility and precision." The flexibility, accuracy, and reliability of the oscilloscope,

with its wide variety of indicator, control, and triggering functions, permits the taking of countless electromagnetic measurements at greatly reduced costs and has transformed the oscilloscope from a simple monitoring tool into a highly precise scientific instrument.

Nobel Prize for Physics. The Royal Swedish Academy of Sciences named three experts on tunneling, a phenomenon of quantum mechanics, as joint recipients of the 1973 Nobel Prize for Physics. They were Leo Esaki, a Japanese consultant with IBM's Thomas J. Watson Research Center; Ivar Giaever, a native of Norway working at General Electric's Research and Development Center; and Brian Josephson of Great Britain, who is an assistant director of research and reader in physics at Cambridge University. Josephson, a theorist, was awarded half of the $122,000 prize; the remainder was shared equally by Esaki and Giaever, both experimentalists.

While working for the Sony Corp. in Tokyo, Esaki developed the tunnel diode, which enables electrical current to pass through normally impassable electronic barriers. He completed his prizewinning research, which subsequently stimulated great advances in the area of tunneling in semiconductors, while doing doctoral studies at the University of Tokyo. After obtaining his Ph.D. in 1959, Esaki left for the U.S. to become a consultant to IBM.

Giaever's contribution was, as he put it, to "marry tunneling to superconductivity." Using a sandwich consisting of an insulated piece of

Leo Esaki (opposite page, left), Ivar Giaever (opposite page, right), Brian Josephson (right).

superconducting metal and a normal one, he achieved new tunneling effects that led to greater understanding of superconductivity. He was a patent engineer for the Norwegian government before emigrating to Canada in 1954. He soon joined General Electric as a mechanical engineer but in 1956 was transferred to the company's Research and Development Center in Schenectady, N.Y., where his interest shifted to physics. His Ph.D. in physics was earned at nearby Rensselaer Polytechnic Institute.

Josephson studied tunneling using superconductors for both sides of the sandwich. The oscillation that occurs between the two sides when electrons tunnel in pairs makes it possible to measure magnetic fields in space as well as laboratory current and provides previously unattainable accuracy. Josephson, a 22-year-old graduate student at the time of his discoveries, earned three degrees at Cambridge before joining the faculty.

Science journalism

AAAS-Westinghouse Science Writing Awards. The American Association for the Advancement of Science annually presents three AAAS-Westinghouse Science Writing awards in different categories, each of which carries with it a $1,000 honorarium. In 1973 the recipients were: (for newspapers with over 100,000 daily circulation) David Brand, a reporter for the *Wall Street Journal,* for articles entitled "Battle for Survival" (Feb. 7, 1973) on protein research, "Catching Sun-

beams" (April 16, 1973) on solar power research, and "I Am a Computer" (June 28, 1973) on artificial intelligence; (for newspapers with under 100,000 daily circulation) Bruce Benson, science writer for the *Honolulu Advertiser*, for his series "The Leeward Islands" (April 8–11, 1973); and (magazine award) Kenneth F. Weaver, assistant editor at *National Geographic* for "The Search for Tomorrow's Power" (November 1972).

Bradford Washburn Award. Boston's Museum of Science annually awards a gold medal and $5,-000 to an individual "who has made an outstanding contribution toward public understanding of science, and appreciation of its fascination and the vital role it plays in all our lives." The 1973 recipient of the Bradford Washburn Award was René Dubos, a microbiologist at Rockefeller University. Dubos, the first person to demonstrate the feasibility of obtaining germ-fighting drugs from microbes, has emphasized in books and speeches the symbiotic relationship between nature and man.

Miscellaneous

Founders Medal. The National Academy of Engineering annually presents the Founders Medal, first awarded in 1965, to honor outstanding contributions by an engineer to both his profession and to society. The 1974 recipient was J. Erik Jonsson, honorary chairman of the board of Texas Instruments, Inc., who was cited for "his utilization of engineering knowledge to improve the quality of

253

life through pioneering efforts in the manufacture of high technology products and innovative use of technology in solving the basic problems of the city" of Dallas.

National Medal of Science. The United States government's highest award for distinguished achievement in science, the National Medal of Science, is presented annually by the president of the U.S. to persons deserving of special recognition by reason of their outstanding contributions to knowledge in the physical, biological, mathematical, or engineering sciences. The 11 recipients of the 1973 gold medals were: (1) Daniel I. Arnon of the University of California at Berkeley "for fundamental research into the mechanism of green plant utilization of light to produce chemical energy and oxygen and for contributions to our understanding of plant nutrition." (2) Carl Djerassi of Stanford University "in recognition of his major contributions to the elucidation of the complex chemistry of the steroid hormones and to the application of these compounds to medicinal chemistry and population control by means of oral contraceptives." (3) Harold E. Edgerton of the Massachusetts Institute of Technology "for his vision and creativity in pioneering in the field of stroboscopic photography and for his many inventions of instruments for exploring the great depths of the oceans." (4) Maurice Ewing of the University of Texas "for extending and improving the methods of geology and geophysics to study the ocean floor and to understand the last remaining unexplored province of the solid earth—that which lies under the sea." (5) Arie Jan Haagen-Smit of the California Institute of Technology "for his unique contributions to the discovery of the chemical nature and source of smog, and for the successful efforts which he has carried through for smog abatement." (6) Vladimir Haensel of Universal Oil Products Co. "for his outstanding research in the catalytic reforming of hydrocarbons which has greatly enhanced the economic value of our petroleum natural resources." (7) Frederick Seitz of Rockefeller University "for his pioneering contributions to the foundations of the modern quantum theory of the solid state of matter, and to the understanding of many phenomena and processes that occur in solids." (8) Earl W. Sutherland, Jr., of the University of Miami "for the discovery that epinephrine and hormones of the pituitary gland occasion their diverse regulatory effects by initiating cellular synthesis of cyclic adenylic acid, now recognized as a universal biological 'second messenger,' which opened a new level of understanding of the subtle mechanisms that integrate the chemical life of the cell while offering hope of entirely new approaches to chemotherapy." (9) John

Wilder Tukey of Bell Laboratories and Princeton University "for his studies in mathematical and theoretical statistics, particularly his pioneering work on broad analysis and synthesis problems of complex systems, and for his outstanding contributions to the applications of statistics to the physical, social, and engineering sciences." (10) Richard T. Whitcomb of the Langley Research Center "for his discoveries and inventions in aerodynamics which have provided and will continue to provide substantial improvements in the speed, range, and payload of a major portion of high-performance aircraft produced throughout the country." (11) Robert Rathbun Wilson of the National Accelerator Laboratory "for unusual ingenuity in designing experiments to explore the fundamental particles of matter and in designing and constructing the machines to produce the particles, culminating in the world's most powerful particle accelerator."

Vladimir K. Zworykin Award. To honor the memory of the inventor of the iconoscope, the National Academy of Engineering in 1974 presented its third annual Zworykin Award to Ivar Giaever, a physicist at the General Electric Research and Development Center in Schenectady, N.Y. Giaever, a native Norwegian and co-winner of the 1973 Nobel Prize for Physics, was given a $5,000 honorarium for "his original contributions to the fields of electronic tunneling, superconductivity, and *in situ* protein detection."

Westinghouse Science Talent Search. Science Service, through its Science Clubs of America, conducts the annual Westinghouse Science Talent Search among high-school students to encourage creative originality in engineering and other branches of science. Winners are determined mainly on the basis of project reports evaluated by experts, but personal data, high-school transcripts, and national test scores are also considered. The top award of 1974 was bestowed on Eric Steven Lander of Stuyvesant High School, New York City, for a mathematical project on quasi-perfect numbers. He received a $10,000 Westinghouse Science Scholarship. Second place prizes were given to Frank Thomson Leighton of Arlington, Va., and Linda Kathryn Bockenstedt of Dayton, O., both of whom received scholarships of $8,000. The next three winners, all receiving $6,000 scholarships, were Emmett Evanoff of Cheyenne, Wyo., Richard Alan Dargan of Palm Bay, Fla., and John Conlin MacGuire of Casper, Wyo. Other major winners, each receiving a $4,000 scholarship, were Edward Harrison Frank of Great Neck, N.Y., Carl Taswell of Rochester, Minn., Jordin T. Kare of Narberth, Pa., and Linda Carol Rabinowitz of New York City.

Information science and technology

The year was a particularly eventful one for the information sciences. New information systems were being built, existing ones were being expanded, information networks were being organized, and governments were expressing concern about the possible effects of computer data banks on society.

New information systems. An Interfile Trade Information Center began operation at the World Trade Center in New York City. This was the first automated library of world trade information sources and was designed to respond to practical questions on international trade and commerce. The computerized data bank contained listings of more than 20,000 trade directories, economic surveys, statistical reports, and other reference documents. The trade centers of Tokyo, Brussels, and London had already contributed regional resource data, and they were expected to join the Interfile system.

A small, specialized information retrieval service on mechanized vibration was announced by the National Science Library of Canada. Called Vibank, this fully computerized system consisted of about 12,000 abstracts of engineering vibration literature classified on the basis of such categories as type of materials, type of vibrations, experimental observations, and applications.

A group of iron and steel companies and related engineering firms from the U.S., Canada, Japan, Australia, and Brazil were supporting the establishment of an Iron Information Center at the Battelle Memorial Institute in Ohio. The center maintained current and comprehensive information on iron processing and iron-ore agglomeration, including blast furnace and electric furnace practice, direct reduction, and pelletizing. Literature written in some 20 languages was scanned by a team of technical experts and translated into English.

The Center for Information and Documentation (CID) of the European Communities, in Luxembourg, undertook the coordination of a multinational program to cover all world literature containing technical and economic information on the manufacture and utilization of ferrous and nonferrous metal products and properties. Each of the member states would be responsible for acquiring, scanning, indexing, and preparing an English-language abstract of documents of interest to metallurgists in their own countries. The CID would amalgamate these contributions and provide for their computerized storage and retrieval.

Specialists use a computer terminal at Interfile, the computerized business information service of the World Trade Information Center in New York City. The Center is a project of the Port Authority of New York and New Jersey.

One of the major problem areas in information science is terminology control. Different words may mean the same thing, and the same word may have different meanings, dependent on context. In computer-based systems, these ambiguities create many problems which are particularly severe in multilingual systems. A new International Information Center for Terminology (Infoterm) was funded to act as a coordinating agency for terminology throughout the world. It would collect terminological publications, disseminate information, give advice on projects, and investigate the possibility of establishing an interconnection among different terminological word banks. Infoterm was located within the Austrian Standards Institute in Vienna.

A pilot program, initiated in New York City, represented the first application of information science technology for providing vital information about fires and other potentially dangerous situations. The system, called Tactical Information About Perilous Situations (TIPS), was designed to provide the officer in command of a serious fire

City officials of Rock Island, Ill., study a map overlay of the city produced from information stored in a computer. The computer maintains a complete inventory of all housing in the city.

with relevant information regarding the building involved, as well as surrounding areas of potential danger.

One of the more unusual information systems was developed in Rock Island, Ill., to provide city planners with an inventory of every parcel of land and the size, type, and condition of all buildings in the city. This information was stored in a computer which kept track of, and cross-referenced, more than 50 factors about housing and land use. One of the analyzed data outputs was a neighborhood profile of housing units by such factors as type, quality, and vacancy rates.

As various computerized information systems were developed for the same service area, managers who had to choose between them were faced with a general lack of knowledge and lack of comparative data about operational capabilities, costs, alternative methods of data input, and other important considerations. These problems had become particularly important for museum cura-

tors. Although computers had been used by museums for only five or six years, about a dozen different systems had already been developed. To coordinate the work being done, a Museum Data Bank Coordinating Committee was developing comparative descriptions of existing computer systems for museum cataloging and for the various data categories and conventions in use. The committee acted as a coordinating agency and disseminated advice and information to all interested parties.

Expanding information systems and networks. While new information storage and retrieval systems continued to be created, a further stage of development had been reached—the integration of on-line computing systems and telecommunications technology to form information networks. A network links people in different cities to a centralized computer installation by means of local telephone access or by leased lines. A typical user installation consists of a cathode-ray tube (CRT) terminal with keyboard and perhaps with a data printer. After dialing and being connected to the computer center, the user types a query to one of several data banks and usually gets an answer on the viewing screen within a few seconds. Many such networks were in existence and more were being established, for they were proving to be both efficient and cost effective.

The U.S. Defense Documentation Center established an on-line systems network linking a number of major Department of Defense facilities to the center's computer installation in Alexandria, Va. The user could begin his search for information by requesting abstracts of technical reports on a carefully defined subject area relating to a defense research or development project. The request was submitted by typing a list of key descriptor terms—contract number, contractor's name, author's name, or other identifying elements—and the retrieved abstracts appeared on the viewer. After having scanned these reports, the user could switch to a different data bank in order to examine documentation on current work or work in progress. A third data base would provide records on planned research and development projects. The two heaviest users of this system were the Air Force Avionics Laboratory at Wright-Patterson Air Force Base, Ohio, and the Redstone Scientific Information Center at Huntsville, Ala.

Having proved its value in the year since its establishment, Toxicon, a computerized information storage and retrieval system containing data on drugs, pesticides, environmental pollutants, and hazardous household and industrial compounds, was being expanded by the National Library of Medicine and made more accessible by means of

Automatic plate embossing machine built for the American Printing House for the Blind is controlled by data from a punched card that has columns corresponding to Braille cells. The embosser produces raised dots on folded metal plates at the rate of a page every 3 min 45 sec.

an on-line network. As an indication of the changes, the program was renamed Toxline.

The New York Times Information Bank, initially designed to provide reporters with computerized access to news items that had appeared in the *Times* and other selected publications over the past few years, was extended to the University of Pittsburgh, Pa. The university's station consisted of a CRT terminal with keyboard and a high-speed printer that could produce a hard copy of information displayed on the screen. Two additional units were a microfiche collection containing the full text of the news documents and a microfilm reader-printer for viewing and making copies. The faculty, students, and staff at the university were able to request information on almost any newsworthy topic by means of the CRT terminal keyboard. They could read abstracts of the retrieved articles on the display screen and the full text of the article on the microfilm reader. If they wished, a printed copy of the abstract or the article could be made.

Social implications. The individual citizen receives many direct benefits from improved information and library services, and the preservation of these services was proving to be of special concern to the U.S. Congress. Sen. Claiborne Pell (Dem., R.I.) introduced Senate Joint Resolution 40, which would authorize and request the president to call a White House conference on library and information services in 1976. The conference would bring together representatives of institutions, agencies, and associations that provide information services to the public and representatives of educational institutions, technologists, and the general public. This body would consider the recommendations of the National Commission on Libraries and Information Science.

The special needs of the blind for current information materials, books, and magazines were being better served as a result of some advances in information technology. A computer program and output printer had been designed that would translate English into Braille at the rate of 12 to 13 Braille pages per minute, compared with the 3 or 4 pages per hour produced manually by experienced Braille encoders. Each book is first keypunched on cards, which are then fed into the computer. The translation program compares each word with its Braille equivalent by means of an 80,000-word dictionary and produces the proper Braille code. If the word is not in the dictionary, the computer signals specialists who supply the appropriate Braille equivalent.

The information science activity that could prove to have the greatest effect on the general public was the release of the U.S. Department of Health, Education, and Welfare report on *Records, Computers, and the Rights of Citizens.* This report was

257

prepared by a 24-member advisory committee chaired by Willis H. Ware of the RAND Corporation. The committee was convened in response to a growing concern over threats to civil liberties resulting from computerized record-keeping operations, and it was charged with the task of analyzing and making recommendations concerning the possible harmful consequences of automated personal data systems.

The report concluded that computerization had made it easier for organizations to collect, integrate, and provide access to many separate records containing different bits and pieces of personal information. At the same time, it had become difficult for individuals to control the content and use of these records. To counteract this threat, the committee recommended the enactment of a federal Code of Fair Information Practice that would provide minimum standards for safeguarding personal files and legal redress if these data were misused. The report also pointed out the implicit dangers in the use of Social Security numbers as all-purpose identifiers and recommended that this trend be curtailed.

—Harold Borko

Marine sciences

Pollution of the oceans and the increasing use of remote sensing from spacecraft again dominated marine sciences during the year. Although it had been precipitated for political reasons, the energy crisis that followed the October 1973 war in the Middle East brought the need for energy conservation and additional energy production into sharp focus, giving rise, in some cases, to conflict with the equally apparent need for pollution abatement. In the field of space technology, instrumentation had become so sophisticated that a suite of instruments on a single spacecraft could be designed to study all the oceans of the world routinely and effectively. The new horizons being opened up as a result would set the pace in marine sciences for many years to come.

One of the most widely publicized oceanographic anomalies of recent years ended as the exceptionally long cycle of the warm El Niño current off the Peruvian coast came to an end. Normally cool and nutrient-rich water upwelling along the coast had provided the basis for the rich Peruvian anchovy fishery, which had been developed into a major source of the world's protein. Approximately every seven years, on the basis of past records, the fish have disappeared as the warm, nutrient-poor waters of El Niño moved close to the coast. The occurrence that began in the spring of 1972

proved unusually severe, however, with resulting worldwide repercussions on grain and livestock markets and supplies. As El Niño receded, the anchovies returned, although the Peruvian government continued to place some restrictions on fishing in order to protect the possibly depleted breeding stock.

Pollution. Plans to drill for oil off the U.S. East Coast and in the Gulf of Alaska led to a series of confrontations between environmental protection groups and the oil industry. Although it had been shown that only about 2% of the total oil pollution in the oceans was caused by accidents at oil drilling rigs and that effects from their routine operation were negligible, the communities near proposed sites strongly opposed drilling. It was hoped that improved technology would make it possible to demonstrate within a few years that the threat of pollution from this source was extremely small.

Among the major sources of oil pollution were the discharging of sludge-oil residues from the tanks of oil tankers at sea, which accounted for 53% of the total oil pollution of the oceans, and accidents at sea, which accounted for 11%. Both might well be eliminated or substantially reduced by about 1980. A giant step in this direction was the draft accord prepared by the Intergovernmental Maritime Consultative Organization and agreed to by the world's maritime nations. When ratified by the required number of signatories, it would establish regulations designed to eliminate completely the intentional discharge of detectable amounts of oil by ships at sea. Its full effect would not be apparent until about 1980, but all new oil tankers would have to meet the requirements and many currently operating tankers would be retrofitted to bring them into conformity. To reduce the hazard of oil spills from accidents, pollution abatement analyses for new designs of very large crude-oil carriers with segregated ballast had been carried out, as described in a recent issue of *Marine Technology*. These designs would greatly reduce the loss of oil should such a tanker be grounded or collide with another ship.

There were signs that another important source of oil pollution in coastal areas might soon be brought under control. This source (16%) was the discharge of used automobile oil into sewers, through which it quickly found its way to nearby bodies of water. Various communities were considering passing laws requiring that this oil be properly collected and reclaimed.

Other sources of marine pollution were being systematically documented as part of the International Decade of Ocean Exploration and in research being carried out by the Marine Ecosystem

Analysis program (MESA) in the New York Bight. In particular, the discharge of sludge in the New York Bight was under active investigation by both government scientists and scientists from the University Institute of Oceanography of the City University of New York.

Scientific meetings. The large number of scientific assemblies held throughout the year, and the correspondingly large number of papers that were presented, covered all aspects of physical oceanography, remote sensing, and ocean engineering. At a meeting of the International Association for the Physical Sciences of the Oceans held in Melbourne, Austr., great interest was shown in the results of studies of the oceans from space made by U.S. scientists and in plans for continuing such research in the future.

At the American Geophysical Union (AGU) meeting in Washington, D.C., studies of internal waves, laser techniques for determining wave spectra, and the computer simulation of oceanic eddies were received with great interest. A paper by Walter Munk demonstrated how the many different procedures for measuring internal waves could be combined into one consistent theory, and subsequent papers illustrated various aspects of the application of his ideas. Research by Denzil Stilwell in 1969 had stimulated a number of studies using, and expanding on, laser techniques for the study of vector wave-number spectra of ocean waves. A computer-based numerical model of the ocean circulation that produced eddies of various sizes was described by W. R. Holland and compared with one of the eddies detected by the Mid-Ocean

Courtesy, NOAA

Instrument that provides rapid continuous measurements of salinity, temperature, and depth is lowered into the New York Bight as part of the Marine Ecosystem Analysis (MESA) research program.

Dynamics Experiment (MODE). A meeting on remote sensing at the University of Michigan also highlighted numerous reports applicable to oceanography.

Oceanography from space. Among U.S. spacecraft currently in operation, the Nimbus 5 meteorological satellite and operational NOAA (National Oceanic and Atmospheric Administration) spacecraft were obtaining data of oceanographic importance. Preliminary analyses of the oceanographic and meteorological experiments conducted by Skylab indicated that the instruments in the earth resources experiment package did provide the desired measurements. As the next step in the scientific study of the oceans from space, the National Aeronautics and Space Administration (NASA) formulated an Earth and Ocean Physics Application Program (EOPAP), which would include the Geos C, Lageos, and Seasat A spacecraft and possibly others in the more distant future.

Nimbus 5, in orbit in 1974, carried a microwave spectrometer that permitted the estimation of water vapor and liquid water in the air over the oceans after correction for the effects of waves and ocean temperature. NOAA 2 routinely provided infrared images of the sea surface in areas where the skies were clear. Techniques had been developed that permitted the routine preparation of sea-surface temperature charts from these data on a day-to-day basis, and Cuddapah Prabhakara and his co-workers described improved procedures that should make the results more accurate. These infrared images were used to locate the boundaries of the Gulf Stream whenever skies are clear, allowing the current to be tracked almost on a routine basis. The images were also used to study sea ice motion along the coast of Labrador, as described by E. Paul McClain, and upwelling in the Gulf of Tehuantepec, as described by Harry Stumpf.

Data from ERTS 1 were coordinated with data from aircraft and conventional sources to study the New York Bight under the MESA program, as described by Robert Charnell. Studies of coastal water turbidity and current circulation in Delaware Bay were made by V. Klemas and his co-workers on the basis of 12 successful ERTS 1 passes over the bay. Other areas were also being studied in detail, and many additional reports on the results were expected.

Measurements of the dynamics of the ocean's circulation will be taken by the instrument being lowered to the ocean floor during the Mid-Ocean Dynamics Experiment (MODE) southwest of Bermuda.

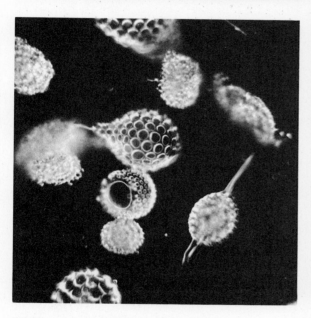

Skeletons of radiolarians, marine protozoans, were recovered from the bottom of the Atlantic and Pacific oceans in cores taken by the Deep-Sea Drilling Project. They are shown here magnified about 100 times.

ERTS 1 also produced a surprise. Its imagery detected internal waves (by means of their effects on the sea surface) along the edges of the continental shelf. When the density discontinuity at which the waves form lies over the shelf, the tides generate pulses at the edge of the continental slope. A train of about five or six internal waves, refracted by the effect of depth, then propagates toward shore. These waves had been known from other mea-surements, but the ERTS photographs clarified their behavior in a remarkable way.

The entire Skylab program, including the Earth Resources Experiment Package (EREP), was in grave trouble a number of times. The loss of the micrometeorite and heat shield and of one full solar panel and the failure of the second panel to unfold when the orbiting laboratory was launched caused great dismay in the scientific community.

261

However, the first Skylab astronauts, Charles Conrad, Joseph Kerwin, and Paul Weitz, succeeded in erecting a sun shield and deploying the stuck solar panel. On the day before the solar panel was deployed, a pass was made near a hurricane in the North Pacific that obtained immensely valuable data with the S193, a combination scanning pencil-beam radar/radiometer, passive microwave radiometer, and altimeter. The second manned mission, with astronauts Alan Bean, Owen Garriott, and Jack Lousma, went fairly smoothly except for a malfunction of the S193. In the third manned mission, astronauts Gerald Carr, Edward Gibson, and William Pogue repaired this complex instrument during a space walk so that it could be used in several of its planned modes. Another difficulty was the loss of one control-moment gyroscope, but the remaining two were sufficient to permit the spacecraft to be maneuvered, and a substantial amount of earth resources data, including oceanographic data, were obtained.

The earth resources data obtained by Skylab included many miles of magnetic tape and many hundreds of photographs. Another year or so would be required to analyze them fully. However, a report on preliminary Skylab results was given at the AGU national meeting in April, with astronaut Garriott chairing the session. Other reports on preliminary results had also been presented.

The S193 proved to be highly successful. The altimeter mode, as studied by J. McGoogan and others, performed as designed. It detected variations in the geoid, the mathematical model of the earth's surface based on a continuation of mean sea level. These included a predicted dip over the Puerto Rico Trench and rises over sea mounts and submarine ridges. Also, features were found that differed from currently available geoids computed from spacecraft orbit and gravity data.

The purpose of the scanning pencil beam-radar/radiometer was to measure radar backscatter and passive microwave temperatures for varying wind conditions and cloud cover to see if winds at the sea surface could be inferred from the measurements. It was believed that this would be possible because the height of the surface capillary waves increases in a well-defined way with wind speed, and the theory was borne out by radar measurements made over a hurricane during the first manned period of Skylab, as reported by R. K. Moore, W. J. Pierson, V. Cardone, and their co-workers. The analysis of additional data, especially that obtained in January 1974, should reinforce this result. The passive temperatures can be used to correct for the attenuation of the radar beam through thick clouds. A study of the high-frequency spectrum of wind-generated waves by Hishasi Mitsuyasu and Tadao Honda of Kyushu University in Japan appeared to provide a logical explanation for the variation of radar backscatter with wind speed, radar wavelength, and zenith angle.

As described by the Special Programs Office of NASA, the goals of EOPAP were "to identify, develop, demonstrate and utilize relevant space measurement techniques that will provide data contributing significantly to our knowledge of earthquake mechanisms, ocean surface conditions and ocean circulation." Important contributions to geodesy would also be made by this program.

Geos C, which was under construction, was designed to carry an improved radar altimeter, based on knowledge gained from the S193, to measure the distance between the spacecraft and the oceans on a routine basis. Data from this spacecraft were expected to shed light on several oceanographic problems. In particular, a number of scientists planned to study the tides on an oceanwide, or even a global, basis.

As described by George Weiffenbach, Lageos (laser geodynamics satellite) was designed to permit direct measurements of crustal motions, polar motion, UT-1 (earth rotation time versus atomic time), solid earth tides, and ocean tidal loadings. The interaction of the ocean tides with the earth's crust under the oceans is an important part of the study of earthquakes and seismic sea waves, or tsunamis.

Seasat A was still in the planning stage, and its final complement of instruments had not yet been decided. Various types of radar would definitely be included. A radar scatterometer, based on results obtained from the S193 and on a program at the NASA Langley Research Center in Virginia, would measure backscatter from the sea surface and hence the winds over the ocean. An improved altimeter would permit a more accurate determination of the position of the sea surface relative to the center of the earth; it was planned to be so accurate that it might be used to study ocean currents and storm surges as well as tides and the geoid. Also, the shape of the return radar pulse would be utilized to determine the height of ocean waves under the spacecraft. Finally, there was a distinct possibility that a synthetic aperture radar would be employed to image ocean-wave patterns and to determine ocean-wave spectra. As John Apel of NOAA stated, Seasat A would be a combination current meter, wave staff, anemometer, and thermometer when used over the oceans, capable of covering the global ocean twice a day on nearly an all-weather basis.

—Willard J. Pierson

Mathematics

The Weil conjectures. Few people would dispute that the most spectacular event in pure mathematics during the past year was the completion, by Pierre Deligne, of the proofs of some celebrated conjectures in arithmetic algebraic geometry made about 25 years ago by André Weil. These "Weil conjectures" have exercised an enormous influence on the prodigious modern development of algebraic geometry, and they have occupied the attention of some of the most brilliant mathematicians of the last two decades. Deligne's affirmation of the last and most resistant of these conjectures—an analogue of the classical Riemann hypothesis—was a combination of the deep originality and technical power that has characterized so much of this 29-year-old mathematician's work. Deftly he used an elaborate arsenal of sophisticated tools forged by many mathematicians, notably Alexandre Grothendieck, Michael Artin, Nicholas M. Katz, and himself.

The Weil conjectures are concerned with questions about "Diophantine equations" that were first raised almost 200 years ago by the German mathematician Carl Friedrich Gauss. Examples of such equations include $x^n + y^n = z^n$, where, for a given exponent n, one asks for integer solutions x, y, z. More generally, one might envisage an equation of the form

(1) $f(x_0, \ldots, x_d) = 0$

where f denotes a polynomial in $(d + 1)$-variables $x_0, \ldots, x_d$ with integer coefficients, and where the object is to study solutions of (1) with integer values of the x_i's. If $x = (x_0, \ldots, x_d)$ is to stand a chance of solving (1), then the integer $f(x)$ must at least be even, since zero is even. When this occurs, a mathematician says x is a solution "modulo 2" of (1), and he writes $f(x) \equiv 0$ (mod 2). More generally, if p is any nonzero integer in place of 2, one writes $f(x) \equiv 0$ (mod p) to signify that p divides $f(x)$; such a relation is called a "congruence modulo p."

A first step toward trying to solve (1) is to treat the easier problem of solving the corresponding congruence modulo p. This problem is easier because it is finite, in the sense that it depends on the various quantities only via the remainders they give upon division by p, and there is only a finite number of possible remainders, the p numbers 0, 1, 2, ..., $p-1$. In the case $p = 12$ one can think of it as being like passing to the arithmetic of hours on a clock except that in this situation the hours are numbered 0, 1, ..., 11, thus identifying 12 with zero (modulo 12). This example has the awkward feature that, whereas 3 and 4 are both nonzero (mod 12), their product is zero (mod 12). To avoid this phenomenon one usually chooses p to be a prime number. Then the arithmetic modulo p has all the customary properties, and in addition it even becomes possible to divide by nonzero numbers modulo p.

The system of p numbers so obtained is denoted

Sidney Harris

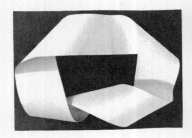

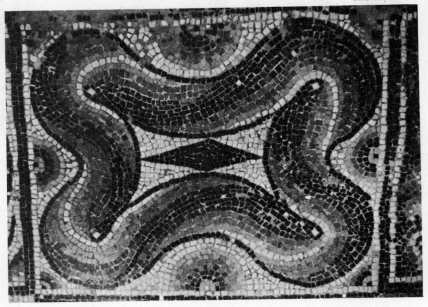

*Möbius strip, formed by giving
a half-twist (180°) to a strip
of paper and then joining
the ends (above), has interested
mathematicians since its
description by A. F. Möbius
in 1865. A Roman mosaic
of the 3rd century A.D. (right),
recently found near Arles, France,
shows that Möbius bands
were known in the ancient world.*

F_p, and called the "field of p elements." It is possible to interpret (1) as an equation over the field F_p, and solutions in F_p of this equation are merely a notational paraphrase for integer solutions of the congruence $f(x) \equiv 0 \pmod p$. One advantage of this point of view is that, for each integer $s > 0$, there is a larger finite field F_{p^s} with p^s elements, containing F_p, and it becomes possible to study the variation, as s increases, of $N_s =$ the (finite) number of solutions of (1) in F_{p^s}. The collective behavior of these numbers N_s is conveniently expressed in terms of the so-called "zeta function"

$$Z(t) = \exp \left(\sum_{s=1}^{\infty} N_s t^s / s \right).$$

Weil's conjectures, made in 1949, claimed, under mild restrictions on the prime p, that:
(I) (Rationality)

$$Z(t) = \frac{P_1(t) \, P_3(t) \ldots P_{2d-1}(t)}{P_0(t) \, P_2(t) \ldots P_{2d}(t)},$$

where the $P_i(t)$ are polynomials with integer coefficients.
(II) (Functional equation)
$$Z(1/p^d t) = \pm \, p^{dc/2} \, t^c Z(t)$$
for a suitable constant c.
(III) (Riemann hypothesis)

$$P_l(t) = \prod_{j=1}^{b_l} (1 - a_{lj} t)$$

where the numbers a_{ij} all have absolute value $p^{i/2}$.

These statements are admittedly somewhat technical in appearance to the nonspecialist, but they express in a striking and elegant way some

profound phenomena underlying the equation (1) modulo p. What is equally remarkable is an interpretation Weil gave to these conjectures, that one can consider the solutions of equation (1) in complex numbers $x_0, \ldots, x_d$. These form a so-called algebraic variety V in the complex $(d + 1)$-dimensional space, and one can attach to V certain algebraic invariants by the methods of topology, a branch of geometry quite remote from the theory of equations over finite fields. Among these invariants are the degrees b_i of the polynomials P_i, called the Betti numbers of V, and the constant c in (II), called the Euler characteristic of V.

Weil's interpretation further suggested imitating such topological methods as the Poincaré duality and the Lefschetz fixed point theorem as an approach to proving the conjectures.

This was indeed the path that was followed. The technical apparatus for carrying out this program was developed by various people, principally Grothendieck. He, Bernard Dwork, and others finally succeeded in proving conjectures (I) and (II), but (III) resisted repeated efforts until Deligne's success. Deligne's work yielded several further deep applications to analytical number theory, and many more were expected to follow.

Serre algebraic problem. In 1955 a French algebraic geometer, J.-P. Serre, called attention to a natural and elementary problem, which, in its simplest form, asks the following: Suppose k is a field (like the rational, real, or complex numbers, or a finite field F_{p^s}). Let $f_0, \ldots, f_r$ be $r + 1$ polynomials in n variables $x_1, \ldots, x_n$ with coefficients in k such that there exist $r + 1$ such polynomials $g_0, \ldots, g_r$ satisfying $g_0 f_0 + \ldots + g_r f_r = 1$.

264

Does there exist a square matrix of polynomials having first row $(f_0, \ldots, f_r)$ and determinant 1? It has been known for ten years that the answer is affirmative if n is at most 2 or if r is at least $n + 1$.

Recently, the first significant new progress was made on this problem simultaneously and independently by M. Roitman (Jerusalem); by M. P. Murthy, Jacob Towber, and R. G. Swan (Chicago); and by A. Suslin (Leningrad) and L. N. Vaserstein (Moscow). They demonstrated that the answer is affirmative always if $r \geq 1 + (n/2)$, in most cases if $n \leq 4$, and if $n = 5$ when k is a finite field with an odd number of elements.

—Hyman Bass

Medicine

Developments in medicine during the past year ranged from two new diagnostic tests for cancer to the establishment of professional review organizations for physicians. An overview of recent advances throughout the medical fields is provided under the title *General Medicine*. Three sections that follow, *Cardiology, Rheumatic Diseases,* and *Dentistry,* concentrate on specific aspects of these general areas. A special section, *Crib Death,* discusses the progress that is being made in dealing with this mysterious killer of infants. Also in this yearbook is a feature article related to medicine, THE LYMPHATIC SYSTEM: MAN'S BIOLOGICAL DEFENSE NETWORK.

General medicine

A vital reason for continuing to advance in medical treatment and research is the inexorable fact that disease also never stands still. Recent advances in combating illness have been matched by previously unrecognized disorders and the resurgence of health threats previously thought conquered.

Immunization. In 1721 the support of Boston civic leader and clergyman Cotton Mather was needed to mount the first mass inoculation campaign, in that case against smallpox. More than 250 years later many Americans still had to be cajoled to take advantage of this means of protection from disease. Despite the availability of proved vaccines, public health officials were alarmed to see a resurgence of old-fashioned measles (rubeola), German measles (rubella), diphtheria, and even polio. Chief victims were preschool children whose parents had not taken advantage of immunization programs. In October and again in May the nation's major medical groups launched immunization awareness campaigns in hopes of vaccinating an estimated five million susceptible children.

The chances for developing a vaccine against serum hepatitis, a major disease of the liver, increased with evidence that a microscopic particle found in patients with the disease is really a virus. Investigators at both Stanford University and the U.S. National Institute of Allergy and Infectious Diseases established the viral identity of the "Dane

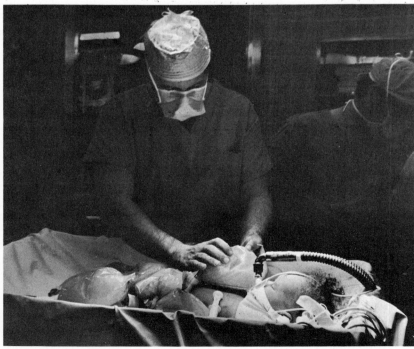

Chicago Tribune

Deep hypothermia, a technique in which the patient's body temperature is drastically lowered and his heart stopped briefly, is used so that surgeons can patch a hole in the wall separating the two ventricles of the heart of a one-year-old boy. Such operations are of great value for children under two, who are generally not ready to undergo conventional open-heart surgery.

Crib death

This year in the United States alone as many as 15,000 infants will die in their sleep, soundlessly, abruptly, from a mysterious malady called crib death. The loss of a child is always tragic; the loss of an infant to crib death is doubly so because of the residual of doubt that weighs heavily upon the parents.

A mother feeds her robust infant at 10 P.M., tucks him in at 11:30, when he is sleeping peacefully; the next morning she finds a cold and lifeless body. The shock is intense because of the suddenness of the tragedy. The magnitude of grief can barely be expressed by the unbelieving parents; a flood of guilt washes over them. Is it something I did (or didn't) do? Could I have saved him if I had gotten him up sooner? Could he have cried and I not have heard him?

The terms crib death or, in Great Britain, cot death are giving way to a more descriptive name for this disease, sudden infant death syndrome (SIDS). A working definition of SIDS is the sudden death of an infant or young child that is both unexpected in light of previous medical history and not adequately demonstrated as to cause by postmortem examination.

SIDS is responsible for 85% of sudden infant deaths in the United States and is probably the leading cause of death in infants over one week and under one year of age. Studies outside the U.S. show similar death rates in Great Britain, Australia, Czechoslovakia, and Denmark. There seems to be a seasonal increase, with more deaths occurring in late autumn, winter, and spring. Clustering of cases tends to suggest localized epidemics. The risk of SIDS is greater among babies of low birth weight, among nonwhites, and among the lower socioeconomic classes.

Medical entity. Apparently SIDS has been victimizing infants since biblical times. It was once called "overlaying," the assumption being that the mother, who usually slept with her infant, rolled over and caused his death. While it is difficult to determine, the incidence of SIDS does not seem to be increasing. It only appears so because of recent widespread publicity.

The complete medical description of SIDS remains to be written. The babies who fall prey to it are seemingly healthy and normal almost until the very hour of death. Most of them, however, do reveal a history of minor respiratory symptoms during the two weeks preceding death. Pulmonary congestion—fluid in the lungs—and inflammation and swelling of the vocal cords are usually present. These and other findings have suggested to some researchers that SIDS may be caused by a violent spasm of the larynx triggered by an overwhelming viral infection of the vocal cords.

Enlightened communites recognize SIDS as a disease entity. Unfortunately, many physicians and community officials continue to list the cause of death in these cases as suffocation and obliquely imply parental negligence or—even worse—child abuse as the ultimate cause of death. In many communities crib deaths are dismissed as due to suffocation or pneumonia, or are listed as undifferentiated accidental deaths—again tacitly implicating the parents. No state legally requires that all sudden deaths be autopsied. This attitude is

particle." Medical researchers hoped that the virus, first detected in 1970 by British scientists led by David M. S. Dane, could be grown in sufficient quantities to help produce a vaccine.

A major source of serum hepatitis infections is transfused blood. By 1974 it had become possible to test for the presence of a hepatitis-associated antigen in blood, but the test was expensive and not entirely foolproof. Much of the infected blood comes from "professional donors," who sell their blood. To combat this the American National Red Cross and 24 other major groups were attempting to set up an all-volunteer blood donor system.

Another viral scourge, spinal meningitis, is the target of a vaccine recently approved by the U.S. Food and Drug Administration (FDA). Since 1970 approximately a half million doses of the vaccine have been administered to U.S. military recruits, resulting in almost total protection. The vaccine, directed at the principal (Type C) variety of spinal meningitis, is expected to be given only where civilian epidemics threaten.

Epidemics: venereal disease and influenza. Unusual developments took place in America's perennial pair of public "epidemics"—venereal disease and influenza. Small but definite decreases in the incidence of syphilis were noted in the wake of nationwide education and screening programs. During the last half of 1973, the Center for Disease Control of the U.S. Public Health Service reported 2.4% fewer cases of syphilis than during the same months of 1972. At the same time, reported cases of gonorrhea among men rose only 1%, the smallest increase in a decade.

In comparison to recent years with their worldwide epidemics of the "Asian flu" and the "Hong Kong" variety, the winter of 1973–74 was not a serious one for influenza. But in the U.S. particu-

particularly distressing since it feeds the self-doubts that parents are already experiencing.

Enough is now known for a physician to make an unqualified determination of death by SIDS. It is not the result of suffocation. Accidental covering of the baby's face and evidence of a struggle have suggested this, but just as many babies have died with face free and without evidence of struggle. The cause remains unknown.

Work being done. In 1972 the U.S. Congress published a report of a special hearing of the Senate Subcommittee on Children and Youth. The objectives were to heighten awareness of SIDS, to explore research efforts, and to inform the public. It was revealed that only one medical research facility in 1971 was engaged in SIDS research, though since that time other research programs have been initiated. A joint congressional resolution introduced by Sen Walter F. Mondale (Dem., Minn.) and 15 other senators called upon the National Institute of Child Health and Human Development to list investigations into SIDS among its top priorities.

The psychological problems arising from the callousness of legal and medical authorities are difficult to manage. Intense grief coupled with self-incrimination assail the parents, and special family counseling is required to handle these considerable emotional burdens. Until 1962 parents confronted with these matters were isolated and shunned. In that year, however, the Mark Addison Roe Foundation was founded to assist the unfortunate parents of SIDS victims and to act as a clearinghouse for information. As of May 1967 its name was changed to the National Foundation for Sudden Infant Death, Inc., with headquarters in New York City but with several local chapters in other cities. Other prominent organizations concerned with SIDS are the International Guild for Infant Survival, Baltimore, Md., and the Andrew Menchell Infant Survival Foundation, New York City.

Medically, there is also some cause for optimism. In April 1974, Richard L. Naeye, chairman of the department of pathology at the Milton S. Hershey Medical Center, Pennsylvania State University, made an important announcement through the National Institutes of Health. Results of his studies indicate that the disease is associated with a chronic lack of oxygen—not a sudden deprivation but a long-standing deficiency for weeks or months after birth. Another finding was that SIDS victims showed a persistence of pigmented fat at ages when such specialized tissue should have been replaced by clear fatty tissue. Eileen G. Hasselmeyer, program director for the NIH's infant mortality program, reported and confirmed Naeye's findings.

These new facts, establishing that the infant victims were predestined to die of SIDS, should relieve anguished parents of their guilt. It may soon be possible to devise a screening test to discover potential SIDS victims and possibly to correct the oxygen deficiency. Work is in progress on a transistorized device that could monitor the body's vital signs. When placed in the crib or near an infant, it would warn of any dangerous departure from normal. Some researchers feel that quick and decisive action at such a critical moment—such as a vigorous awakening—might turn the tide.

—Richard C. York

larly alarming clusters of a rare complication appeared. Between mid-December and mid-March, about 150 children developed a degeneration of brain and blood known as "Reye's syndrome"; this always occurred after a common viral infection, usually influenza. Some 50 of the victims died, and public health spokesmen were unable to explain why this rare complication was taking place with greater frequency and only in certain areas, principally the Middle West. By spring the outbreak had subsided, but by then more than 300 cases had been reported.

The overwhelming number of flu infections in 1973–74 were of the comparatively mild Type B variety; however, as in all types of flu each individual case differed from all others because the influenza virus changes subtly with every human transmission. For this reason, past flu vaccines have been partly or almost entirely ineffective. Late in 1974 Edwin D. Kilbourne and his colleagues at Mount Sinai School of Medicine, New York City, planned to vaccinate about 2,000 children with a new "hybrid" vaccine made from minor components of the viruses. They hoped that it would provide partial but overall protection, inhibiting the impact of the virus rather than trying to block it completely. The key mechanism of the vaccine would be to limit viral replication by the use of neuraminidase, a protein component of all "flu bugs."

Immunology. A substance from within the bones, the marrow, is being used successfully to treat aplastic anemia, leukemia, and other diseases of the immune system. Although survival results have not been uniformly good, dozens of patients are surviving as the result of marrow transplants, usually donated by close relatives. Grafts of fetal tissue also are being employed to boost the capacities of immune systems left defi-

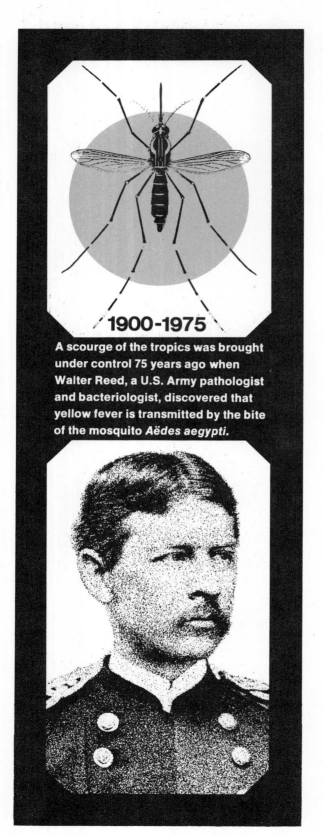

1900-1975

A scourge of the tropics was brought under control 75 years ago when Walter Reed, a U.S. Army pathologist and bacteriologist, discovered that yellow fever is transmitted by the bite of the mosquito *Aëdes aegypti*.

cient by a variety of birth defects. Cooperative bone marrow and fetal tissue teams were being planned or were operational in Chicago, Seattle, Boston, and other major medical centers. Fritz H. Bach of the University of Wisconsin predicted at an international conference in April that marrow "banks" containing registries of immunologically compatible potential donors such as those already established in the blood bank system were not more than five years from reality.

Two other components of man's immunologic system, interferon and transfer factor, were also in clinical use. A natural antiviral substance, interferon has been used at both Stanford University and Salisbury, Eng., against illnesses ranging from the common cold to chicken pox and shingles. Transfer factor seemed to increase the period of survival in certain patients with bone cancer, kidney tumors, and the Wiskott-Aldrich syndrome. As of 1974 both substances had been incompletely evaluated and required sophisticated immunologists to administer.

Bioengineering. Man's natural resources were being assisted by machinery in several important clinical areas. For example, several hundred hospitals possessed the equipment and expertise to perform autotransfusion of human blood. After massive trauma or during surgery, the blood lost is collected, filtered, and processed, and then introduced back into the patient's circulatory system. Where it can be used, autotransfusion eliminates the need for blood cross-matching, assures blood for those with rare types, and drastically cuts the cost of transfusion.

In another application, a surgical team at Boston University Medical Center adapted a perfusion machine, ordinarily used in transplantation surgery, to restore to a man the use of his kidney. In a seven-hour operation, the kidney was taken out and the tumors in and through it removed while the pump continued to perfuse it. It was then reimplanted into the patient.

Where man cannot provide help, bioengineering sometimes can step in completely. U.S. Navy physicians placed three seriously anemic patients, whose religious beliefs prevented transfusions, in a hyperbaric chamber until their blood supplies regenerated, thereby forcing oxygen into the body during the critical period. (In a hyperbaric chamber the pressure of oxygen is greater than normal.) The first artificial shoulders, complete with movable plastic/metal joints, were implanted in patients by surgeons at Chicago's Michael Reese Hospital. And an artificial kidney that weighs less than five pounds and can be worn on the body was being tested at the University of Utah Medical Center, Salt Lake City.

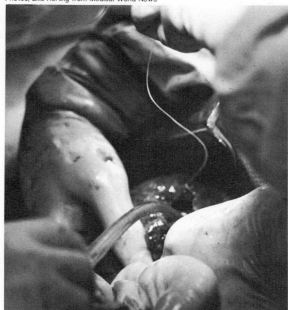
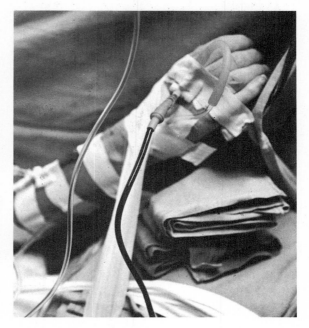

Blood is sucked up from the area of surgery (left) into a reservoir by autotransfusion device developed at Massachusetts General Hospital. The patient's own blood is then reinfused (right) into his body from the other reservoir. In one such operation the patient supplied 80% of the blood needed.

The atomic pacemaker passed the first anniversary of its introduction in the U.S. in April with all of the 15 pioneer recipients doing well. Approximately 200 of the plutonium-powered pacers were in use in the U.S., and about 500 others had been implanted in heart patients elsewhere in the world.

Perhaps the most sophisticated bioengineering feat, the use of a satellite to transmit medical information, would soon become commonplace with the orbiting of a special vehicle for that purpose in mid-1974. Transmissions of X rays, electrocardiograms, and other images had already taken place, even from ships such as the "SS Hope" and the "Queen Elizabeth II."

Cancer. The major focus for research and detection efforts during the year was a dual one: cancer and heart disease. Of the $2.6 billion in federal funds to be spent on biomedical research during fiscal 1974–75, more than $600 million were to go to investigate cancer and $334 million for diseases of the heart, blood, and lungs (see *Cardiology,* below). Although the overall federal funds for research declined, that segment devoted to cancer rose by more than $73 million over fiscal 1973–74.

Of the 655,000 persons in the U.S. expected to be diagnosed as having cancer in 1974–75, about 355,000 will die from the disease. Only early detection, before the malignancy spreads (metastasizes), can reduce that toll, experts believe. At least two immunologic tests, one for bladder and

the other for intestinal cancer, did not require tissue samples. Teams at Northeastern and Harvard universities were developing the bladder test, which relies on antibody reactions in urine. The intestinal malignancy assay, the first actually licensed by the FDA, was developed by investigators at Montreal General Hospital and Hoffmann-La Roche Inc. In this test physicians used serum samples from the patient's blood to look for the presence of carcinoembryonic antigen, a protein produced by tumors that is only present otherwise during fetal life. The antigen test was expected to be particularly helpful in monitoring the results of therapy after cancer is discovered.

A formidable battery of diagnostic tests was being used against breast cancer, the leading killer among women and third-ranking overall (lung is the leader, followed by cancer of the colon). Mammography, thermography, and ultrasound devices were found valuable. But, according to the American Cancer Society, the earliest and most frequent diagnoses are still made by women conducting the recommended self-examination on a monthly basis.

One of the rarest types of cancer, angiosarcoma of the liver, occurs only about 25 times a year in this country. Industrial physicians were, therefore, startled when a search of records revealed five deaths from this cancer in only a few years among workers in a Kentucky factory. The cause established was chronic exposure to vinyl chloride monomer, used in synthesizing polyvinyl chloride for a wide variety of plastic products. About 20% of the workers at a single plastics manufacturing plant showed liver abnormalities during subsequent screening. The U.S. Occupational Safety and Health Administration quickly issued tighter guidelines for using vinyl chloride monomer and launched an investigation of more than 400 other toxic substances used in industry.

Malpractice and fraud. This century's second "mercy killing" trial of a U.S. physician ended with the doctor acquitted. Vincent A. Montemarano, a Long Island physician, was charged with causing the death of a terminally ill cancer patient by injecting a lethal dose of potassium chloride. In the face of conflicting medical testimony he was found not guilty by a civil jury. (In 1949 a New Hampshire physician also was acquitted of a similar charge.)

Deliberate falsification of his research results was charged in May against William T. Summerlin, chief of transplant immunology at the Memorial Sloan-Kettering Cancer Center in New York City. Summerlin had claimed that he had grafted cultured skin from one animal onto another with no subsequent rejection of the graft. It therefore appeared that Summerlin might have discovered a method of overcoming transplant rejection, an achievement that would have great implications for medicine. However, other researchers discovered that black patches Summerlin claimed to have grafted onto white mice were not grafts but had simply been painted on. Summerlin later admitted that he had done this. An investigating committee placed part of the blame for the scandal on the general prevailing conditions of research, in which institutions must compete vigorously for scarce funds.

In perhaps the largest malpractice judgment against an individual physician to date, California orthopedic surgeon John G. Nork was ordered to pay a total of $3.7 million to a patient. The California Board of Medical Examiners later revoked Nork's license on the basis of 13 charges of negligent or unnecessary surgery.

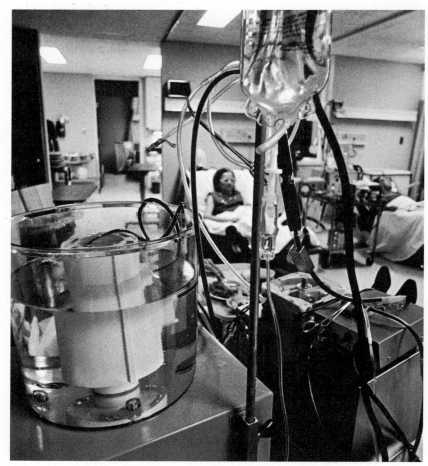

Dialysis unit substitutes for inoperative kidneys, cleansing the blood of patients suffering from kidney disease. As of July 1, 1973, Medicare paid the dialysis treatment costs for all persons in the U.S. covered by Social Security.

Wide World

Government programs. The California case heightened an already lively controversy about medical review mechanisms and their effectiveness. At the center is the federal program, mandated by act of Congress, to establish Professional Standards Review Organizations (PSROs) overseeing all medical care funded by tax moneys. Although the American Medical Association (AMA) took a position of compromise on PSROs, a number of state medical societies and national groups vowed to resist them. The system, scheduled to be operational in early 1976, divides the country into 203 PSRO units. Although peer review has in the past been a fixture of some clinical practices, this will mark the first time that physicians will be required to submit to it if they are to receive fees from such federal programs as Medicare and Medicaid.

The burgeoning issue of national health insurance created a thicket of proposals in the U.S. Congress. The AMA came forth with a plan called Medicredit, while the Nixon administration supported a bill calling for a Comprehensive Health Insurance Plan (CHIP). Sen. Edward M. Kennedy (Dem., Mass.) and Rep. Wilbur D. Mills (Dem., Ark.) joined forces to support yet another, untitled plan. Passage of a bipartisan compromise was expected in 1975. Thus far, all proposals would be funded at least in part from Social Security revenues.

An important impetus toward both the peer review and national health programs was consumerism. Nowhere has the voice of the taxpayer been stronger than where he is also the patient. In Maryland an affiliate organization of consumer advocate Ralph Nader published a controversial directory of local physicians (Prince Georges County), listing such details as the doctor's qualifications, average waiting time in his outer office, fees, and whether or not he makes house calls. Similar directories were being planned by groups in 32 other states, plus the District of Columbia, according to the Nader group. In Minnesota a new state law required all hospitals and nursing homes to post prominently a patient's "list of rights." Among these rights were privacy and being informed of the name and specialty of the doctors

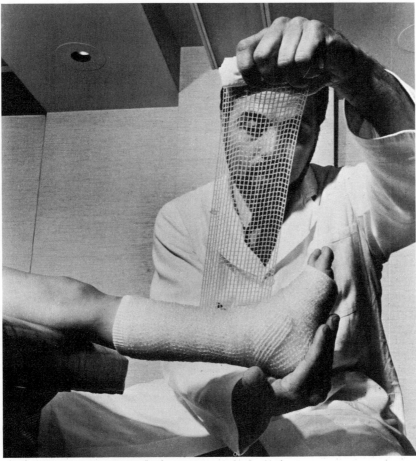

Courtesy, Merck Sharp & Dohme Orthopedics Company, Inc.

Newly developed cast, consisting of a stocking made of polypropylene and an open-weave fiberglass wrapping tape, is one-third to one-half the weight of conventional plaster-of-paris casts. The pliable tape conforms to body contours, and the cast can be immersed in water, allows for circulation of air, and is resistant to breakage.

Courtesy, Drs. L. S. Lessin, W. N. Jensen, and Panpit Klug, George Washington University

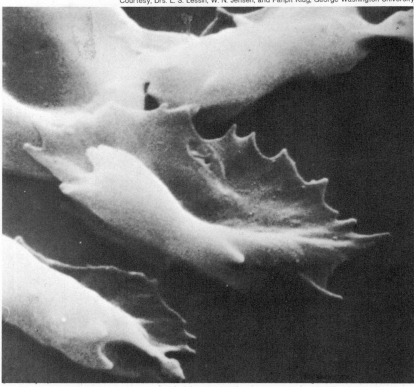

Sickled cell with its typical pattern of polymerized hemoglobin S is revealed in an electron micrograph. An intensive campaign against sickle-cell anemia, a crippler and killer mainly of black people, was under way.

coordinating one's care. The American Hospital Association urged all its member institutions to post a similar declaration.

In clinical research, new guidelines by the Department of Health, Education, and Welfare sharply limited participation in research experiments of prisoners, children, and infirm individuals without strict peer review. Under the HEW rule, protection committees would have to be set up regardless of the location and auspices of the research provided it was funded with tax dollars.

Physician shortage? A major question in any assessment of health care in the U.S. is how many and what kinds of doctors there should be. The "physician shortage" noted a few years ago has been declared by both the government and the AMA to be really a problem of maldistribution.

U.S. Pres. Richard Nixon proposed in his 1974 health message to correct the shortage of physicians in ghetto and rural areas by requiring new medical school graduates to serve a period of time in those areas of need. In return for this service the individual medical student's academic career would be directly backed by the government. In another approach a special panel of the National Board of Medical Examiners called for a two-stage medical licensure system; the physician would first receive a preliminary license and would become fully qualified only after serving an ap-

proved residency. Physicians would no longer be able to declare themselves "specialists" without full certification and training. The board also hoped that these more stringent requirements for specialization would convince more new doctors to go into general practice.

The American Academy of Family Practice, a specialized organization for the nation's general practitioners, set a goal at its October meeting of recruiting one-fourth of U.S. medical school graduates into their undermanned specialty. The AMA, in turn, called for half of the new M.D.s to go into "primary care" specialties—general practice, internal medicine, pediatrics, and obstetrics-gynecology. A Rand Corp. study estimated that the physician population would grow from its present 360,000 to 436,000 by 1980. Current pressures are toward directing most of this increase away from the large, urban medical centers.

Many of the new generation of physicians will be women. Although only 11% of medical school freshmen were females as recently as 1970, the Association of American Medical Colleges estimated that in 1974 almost 20% of the entering students would be women. By the 1980s the proportion of women in medical schools was expected to reach nearly one-third. But while the professional barriers to women were breaking down, public resistance remained strong. A Columbia University survey of 500 patients showed that four of five male patients and three of four women preferred to be seen by a male doctor. Those who had been treated previously by women were more likely to be positive toward them.

Other developments. The severe pain caused by the bony destruction in Paget's disease can be considerably relieved by injections of a substance extracted from salmon, U.S. Air Force physicians found. Calcitonin, a natural substance that also occurs in humans, is abundant in the fish and is thought to inhibit the turnover of calcium in the body. If obtained or synthesized in sufficient amounts, it may be used to reverse the effects of several bone diseases.

The ancient anesthetic technique of acupuncture underwent its initial year of systematic evaluation by skeptical U.S. clinicians. Some findings were cautiously confident. For example, an Ohio State University specialist in rehabilitation medicine reported pain relief in close to half his patients by using electrically stimulated needles. However, reports from a number of patients indicated disappointment in this therapy.

Among the most famous and perhaps best-examined patients during the year were the more than 500 American POWs released from Vietnam and Cambodian prison camps in February 1973. Still the subject of intensive periodic health examinations, the group had not developed the psychiatric difficulties, principally psychoses and criminal violence, expected on the basis of World War II and Korean internees. The severe malnutrition of their confinement caused a number to develop vision problems, including amblyopia. Although most had intestinal parasites and many had malaria, all seemed to have recovered almost fully.

The most dreaded disease among blacks, sickle-cell anemia, can be detected at birth by an inexpensive analysis of umbilical cord blood specimens, according to Yale University investigators. In this way major complications of the blood defect may be avoided later in life.

Cold baths can be used in some instances to correct the low sperm counts of formerly "sterile" men, according to a Salt Lake City study. After two weeks of daily cold baths both the counts and the motility of sperm in the men were increased. Most had previously been partial to long, hot baths.

Extreme sensitivity to sunlight, known as porphyria, was treated successfully with large doses of beta-carotene, a pigment abundant in yellow and green fruits and vegetables. In the body the substance is converted to vitamin A. Most of the 33 patients so treated became able to develop a suntan, Harvard University investigators report.

—Byron T. Scott

Cardiology

Each year in the United States approximately one million people die of heart disease. In 75% of these cases the deposition of cholesterol in the linings of the coronary arteries (atherosclerosis) is either the primary or secondary cause of death. Of the one million cardiac deaths, 260,000 occur in people under the age of 65.

While coronary heart disease is the leading cause of death in Western society, rheumatic heart disease remains the major cardiac problem in the less developed countries of the world. The infrequency of rheumatic fever in the U.S. today is the result of the widespread use of antibiotics. In the less developed countries the large numbers of young children with advanced heart failure caused by rheumatic valvular heart disease on hospital wards documents the urgency for adequate worldwide measures to prevent this crippling disease. (See *Rheumatic Diseases,* below.)

In 1973–74 the mystery of coronary artery disease due to atherosclerosis remained unsolved, but physicians were becoming increasingly discontented with the traditional concepts concerning its causation. Statistically, it has long been recognized that such factors as heredity, a diet high in fat content, an elevated cholesterol level, emotional stress, cigarette smoking, lack of exercise, high blood pressure, diabetes, and obesity are associated with an increased prevalence of heart attacks in most populations throughout the world. Although the greatest emphasis has been placed on the importance of the level of blood cholesterol, susceptibility to atherosclerosis varies widely among individuals with identical levels. Similarly, coronary heart disease is found in persons manifesting none of the factors currently held suspect in the causation of this disorder. Therefore, it seems evident that while risk factors are significant, susceptibility must also be related to differences in structure or function among individual arterial systems. Such susceptibility could be genetically endowed or acquired by environmental factors or unhygienic living habits.

Development of coronary atherosclerosis. The intriguing studies of Takio Shimamoto and associates in Tokyo not only offered a new and plausible concept of the genesis of cholesterol deposition and clot formation in the arterial system but also indicated a promising approach to their prevention. Using a powerful electron microscope, these investigators showed that the cells lining the arteries in the human body are normally in such close apposition that only small amounts of cholesterol can enter the intercellular spaces and thereby penetrate the wall of the vessel. In sharp contrast,

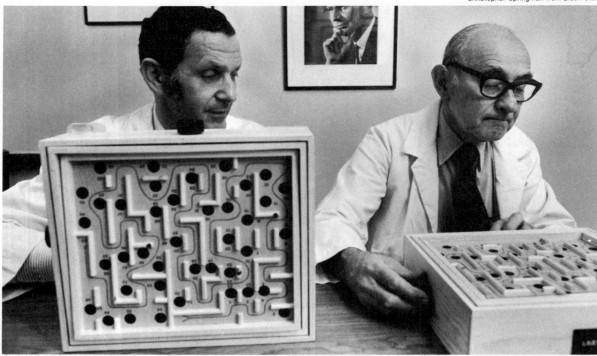

Maze-like games were used by physicians Meyer Friedman (left) and Ray Rosenman to determine frustration levels. The easily frustrated, hard-driving "Type A" people were those most likely to suffer heart attacks.

under conditions of stress (physical, emotional, chemical, dietary) the cells lining the blood vessels contract and swell, changing their form and increasing the size of the intercellular spaces. As a consequence increased quantities of fatty material enter the enlarged "gateways" and become dammed up in the arterial wall, providing the potential for development of cholesterol plaques. Accompanying these changes, blood platelets become more "sticky"; this causes them to adhere to each other and to the endothelial cells and thus affords a stimulus for clot formation.

All of these changes were shown to be prevented in rabbits and rhesus monkeys by administration of pyridinyl carbamate (Anginin) and other related agents. Whether or not similar therapy in man will ultimately prove of value in arresting the atherosclerotic process remained to be determined. In 1974 clinical studies were under way to provide an answer to this important question.

The candidate for a coronary attack. In a long-term study of the prevalence of coronary heart disease in Evans County, Ga., Curtis Hames observed that there are genetic differences in the capacity of individuals to mount a physiologic response to challenge (the stresses of life) and that these differences may account for the varied susceptibility to coronary attacks. The demands and respon-

sibilities of ownership, achievement, and "keeping up with the Joneses" cause responses in the human organism which, in excess, may lead to premature atherosclerosis. Like all animals, man responds to stress, actual or symbolic, by secreting adrenal hormones, increasing the level of cholesterol in his blood, and accelerating the coagulability of the blood. When such responses are excessive, as they are in some individuals, there is a predisposition to cholesterol deposition and clot formation in the arteries of the body, notably the coronary vessels. Significantly, Hames demonstrated that coronary disease is far more prevalent today than it was 20 years ago, even in the lowest socioeconomic groups, quite possibly because of the increasing pressures of modern society.

This concept is in keeping with the fate of the "Type A" personality, described by Meyer Friedman and Ray Rosenman. The individual with A-type behavior characteristics is a "hyperreactor" to stress who is highly competitive, compulsive about time, and deadline-oriented. Friedman and Rosenman found that persons so endowed possess a marked predisposition to coronary disease. On the other hand, Henry Russek showed that there is a significant correlation between the stressfulness of employment and the prevalence of heart attacks. Thus, it would seem that the response to challenge, whether genetically conditioned or environmentally augmented, may be a potent influence in the causation of coronary heart disease. These observations appear to point up the fallacy of our continuing preoccupation with "symptoms" (cholesterol levels, smoking, obesity, lack of exercise) rather than with causes.

Treatment. The ultimate treatment of coronary artery disease is prevention. Just as the respirator centers once used for the victims of polio are today viewed only as a bad dream, the coronary care units of the present era hopefully will one day be considered in the same light.

Drugs. During the past year, the U.S. Food and Drug Administration approved the use of a new agent, propranolol (Inderal), for the treatment of angina pectoris, a disorder characterized by recurring discomfort in the chest on effort (resulting from narrowing of the coronary arteries due to cholesterol deposition). This drug not only acts like a "governor" on an automobile by reducing the "speed" of the heart's performance but also diminishes the heart's requirements for oxygen. It can accomplish the latter because of its ability to interfere with the action of adrenal hormones, which whip up the action of the heart. Even more importantly, when used in combination with the standard nitrate drugs such as nitroglycerin or Isordil, propranolol produces remarkable responses. As a result of such combination therapy, most patients, even those with severe angina pectoris, have responded to an appropriate medical regimen of treatment.

Surgery. By 1974 the coronary bypass operation was being performed more frequently than any other form of radical major surgery in the U.S. During the past year surgeons opened the chests of more than 25,000 heart disease victims and implanted there one or more of the patient's own veins or arteries so that they would carry blood beyond obstructions in the coronary arteries that feed the heart muscle. This procedure has been credited with saving at least 60,000 lives and restoring many cardiac cripples to productive existence. However, its popularity and the enthusiastic support given to it by the medical profession unfortunately resulted in surgery being performed before an adequate trial of medical therapy was undertaken.

Recent observations seem to indicate that the coronary bypass operation might not be as safe as its advocates claim nor the results as beneficial. The mortality rate from surgery varies from less than 5 to 20% in different institutions. Complications such as myocardial infarction (the classic heart attack), brain damage, hemorrhage, kidney

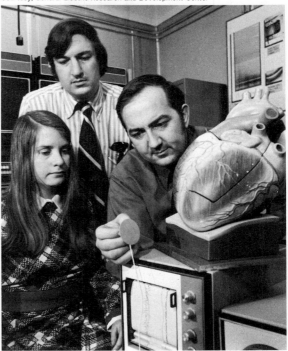

Researchers test a device that combines a minicomputer with a highly sensitive electronic "ear" in order to detect heart defects not found during routine electrocardiogram examinations.

failure, or closure of the bypass are not uncommon. In recent months mounting evidence suggests that the operation may not prevent heart attacks or prolong life but is effective in improving the quality of life in those who have remained unresponsive to "optimal" medical care. The need is therefore evident to ensure that all patients receive optimal care before surgery is even contemplated. Studies under way in 1974 were expected to determine the true indications for this form of intervention.

Mobile coronary care unit. The coronary care unit (CCU) has provided an effective means of preventing and treating life-threatening irregularities of the cardiac rhythm after a coronary attack. By providing trained personnel, constant patient monitoring, and immediate treatment when needed, such units have reduced substantially the hospital mortality rate from coronary artery disease. But because the majority of deaths from coronary disease occur outside the hospital, and in most instances without medical attendance, further reductions in mortality must require early coronary care in the pre-hospital phase of the acute heart attack. It has been estimated that at least two-thirds of sudden deaths are due to un-

275

recognized and untreated disorders of the cardiac rhythm. To prevent such disasters, it seemed logical to bring the trained personnel and equipment of the CCU to the patient outside the hospital as quickly as possible. This is now being accomplished in many cities throughout the United States and other areas of the world by means of mobile coronary care unit vehicles (MCCU) staffed by residents-in-training, nurses, paramedical personnel, or trained fire rescue squads.

Authorities have recognized that the degree of success of these efforts depends in large measure on increasing the public's awareness of the significance of chest pain and the lethal risks of delay in seeking skilled assistance. There are many persons who are alive today only because of the availability of the MCCU. Mortality has also been reduced by stationary life-support stations in factories, airports, office buildings, and athletic stadiums. In the future there will almost certainly be more of these in public places where large groups assemble.

Hypertension. During the past year a campaign was waged to contain the ravages of hypertension (a blood pressure of 160/90 or higher) by identifying its victims and instituting effective therapy before it can lead to such major complications as stroke, kidney failure, heart failure, and artery disease. The physical impairment and loss of life from hypertension need not occur considering the availability of potent drugs that are capable of preventing most of the major complications. Yet, surveys in the U.S. indicated that half of those with hypertension did not know they had the disorder; of those who did know, only 50% were under treatment and only half of them were being treated effectively. In fact, most surveys showed that the percentage of hypertensives under adequate treatment was no more than 10–20%. It was clear, therefore, that not only was treatment being ineptly applied but many victims were not even being identified.

Most affected by these inadequacies in the U.S. was the adult black population, of whom about 27% were found to be suffering from hypertension. Among blacks, hypertension is the leading cause of death and its importance may be better understood when it is realized that for every black person who dies of sickle-cell anemia, 100 others die from the complications of hypertension.

The present depth of concern was being manifested by the large-scale screening programs launched by medical school groups, pharmaceutical companies, and local heart associations to identify unrecognized hypertensives. These programs made it possible to screen large numbers of people at low cost over a relatively brief period.

Still to be worked out are the problems of delivery of appropriate medical care, patient education, and maintainence of patient motivation.

Preventive measures in childhood. One of the newer important issues in heart research is concerned with initiating preventive measures against artery disease among children and teenagers. Although many believe that artery disease begins as a pediatric disorder and should be dealt with in the community at large, most authorities agree that preventive measures are justified at present only for the child or teenager who demonstrates an exceptionally high possibility of developing premature coronary heart disease. By seeking out and counseling young people with elevated levels of cholesterol or blood pressure or with unequivocal family histories of these abnormalities, researchers believe that it may be possible to alter the familial pattern of the disease in later life.

Although the dangers of the typical American diet have frequently been emphasized in relation to the growing child, it seems remarkable that the "stresses" to which a student is exposed during the many years in educational institutions have not also been considered as possible detriments to health. Time-consuming assignments extending over holidays or the scheduling of tests immediately thereafter often destroy both the opportunity and the inclination for beneficial recreation and exercise during these vacation periods. Compounding the effects of these pressures, unrelenting competition for high scholastic standing and college placement makes formalized education in the U.S. a highly traumatic experience for many students.

Treatment today is largely focused upon the end result of a disease process rather than upon its prevention. Therapeutic interventions may someday block the dangers of a high-fat diet, high levels of blood cholesterol, and emotional stress, but until then hygienic measures and preventive medicine must begin in childhood and continue throughout life if any real impact is to be expected on the mortality risk from coronary heart disease. It has long been evident that the mental, emotional, physical, recreational, and spiritual needs of the young are being increasingly unfulfilled, both in the academic and home environments, despite unprecedented affluence. The resultant stresses almost undoubtedly are responsible for the atherosclerosis observed in young people. The educational institutions of this country can play a crucial role in preventive cardiology. By teaching and encouraging prevention measures both at school and at home, they may do much to help curb this "black plague" of the 20th century.

—Henry I. Russek

Rheumatic diseases

Rheumatism, or inflammation of the joints (arthritis), is one of man's oldest medical enemies and one of his newest research frontiers. The varied and seemingly mystical causes of joint inflammation have confused physicians for centuries. Until the modern era of medicine, all rheumatism was simply called "the gout." However, the fevers that swept Europe were often accompanied by acute arthritis. Acute rheumatism and infectious diseases were therefore soon recognized to be related and to be clearly distinguished from gout.

The rise of modern microbiology in the 19th century still left much of arthritis a puzzle. True, many cases of acute arthritis could be attributed to joint inflammation that was shown to be due to specific bacterial agents, particularly following severe septic states in which bloodstream infections (septicemias) seeded organisms in the joints. But the majority of the rheumatic diseases remained of mysterious origin. The rise of the science of immunology in the 20th century soon made it clear that many of the rheumatic diseases fell into a classification of hyperimmune states; that is, these diseases demonstrated features that were common to patients who were allergic or hypersensitive to foreign proteins, serums, drugs, inhalants, or other antigens.

By the 1970s there was intense study of the body's defenses against the invasion of foreign agents of all kinds—viruses, cancer cells, grafted tissues, drugs, inhalants, contactants, and pollutants. Possible relationships between the mechanisms of immunology and the varied causes of arthritis were being considered. This survey of current research in the rheumatic diseases emphasizes the role of infection and touches upon only a few highlights to illustrate some major new insights into various rheumatic processes.

Viruses. Two viral diseases have recently received particular attention because of the frequency with which they cause arthritis and because of interest in the mechanism by which they may do so. These are hepatitis B and rubella.

Hepatitis B, or what until recently was called "serum hepatitis" virus, is an infection about which a great amount has been learned since a serologic test was described to detect the presence of the virus—the so-called Australian antigen—or at least a part of it in the blood. It had long been known that some patients early in the course of viral hepatitis develop a syndrome similar to that of serum sickness: fever, hives and itchy rashes, joint swelling and pain, swollen lymph nodes, and even acute glomerulonephritis (an inflammatory disease of the kidney). The recent spread of hepatitis B virus in the United States has been alarming, and the extent of exposure to the disease can now be shown by blood tests that measure not only circulating viral antigen but, more sensitively, antibodies to the Australian antigen. The presence of these antibodies proves prior infection with hepatitis B. Like many other viral

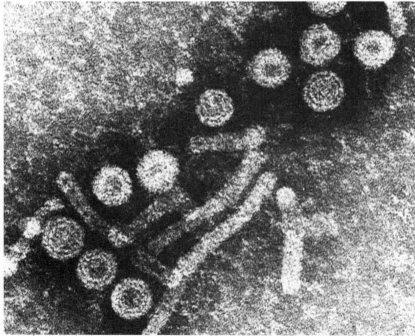

Courtesy, National Institutes of Health

Hepatitis B antigen, seen here magnified 150,000 times, reveals filamentous structure in association with round Dane particles.

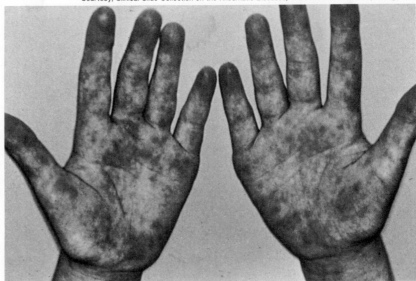

Rash and lesions on the hands are associated with systemic lupus erythematosus (SLE), a rheumatic disease in which antibodies to the host's own tissues and cellular components are generated.

diseases, hepatitis B frequently infects without producing symptoms, or if illness does occur it may not be associated with jaundice or other symptoms that would alert one to the diagnosis of hepatitis.

Because hepatitis B's peculiarity as an infection is the tendency for the virus to persist in the blood for a prolonged period, the antigen is still present when antibodies to it appear. This situation results in the circulation of large amounts of antigen-antibody complexes, setting up the conditions for what has come to be recognized as one kind of acute arthritis, "immune complex disease," and also for what once were classified as "serum-sickness-like" illnesses. Antigen-antibody complexes bind and activate complement, a generic name for a system of enzymes in the blood that produces many of the tissue responses of acute inflammation. Many of the lesions associated with serum sickness or immune complex disease have demonstrated that complement and antibodies are bound to the tissues and deposited in blood vessels, but seldom has the actual antigen causing the reaction been identified. Hepatitis B infection accounts for at least some cases of arthritis of the immune complex type. Furthermore, this virus has close relatives, such as hepatitis A, Epstein-Barr virus (the agent of infectious mononucleosis), and cytomegalovirus, all of which can produce the same kind of rheumatic disease.

German measles (rubella) is another viral infection that has long been known to produce arthritis, most frequently in older children, adolescents, and adults, and apparently more often in females. Because more people than in past years reach adolescence and adult life without exposure to rubella, it is more often accompanied by arthritis when it does strike.

A dramatic focus on the disease was provided by the relatively recent development of the live rubella virus vaccine, which has been used extensively in the past few years. Approximately 5% of those receiving the live vaccine may develop an acute arthritis, and this should permit considerable study of the mechanism by which the disease is produced. The latter is worth studying because the clinical manifestations of rubella arthritis may be strikingly similar to those at the onset of acute rheumatoid arthritis (*see below*). Sudden involvement of many small joints of the hands and feet may appear about two weeks after the vaccine is given or after the early symptoms of German measles have come and gone. But because these symptoms are often mild or even absent, there may be no clue to indicate the viral origin of the arthritis. Moreover, some investigators suspect that the joint reaction may represent an intense allergy to the virus.

Rheumatic fever. One of the great sagas in the history of the rheumatic diseases has been the dramatic decline in the incidence and severity of rheumatic fever in most countries of the Western world. Only a generation ago (in 1942) this disease caused more deaths in school-age children (5 to 19) than did all others combined. Rheumatic fever is terrifying not only because of the fear of death from acute inflammation of the heart during the attack but because of the high likelihood of permanent scarring of the heart valves during healing, leading to a crippled heart and premature death from heart failure.

More than 50 years of intensive research were

required to establish that rheumatic fever occurred as a late complication of the commonest bacterial cause of simple sore throat, the group A streptococcus. The mystery of rheumatic fever is the mechanism by which the streptococcus produces the disease. The manifestations of rheumatic fever begin a few weeks or longer after all traces of the streptococcus sore throat are gone. This complication, however, occurs in only a few percent, at the most, of those who suffer an untreated streptococcal ("strep") sore throat.

Prompt penicillin therapy of strep throat has prevented initial rheumatic attacks. Furthermore, the continuous use of small doses of penicillin or sulfonamides to prevent streptococcal infections in those who have once had rheumatic fever will prevent recurrences of the disease.

By 1974 the major remaining challenges to researchers in rheumatic fever were to discover the precise process by which the disease is caused and to find an effective vaccine against the group A streptococcus. In regard to the first, nobody has yet found a single streptococcal toxin or antigen among the dozens studied that can cuase the process of rheumatic fever, but the following recently established facts are of special interest: (1) Group A streptococci must infect the throat to produce the disease. Strains that cause skin infections or infections at other sites do not cause rheumatic fever even though some of these can cause acute glomerulonephritis by producing immune complexes of streptococcal antigens and antibodies.

(2) Rheumatic fever becomes much less frequent when the strains infecting the throat are attenuated or less virulent. (3) Group A streptococci contain antigens that are very similar to those found in human tissues, particularly the membranes of heart muscle bundles and the fibers of the heart valves.

Regardless of race, climate, or geographical location, the frequency of acute rheumatic fever is related to the conditions of crowding that promote intimate contact and pharyngeal passage of virulent group A streptococci. For these reasons, slums and military barracks during the rapid mobilization of recruits have bred the greatest epidemics of streptococcal sore throat and rheumatic fever. By contrast, the disease has virtually disappeared in affluent communities that have good housing and physicians who can diagnose and treat streptococcal sore throat.

In 1974 streptococcal vaccines were receiving intensive study by several research groups. Attention was focused upon one particular surface protein of the streptococcus, the M protein, which seems to be concentrated in small hairlike projections on the surface of the bacterium's cell walls. These projections are visible only by electron microscopy. M protein appears to be responsible for the difficulty white blood cells have in recognizing group A streptococci. The virulent strains of streptococci escape phagocytosis (ingestion) by white blood cells unless specific antibodies have been produced against the streptococcus' particular

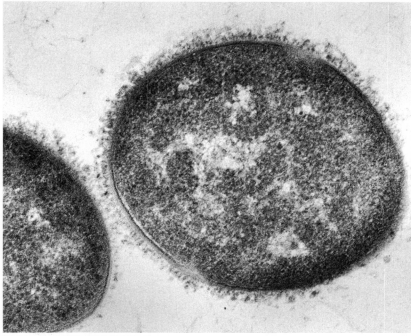

Ultrathin sections of group A streptococci as seen in electron micrographs reveal hairlike projections, indicating the presence of M protein. This protein is believed to be responsible for the difficulty that white blood cells have in recognizing group A streptococci, which cause rheumatic fever.

type of M protein. Such antibodies will combine with the surface M protein, activate complement (*see* above), and stimulate rapid phagocytosis and destruction of streptococci. Therefore, the production of such antibodies in response to M protein vaccine should confer immunity against the streptococcus.

Two major problems must be overcome: (1) There are several virulent strains of group A streptococci, each with its own specific M protein, so that vaccines must be produced against multiple types of M antigen. (2) More important, vaccines of purified M protein must be shown to be free of contaminating proteins that may cause adverse reactions with the heart, kidney, or other tissues. These problems are now being vigorously attacked. Preliminary trials of M protein vaccine in man have been encouraging. The methods of purifying M protein seem to improve each year, and the complexity of its structure is being unraveled.

Rheumatic diseases of unknown cause. Sir William Osler early in this century described syphilis as "the great imitator" because of the great variety of lesions and immunologic reactions it could produce. Since the decline of syphilis, the most florid immunologic disease has been systemic lupus erythematosus (SLE). Because almost all patients at some time in the course of this disease develop arthritis, SLE is a rheumatic disease, but its great variety of skin, blood cell, kidney, lung, heart, and liver lesions have involved many medical specialists in its investigation.

The immunologic phenomenon most characteristic of SLE is the development of autoantibodies, that is, antibodies to a great variety of the host's own tissues and cellular components. Most characteristic of these is antibody to DNA (deoxyribonucleic acid), the genetic material of the cell's own nucleus. Anti-DNA antibodies are the hallmark of this disease, and such antibodies are demonstrable in virtually all cases. Circulating immune complexes of DNA–anti-DNA undoubtedly cause some of the lesions of acute nephritis and vasculitis characteristic of SLE. In addition to antibody-forming lymphocytes (B lymphocytes), however, T lymphocytes and cell-mediated immune responses seem also to be common. Thus, SLE patients may show tuberculin-like skin reactions to tests with DNA, and may develop spontaneous skin lesions that are usually seen in other diseases associated with cellular hyperimmunity.

An exciting development of the past few years in SLE research was the finding of a viruslike structure in the cells that form the lining of the capillaries. The search for a viral agent led several investigators to examine the lesions of SLE with an electron microscope. Almost simultaneously,

workers in several laboratories began to notice structures that strongly resembled virus inclusion bodies within the cytoplasm of the lining cells of the capillaries in the kidney and in the skin; this is where one might expect a causative virus to be, since capillary lesions are basic to the pathology of SLE. The problem is that as of mid-1974 nobody had been able to grow a whole, live virus from these lesions, and it was therefore not clear whether these inclusion bodies represented an incomplete virus or some other alteration of the cell structure.

Of all the rheumatic diseases, rheumatoid arthritis remains the greatest challenge. Its frequency (estimated to be 2–5% of the adult population), its chronicity (virtually lifelong in some patients), and its characteristic tendency to deform the joints make it one of the most dreaded forms of arthritis. Although attacked with all the modern tools of microbiology and immunology, rheumatoid arthritis has been stubborn in yielding the secrets of its causation.

Most characteristic of the disease are its immunologic features. For the past two decades, great attention has been given to a characteristic antibody present in the serum of rheumatoid patients known as "rheumatoid factor." This is usually an "IgM," an antibody of large molecular weight that reacts with other antibodies in the blood, the common "gamma globulin" that we all possess. Rheumatoid IgM forms complexes with slightly altered IgG (gamma globulin), and these complexes bind and activate complement. This reaction causes the ingestion of these immune complexes by white blood cells in the joints and undoubtedly contributes thereby to the inflammatory reaction of rheumatoid arthritis.

Despite the many immunologic reactions of rheumatoid factor, no clear evidence incriminates it as the primary event in the production of the joint lesions. Instead, it seems to be a by-product of intense antigenic stimulation of the patient by some other agent or factor. Rheumatoid factor of very high concentration can be produced, for example, by patients who have chronic bacterial endocarditis, a heart valve infection caused by the common bacteria of the mouth and gut. In addition, some other rheumatic diseases (but not rheumatic fever) also may have rheumatoid factor in low concentration in the blood. Therefore, the factor is not entirely specific for rheumatoid arthritis.

What remains is the frustrating quest for an agent that can create the potent hyperimmune reactions observed in this disease. The recent interest in rubella arthritis (*see* above) stimulated investigators because German measles can cause reactions that look like rheumatoid arthritis in its early form, before progressive and destructive joint disease begins. Some reasoned by analogy with viral hepatitis that such specific infections occasionally cause sufficient damage to an organ so as to liberate tissue antigens that can sensitize an individual to his own tissues or cells. Such tissue injury could set in motion "autoimmune reactions" by means of which the host then destroys his own tissues. In "chronic aggressive hepatitis" the liver seems to be destroyed by such autoimmunity, which may at times be a consequence of some attacks of viral hepatitis; in progressive glomerulonephritis, the kidney is destroyed by autoimmune reactions following some forms of immune complex disease. Perhaps rheumatoid arthritis is a complication of an acute arthritis-producing infection which in the great majority of cases heals but which sometimes may go on to produce an "autoimmune disease." In future studies of early forms of rheumatoid arthritis, researchers hope to identify, at the earliest phase of the disease, the factors that may have initiated the intense immunologic process.

—Gene H. Stollerman

Dentistry

The dental profession's role in a future national health insurance program continued to be a key challenge for dentists in the U.S. A number of health insurance bills had been introduced, and in testimony before Congress in May 1974, the American Dental Association (ADA) pledged its continued active participation in the design and support of a dental component that would serve the needs of all people. However, the ADA reiterated its long-standing opposition to any program that would use public funds to provide health care services for persons who could afford to pay for them or that would impose a federalized structure on the health care delivery system.

Among the major provisions that the dental profession wished to see incorporated into a national health insurance dental component were: comprehensive dental services for children; emergency care for all; use of the usual, customary, and reasonable fee concept in reimbursement for services; preference for private carriers as opposed to governmental agencies; and control by the dental profession of any peer review mechanism associated with the program.

Dental insurance, the fastest growing health benefit in labor-management negotiations, was proving to be a highly effective tool for bringing dental care to more people. Inclusion of dental coverage in the health insurance policies provided in such industries as steel and automobile manu-

(Top) Wide World; (bottom) adapted from NEWSWEEK magazine

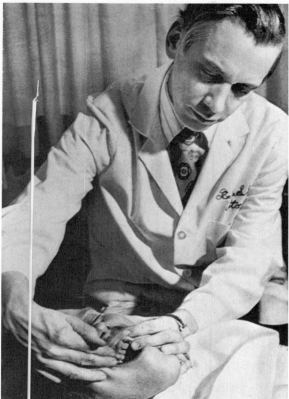

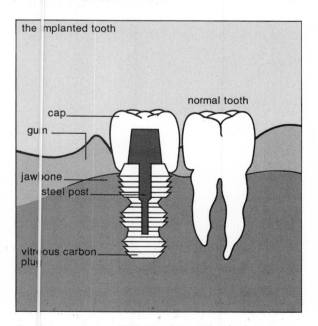

the implanted tooth

cap

gum

normal tooth

jawbone

steel post

vitreous carbon plug

Dentist examines artificial tooth root that he has implanted in a patient (top). The root is made of vitreous carbon, a hard, glassy substance that will not corrode or decompose. As seen in the diagram, a stainless steel post is fitted into the root and an artificial cap placed on it.

Courtesy, U.S. Department of Health, Education, and Welfare

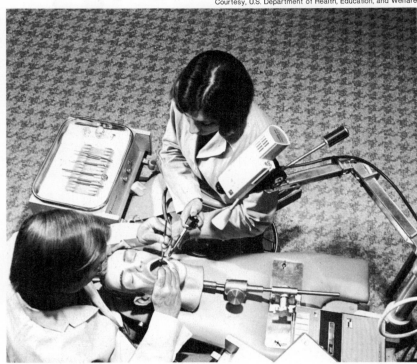

Robot patient is used to help train dental students in Louisville, Ky. If a student makes a mistake with the drill, the robot's head may jerk or its gums may simulate bleeding.

facturing would give added impetus to this rapid expansion. By 1974 some 20 million Americans were covered by dental care policies, compared with only a handful of people 15 years earlier.

The prospect of a national health insurance program focused concern on the issue of meeting growing dental manpower needs. At the request of the ADA the Leonard Davis Institute of the University of Pennsylvania embarked on a two-year study of future dental manpower requirements. Already, dental leaders noted, the expanding use of dental auxiliaries was enabling many dentists to treat more patients.

Diet and dental health. The importance of proper nutrition in maintaining good oral health was again cited by the dental profession at the ADA's annual session in Houston, Tex., in the fall of 1973. The association called for amendment of the National School Lunch Act to prohibit the sale of nonnutritious and decay-causing foods under the federal reimbursement program. It also reaffirmed its long-standing policy of urging school administrators and local government officials to ban the sale of sweetened beverages and sugar-rich products in competition with nutritional foods provided in school food programs. Sugar has long been identified as a major factor contributing to tooth decay in children.

Studies conducted by the University of Texas Dental Science Institute at Houston, in coopera-

tion with the Lyndon B. Johnson Space Center, indicated that astronauts may be especially prone to oral health problems because of their special carbohydrate-enriched diet. Regularly scheduled samples of oral bacteria were taken from the members of the prime and backup crews for the three Skylab missions before and after the astronauts were placed on the space diet. They followed routine oral hygiene procedures during the entire mission, including toothbrushing for two minutes twice daily and using dental floss once a day. Samplings were taken as early as 57 days before launch and resumed 4 days after the mission ended. The results showed increases in various types of oral bacteria. Since most of the changes were observed in both the prime and backup crews, they apeared to be related more to diet than to space flight per se.

Periodontal disease. Edward R. Loftus of West Roxbury, Mass., reported further evidence indicating that periodontal disease is more prevalent among smokers than among nonsmokers. His research also reconfirmed the relationship between smoking and leukoplakia, a type of oral tissue lesion. The study, conducted at the Massachusetts Veterans Administration Hospital, West Roxbury, showed that among patients of comparable age with similar oral hygiene habits, periodontal disease was more advanced in smokers than in nonsmokers and smokers exhibited more bone loss

R. F. Boehm of the University of Utah demonstrates laser which he plans to use in order to fuse caps of ceramic material onto cavity-prone molars, thereby protecting them from decay and making them easier to clean.

and tooth mobility. The link between smoking and periodontal disease was even more marked in patients with removable partial dentures.

In other research directed at controlling periodontal disease—still the major cause of tooth loss in adults—scientists from the National Institute of Dental Research, Bethesda, Md., discovered evidence that the body's immune system can be stimulated by dental plaque products and could be involved in the gum inflammation characteristic of the disorder. Plaque is a sticky substance that forms on teeth and is considered to be a major contributing factor in the development of both tooth decay and periodontal disease.

The body's natural defenses against disease include two types of "immune reaction," the formation of antibodies and the development of what is called cellular immunity. Antibodies are manufactured by certain white blood cells (lymphocytes) called B cells, which are derived from the bone marrow, while cellular immunity is controlled by another class of lymphocytes, called T cells, which are processed by the thymus gland. The investigators removed B and T cells from patients with severe periodontal disease, cultured them separately, and then introduced cell-free extracts of dental plaque into the cultures. Both types of cells divided actively. On the other hand, similar cells from individuals without periodontal disease gave no indication of being sensitized and did not mul-

tiply unduly, leading the researchers to speculate that the reactions of both B and T cells may aggravate chronic gum tissue inflammation.

Other developments. A Georgetown University (Washington, D.C.) study showed that aspirin was more effective than the usually prescribed codeine in relieving discomfort following oral surgery. The study involved 128 patients, aged 16 to 35, with impacted wisdom teeth. "We found aspirin to be extremely effective," said Steven A. Cooper of Washington. "In controlling . . . post-operative pain, we found most effectiveness came from the aspirin, not the narcotic. Its effectiveness against pain and swelling was probably due in large part to its anti-inflammatory properties."

At the annual meeting of the International Association for Dental Research, Wade B. Hammer of the Medical College of Georgia reported on the possibility of implanting "ceramic jawbones" to help stabilize lower dentures. "Frequently, we find that the bone base on which the denture rests— the alveolar ridges—shrinks. Sometimes this resorption of bone by the body is so marked that the ridge disappears entirely. This leaves no support for the denture." Augmentation of the ridge involves the implantation of a ceramic calcium aluminate material. This ceramic is lightweight and porous and allows bone and other surrounding tissue to grow into it, thus further increasing its stability.

283

Dental appliances such as bridges made of nonprecious metals were successfully welded together with laser beams. J. D. Preston, a dentist at the Veterans Administration Hospital-Wadsworth, Los Angeles, discovered that controlled application of laser beams to two pieces of base metal of the kind used for dental appliances resulted in a weld almost as strong as the metal itself. His findings were supported when the joints were ruptured in a laboratory test unit and the ruptured surfaces examined with a scanning electron microscope.

Dental scientists at Washington University in St. Louis, Mo., uncovered evidence suggesting that certain wooden chewing sticks, widely used in Africa for cleaning teeth, may contain substances that inhibit tooth decay. The chewing sticks are taken from over 100 varieties of plants—though most commonly from those in the citrus, ebony, and coffee families—and are generally about eight inches long and the width of a finger. The stick is vigorously moved sideways and vertically over the tooth surfaces and, once the cleaning is completed, is often sucked for several hours more. Memory Elvin-Lewis, who headed the study team, said tests showed that *Massularia acuminate,* a hardwood member of the coffee family used in southern Nigeria, apparently contains substances that prevent the growth of oral bacteria implicated in the development of decay. This was also true, although to a lesser degree, of *Fegara zanthoxyloides*, commonly called candlewood. Companies in England and India began manufacturing toothpastes that utilize substances taken from these chewing sticks, and limited clinical studies were being conducted at Washington University.

—Lou Joseph

Microbiology

The year was a characteristically active one for microbiologists. While there were no breakthroughs of any great magnitude, there was an abundance of high-level research. Environmental problems continued to receive attention, but there appeared to be a shift of emphasis as increasing attention was directed toward problems involving energy and other practical applications.

Environmental microbiology. For environmental microbiology, the year was a time of consolidation of prior findings. Nevertheless, some interesting new data were reported. There was increasing evidence that nitrogen, not phosphorus, is the growth-limiting nutrient for phytoplankton in aquatic environments. As a result, considerable emphasis was being placed on finding ways to

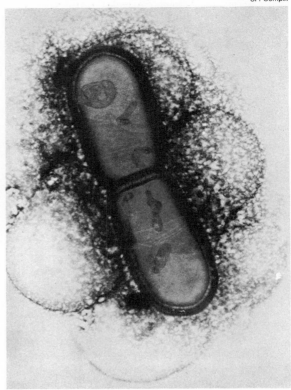

UPI Compix

Rare unnamed microorganism of northern California lives in an environment ten times more alkaline than believed possible to support life. The planet Jupiter is thought to have a similar environment.

remove nitrogen from wastewaters in order to control and prevent excessive growth of organisms in aquatic systems.

Further evidence was reported indicating that microbial utilization of nutrients is extremely slow on the deep-sea floor. An enlarged gut was found in deep-sea mollusks, providing support for the hypothesis that in the deep oceans the microorganisms may be active only in the intestinal tracts of animals. The resulting conditions of enriched nutrients in the intestinal tract would enable microorganisms to decompose refractory materials, such as chitin or cellulose. This indicates that there may be a fundamental difference between the role of microorganisms in the turnover of organic matter in deep-sea sediments and that of microorganisms in shallow-water sediments or soil.

Fluorescent antibodies were being used to detect specific bacterial species in various ecosystems. When bacteria are injected into an animal, proteins called antibodies are produced in the bloodstream. These antibodies can be "harvested" from the bloodstream and chemically coupled to a dye that will fluoresce when viewed microscopically under ultraviolet light. When the fluorescein-labeled antibodies combine with a bacterium, the organism itself appears luminescent when viewed by fluorescent microscopy. Antibody reactions are highly specific, since antibodies react only with the bacterium that stimulated their production. Thus, known fluorescein-labeled antibodies can be used to detect specific bacterial species occurring in ecosystems under natural conditions.

Bacterial production of fuels. Fuel production by bacteria was already being practiced, for example, by the fermentative production of industrial alcohol, acetone, and butanol. Costs in these processes were high, however; the starting materials were more valuable for purposes other than fuel production, and only a relatively small amount of fuel could be provided. A better prospect was methane, which is a prime constituent of "natural" gas.

Bacteria have long been known to produce methane. For example, it is produced by bacteria in municipal sewage-disposal plants and by microorganisms in the stomachs of cattle and sheep. Commercial production of methane by processes analogous to present industrial fermentative practices, however, would involve staggering bacteriological, economic, and engineering problems. A possible solution lies in the fact that methane can be produced by bacteria from common agricultural waste materials such as cellulose. To avoid the prohibitive expense of collecting and transporting the cellulose, a practical approach would

be to produce methane where large quantities of fermentable material are already gathered, as in the huge feedlots where cattle are fattened. This would have the additional advantage of reducing a pollution hazard. Since supplies of methane obtainable from these sources would be relatively small, their use would necessitate thinking in terms of many small packets of fuel rather than in terms of huge quantities produced in a few fermentation plants.

Other bacterial products could also be used as fuels. A blend of 10 to 30% ethyl alcohol added to gasoline was used successfully to help to alleviate gasoline shortages in many European countries during the 1930s and World War II, and a similar practice was followed in the Philippines. There are many other bacterial products that would readily lend themselves to use as fuels, either directly or after chemical modification.

Food microbiology and hygiene. The incidence and control of salmonellosis continued to be of great concern. The term usually refers to cases of acute gastroenteritis with diarrhea, but it may also include typhoid and typhoid-like fevers. These diseases are caused by members of the genus *Salmonella,* all of which are intestinal organisms of man and animals.

At the eighth international symposium of the Committee on Food Microbiology and Hygiene of the International Association of Microbiological Societies, held in the U.K., it was generally agreed that the salmonellosis problem was growing in

John Pearson from Popular Mechanics

Methane, produced by the action of bacteria in a mixture of cow manure and water, is pumped from a generator (left) through a garden hose to operate a Chevrolet engine.

spite of improved surveillance efforts. Two factors were thought to be responsible: increasing environmental pollution, and contamination of food eaten by both man and animals. Contaminated feed leads to infected animals and to the infection of man when he eats them as meat. Control measures would involve not only sterilization of animal feed but also more effective treatment of wastes from slaughterhouses and meat-processing plants and better sanitation in animal quarters, packing plant operations, and food service operations.

High-nutrient industrial wastes pose additional health-related problems. Salmonella and other intestinal bacteria have been shown to persist for long periods of time, and even to multiply, in streams and lakes into which nutrient-rich industrial wastes are added. Thus, considerable interest was being shown in the chemical and bacteriological quality of wastewater effluents from a variety of sources, including sewage-purification plants, cattle feedlots and stockyards, food-processing plants, and paper and textile mills.

Seven years earlier a food-borne disease caused by *Vibrio parahaemolyticus* had been known only in Japan, where it accounted for about half the reported food-borne diseases annually. By 1974 the organism had been found in coastal waters throughout the world, and outbreaks of *V. parahaemolyticus* food poisoning were being reported in the adjoining countries. As in Japan, the disease was transmitted by seafood or seafood-containing food.

Plasmids, cellular elements that contain extrachromosomal genetic material, are found in the cytoplasm of certain bacterial cells. These pieces of DNA carry extra genes that are not essential for the bacteria but frequently confer a selective advantage upon the cells they inhabit. One class of plasmids, known as R factors, confers resistance to a number of antibiotics. Harmless bacteria of animal origin may acquire resistance to drugs via R factors as a result of exposure to antibiotics in animal feed, and it has been shown that these bacteria can transfer R factors to intestinal bacteria, thus conferring antibiotic resistance upon the latter. The implications are disturbing. Some intestinal bacteria cause disease in man, and they may be transmitted to man as a result of direct contact with animals or through the ingestion of improperly processed animal products. If drug-resistant bacteria can transfer this resistance to other, non-related bacteria, the chain can stretch very far. This finding provided further reason for concern about the indiscriminate use of antibiotics as feed additives and for prophylaxis in animal farming.

Other developments. Herpes viruses are extremely common, and most men and animals harbor them. They cause fever blisters, or cold sores, and genital sores. The herpes viruses have been the focus of attention of virologists around the world for some years, not only because of the diseases they cause but because of an intriguing characteristic —their ability to enter into a latent state in individuals who have recovered from the clinical symptoms of disease and then to cause recurrent infection. Although conclusive proof was lacking, there was a considerable body of circumstantial evidence indicating that herpes viruses may play a role in causing certain cancers.

Previously, herpes-type viruses had been detected only in humans and animals. Recently, however, a marine fungus was discovered that was

Newcastle viruses (V) interact with chicken red blood cells (RBC) and cause them to clump. This occurs because the viruses are adsorbed onto the cells, and the cell-virus aggregates then agglutinate.

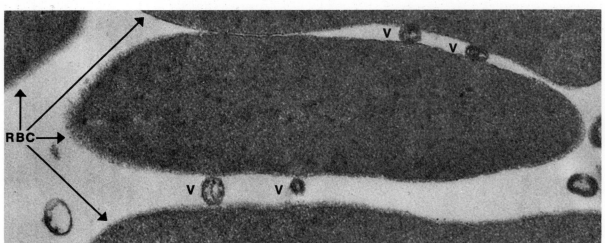

Courtesy, Pedro Villegas, S. H. Kleven and I. L. Roth

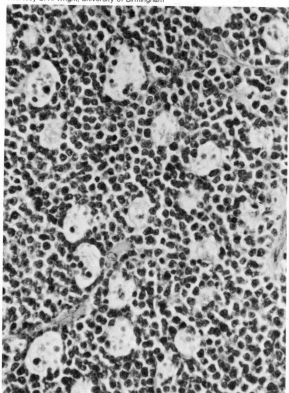 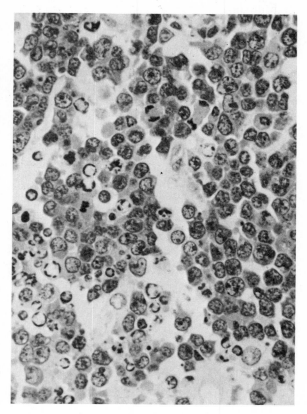

Section of Burkitt's lymphoma tumor displays in photomicrograph the "starry sky" effect of such tumors (left). The white "stars" are large histiocytes, cells that engulf debris from dying cells. At higher magnification (right) can be seen empty-looking circular structures characteristic of herpes infections.

infected by a probable herpes-type virus. The tentative identification of the virus was based on its appearance and on its replicative cycle. If this identification proved correct, interesting new research possibilities would be opened. Scientists would then be able to study a herpes virus in a relatively simple organism that could be grown under simple cultural conditions.

A bacterium that causes citrus stubborn disease was cultured from diseased plants and was shown to be related to the bacterium causing corn stunt disease. An interesting characteristic of the latter bacterium is that it affects both the plant and the insect that transmits the disease. The discovery of a related organism suggested the existence of a unique group of microorganisms capable of infecting both plants and insects.

A sterilization process developed for interplanetary spacecraft offered new possibilities for the sterilization of medical implements that are damaged by traditional means. The method involves the application of inert gas plasmas that kill microorganisms without the use of heat, chemicals, or moisture. When exposed to an electrical field, a gas such as nitrogen, argon, or helium loses electrons and becomes a plasma. The object to be sterilized is exposed to the gas plasma, which is then vented off. The mechanism whereby the gas plasmas kill microorganisms was not yet known.

Microbiology and the future. As man turns more and more to microbiology for the solution of some of his difficult technological problems, there will be more urgent need for applications in microbial genetics. Examples include the selection of genetic strains for the production of useful substances; the conversion of biologically active substances to more active or more useful forms; and means for degrading organic pollutants that are presently resistant to microbial action.

Increased surveillance of foods will be required to avoid microbial contamination. More and better food microbiology guidelines for the processing and preservation of foods will be required. Food microbiologists will be called upon to develop novel types of food, and to develop foods that may be made more nutritious, digestible, or attractive by the use of microbial processes.

In the future, man must depend upon the mi-

287

crobial process for disposition of the pollutants and waste of modern industrial society. Organic pollutants must be disposed of or stabilized by the microbial process. The bioconversion of organic wastes to methane also has great potential. At present, researchers have not even begun to investigate and explore the broad spectrum of microbiological energy sources. Moreover, microbiological processes for extracting fuels from oil shales may be possible.

Finally, there is still need for new antibiotics to replace those that have lost efficiency, to supplant those that are excessively toxic, and to treat diseases for which there presently are no effective antibiotic agents.

—Robert G. Eagon

Molecular biology

The whole cell provided the subject of much of the research in molecular biology during the year. Geneticists concentrated on somatic mammalian cells, while biochemists studied the mechanisms by which cells are regulated in response to their environment. A new instrument, the proton scanning microscope, held great promise for research in biophysics.

Biochemistry

Although biochemistry traditionally has dealt with the properties of isolated biological molecules in solution, it is turning more and more to the study of mechanisms by which the functions of whole cells are regulated in response to their environment. Collectively, these mechanisms are concerned with what has been called "the social life of cells." This phrase encompasses those processes in which occurs some form of cellular recognition that is mediated by molecules located on the cell surface. From the experimenters' viewpoint, the difficulty with such research is that the phenomenon under investigation is a property of the intact organism, whereas the biochemical process that underlies it can be fully understood only by the separation of the individual molecular components and their study in a reconstituted system. Intercellular adhesion, chemotaxis (orientation or movement of cells in relation to chemical agents), and hormone action are examples of processes that involve recognition between molecules on cell surfaces, and important advances were made during the year in understanding these phenomena.

Cell-cell recognition and adhesion. When squeezed through a fine mesh cloth, common ma-

Spider-shaped fibrous molecules in Microciona parthena, *a variety of marine sponge, are protein-polysaccharide complexes that cause cells in suspension to reaggregate only with their own kind.*

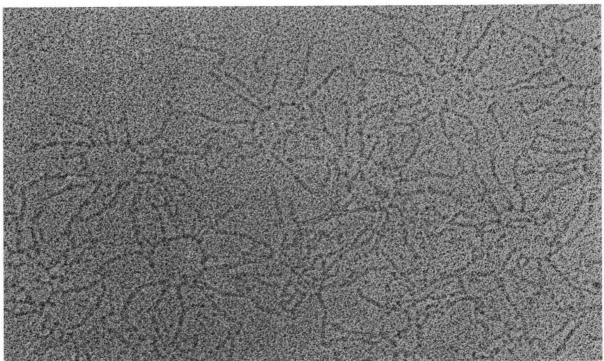

rine sponges yield suspensions of single cells that readily reaggregate. In 1910 H. V. Wilson showed this to be a species-specific reaction when he observed that mixed suspensions of cells from different kinds of sponges reaggregate only with their own kind. Such "sorting out" can be followed easily with the two genera *Microciona* and *Cliona* because one is orange in color and the other is yellow. The multicellular aggregates that reform from a suspension of the two kinds of cells are either orange or yellow, not a mixture of the two. In recent months researchers purified an aggregation factor dissociated from the cell surface of *Microciona parthena* and showed it to be an unusual proteoglycan, a protein-polysaccharide complex. An electron micrograph of a preparation of this factor shadowed with a heavy metal revealed the presence of spider-shaped fibrous molecules composed of a ring about 800 Å in diameter to which were attached about a dozen arms, 1100 Å in length, that radiated from the circle. In the presence of calcium ions the individual macromolecules aggregate to form a gel, and it is proposed that the biological activity of the factor is related to its ability to associate by means of noncovalent forces (that is, weak associations that do not involve bonds formed by shared electrons) with the surfaces of two adjacent cells, thus forming a cohesive cross-link.

Cell-cell recognition is not a static process, but rather it may show temporal changes as cells differentiate from one form to another. Thus, whereas common bakers' yeast may grow by vegetative budding in either the haploid or diploid forms (forms the cells of which have either one or two complete sets of chromosomes, respectively), the mixing of two haploid forms of opposite mating type may produce a specific cell-cell agglutination as a prelude to the sexual fusion of the cells to form a zygote. This agglutination is both species-specific and sex-specific, and it involves recognition between complementary glycoproteins that are located in the cell wall and on the cell surface of the haploid yeasts. Following zygote formation, a new diploid cell buds off that no longer produces the sexual agglutination factors. Thus, the biosynthesis of these recognition factors is apparently shut off in diploid cells.

Similar time-dependent syntheses of recognition factors are also involved in differentiation of animal cells as the fertilized egg divides to give a multicellular organism. Again, tissue-specific

To track bacteria a cuvette, filled with bacteria in a defined gradient, is observed through a microscope. Moving the cuvette in the x-z directions with a "joystick" (upper left) and in the y direction with a foot pedal keeps an individual bacterium in focus. These movements activate motors connected to a punch-tape apparatus that records data for calculating the coordinates.

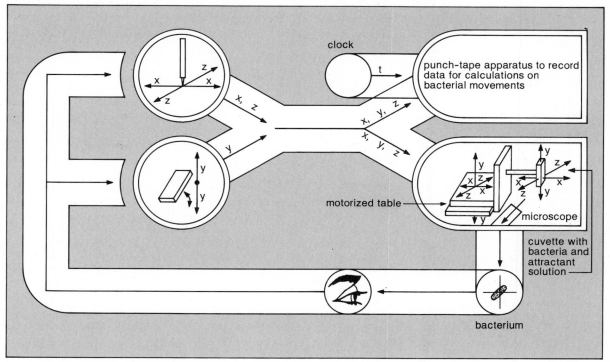

reassociation of dispersed single cells can be demonstrated, but the biochemical basis of the recognition and reaggregation processes was not defined as of 1974. Part of the difficulty stemmed from the lack of reliable quantitative assays for use in the purification of the presumed adhesion factors. Significant advances were made in developing such assay procedures by Saul Roseman and co-workers at Johns Hopkins University, and it seemed probable that such factors soon would be obtained from human cells and their biochemical nature determined.

Transformation of cell-surface antigens. Every cell has on its surface an assortment of antigens that give the cell its individuality, and these antigens may undergo change as the cell interacts with its environment. In the case of the enteric bacterium *Salmonella,* which is responsible for such human diseases as typhoid fever and gastroenteritis, the outer membrane lipopolysaccharide that coats the cell is an important antigen. The typing, or identification, of the strain of *Salmonella* involved in an infection has long been facilitated by the Kauffmann-White classification, which is based primarily on differences in the antigenic structure of the O-polysaccharide chain of the lipopolysaccharide. The O-antigen may also serve as the receptor for attachment of bacterial viruses that are able to infect the bacterium.

It was shown during the last decade that subsequent to viral infection the surface antigens of the bacterium may undergo a change in chemical structure. Researchers demonstrated that this transformation results from the new genetic information introduced with the viral genome (haploid set of chromosomes with the genes they contain) into the bacterium. In one example, the infection of *S. anatum* with the bacterial virus ϵ^{15}, the structure of the O-antigen is altered as the result of three new proteins made from the template of the viral genome, these being an inhibitor of a bacterial enzyme, a repressor of the synthesis of a second bacterial enzyme, and a new enzyme that is involved in synthesizing the altered lipopolysaccharide. Analogous transformation of animal cells following infection with certain viruses stimulated great excitement among scientists who were seeking a general mechanism for the cause of cancer. Such virus-infected animal cells often show a loss of density-dependent inhibition of growth that mimics the uncontrolled growth of neoplasms.

Chemotaxis. Cells that are motile (capable of movement) can respond to diffusible substances in the environment by moving either toward the source of the substance (positive chemotaxis) or away from it (negative chemotaxis). In the human body, sites of infection are immediately attacked by macrophages and lymphocytes that converge on the damaged area to isolate it and to eliminate the foreign agent by phagocytosis. This immune reaction is based on a chemotactic response to substances released near the damaged tissue.

The biochemical basis of chemotaxis is slowly being elucidated, and important advances have occurred recently. Through studies with the flagellated unicellular organism *Escherichia coli*, some of the fundamental aspects of chemotaxis have been clarified. (A flagellum is a threadlike extension from the membrane of a cell that enables the cell to swim.) This bacterium has been shown to contain in its plasma membrane many different receptor proteins that bind substances which stimulate the chemotactic response, and Julius Adler at the University of Wisconsin obtained mutants that fail to respond either because they lacked a good receptor protein or because some other defect occurred. Most past efforts to follow chemotaxis used an extremely simple assay procedure in which a capillary tube filled with a solution of an attractant is immersed in a suspension of bacteria. If the bacteria are attracted, they enter the end of the capillary tube from which the solute is diffusing and move up the concentration gradient. The capillary tube can then be withdrawn and the number of trapped bacteria counted. Bacteria move up or down a gradient by a "random walk" mechanism; that is, they tumble erratically and then glide off in a new random direction. If the new direction happens to be up the gradient of an attractant, the frequency of tumbling is lower than if the new direction is down the gradient. Thus, the net effect is to move toward the source of an attractant or away from a repellent.

Peter Lovely, F. W. Dahlquist, and D. E. Koshland at the University of California devised an ingenious tracking apparatus that allows an experimenter to follow the three-dimensional movements of bacteria in the presence and absence of gradients. With this apparatus it is possible to study the quantitative response of bacteria both as a population and as individuals, and it was shown that they detect gradients by comparing ratios of concentrations rather than absolute concentrations. The tracking apparatus also allows an estimate of the average rate of movement of an individual bacterium and the mean free path or net velocity toward an attractant.

Thus, the chemotactic response of bacteria can be broken down into several biochemically solvable problems. First is the identification of the specific receptor molecules, proteins that are usually to be found in the plasma membrane. Second is the identification of the change occurring in the membrane that, in some way, imparts motion to the flagellum embedded in the cell envelope. Finally there is a sensing mechanism by means of which the cell alters its response as a function of the changing concentration of the stimulant. It is proposed that the detection of a gradient involves a "memory" in that the bacteria utilize a time-dependent response that compares past and present environments. Such an experimental model may be useful for future studies aimed at understanding the memory function of neural systems in higher organisms.

Hormone receptors. Individual cells in a multicellular organism are subject not only to the influence of neighboring cells exerted through direct contact but also to the action of diffusible substances called hormones. Some hormones, such as insulin, appear to initiate their effects through an interaction with the plasma membrane of the cell to which they bind, whereas the steroid hormones pass through the membrane to the nucleus of the cell, where their primary action is exerted.

The peptide hormones, such as insulin, exert their primary effect on the adenylate cyclase system in the cell membrane, altering the rate of formation or degradation of a cyclic nucleotide. This nucleotide serves as a second messenger to initiate a specific action elsewhere in the cell, such as modification of the transport of glucose and amino acids. In contrast, considerable evidence indicates that steroid hormones, such as the estrogens, function by a different two-step mechanism in which an extranuclear (cytoplasmic) receptor protein, activated by specific binding with the steroid, apparently delivers the regulatory message to the cell nucleus. At least one consequence of the interaction of the hormone-receptor complex with the nucleus may be a regulatory effect on ribonucleic acid (RNA) synthesis.

One controversial point that surfaced during the year concerned whether insulin registered its effects by interaction with the plasma membrane surface or whether the hormone must penetrate the membrane to act. A previous attempt by Pedro Cuatrecasas at Johns Hopkins University to answer this question involved attaching insulin covalently to insoluble agarose or polyacrylamide beads that are apparently unable to enter the cell. Such "immobilized" insulin still initiated a typical response of free insulin. This result suggested that all of the insulin receptors are located on the outer surface of the cell membrane and that the interaction of insulin with these surface receptors triggers some change in the membrane that initiates a cascade of effects as yet poorly understood.

The above experiment, and others like it, were criticized on the basis that some of the insulin molecules could come free from the insoluble carrier and penetrate the membrane to stimulate the observed response. Moreover, nonspecific effects resulting from interaction of the beads with the membrane may also complicate interpretation of the observations.

—C. E. Ballou

Biophysics

Coenzyme-binding sites. The three-dimensional structure of the enzyme glyceraldehyde-3-phosphate dehydrogenase (GPDH) was determined to a resolution of 3 Å by Michael Rossmann and his colleagues at Purdue University during the year. This became the fourth enzyme that utilizes the coenzyme nicotinamide adenine dinucleotide (NAD) to have its structure elucidated. (A coenzyme is a substance accompanying, and essential to, the activity of an enzyme.) Comparisons of these structures, particularly the NAD-binding portions, indicated similarities that may be of considerable evolutionary significance. Figure 1 shows a comparison of the conformation of the NAD-binding portion of GPDH and the corresponding part of another enzyme, lactate dehydrogenase (LDH). The similarity in the disposition of α-helical segments (cylinders in the figure) and parallel β-pleated sheet elements (fat arrows) around the coenzyme-binding site is remarkable.

Although the amino acid sequences of GPDH and LDH are very different, their three-dimensional structures appear to be closely related. Both contain four identical protein chains. The four subunits of these molecules are spatially related to one another by three mutually perpendicular 2-fold axes (labeled P, Q, and R in figure 2). The figure also shows that, although the coenzyme-binding sites of the two enzymes are similar, the way the subunits form tetramers differs: in LDH the NAD sites are on the outside of the molecule, while in GPDH they are close to the subunit interfaces.

The NAD-binding site shown in figure 1 is structurally similar to the NAD-binding sites of two dimeric enzymes: malate dehydrogenase and liver alcohol dehydrogenase. (A dimer is a molecular species formed by the union of two like molecules.) Finally, the monomeric protein flavodoxin contains a site that binds flavin mononucleotide (another adenine-containing coenzyme), which is structurally similar to the NAD-binding sites of GPDH, LDH, and the two dimeric enzymes. Thus, Rossmann and his colleagues suggested that the basic protein fold that forms the nucleotide-binding portions of these very different proteins evolved very early and has been conserved

291

Proton scanning microscope

A proton scanning microscope, constructed at the Enrico Fermi Institute of the University of Chicago by Riccardo Levi Setti and two graduate students, William Escovitz and Timothy Fox, yielded its first pictures at the beginning of January. This new instrument, still in the early stage of development, held promise for applications in many areas of biology and biophysics, complementary to the uses of the electron microscope. In the prototype form, schematized below, a two-electrode lens accelerates and focuses the hydrogen ions produced by a field ionization source into a probe about 2000 Å in size. A more advanced design envisages an optical system capable of reaching a resolution in the 5–10 Å region.

With an anode consisting of a fine tungsten tip cooled to the temperature of liquid nitrogen (about −195° C) in a hydrogen atmosphere, the field ionization source is a modification of E. W. Müller's field ion projection microscope. Due to its small size and ion yield, such a source is by far the most intense ever used in a particle accelerator. One of the first pictures of biological material obtained with the proton probe is shown in the micrograph below.

The features that make the proton scanning microscope particularly attractive for biological applications rely on the phenomena that are characteristic of the interaction of ions with matter. Thus, it will be possible to obtain high-contrast microradiographs of unstained sections of organic matter, using the well-defined proton range as an image-contrast mechanism. Protons will be transmitted only by areas of lower density and will be absorbed by the denser material, in a process called critical range absorption. The micrograph is an extreme example of this approach.

Another unique source of image contrast is anticipated to arise from the process of electron pick-up, neutralization of a portion of the ion beam after traversal of a thin specimen. The detection of the neutralized beam component, hydrogen atoms, should in principle reveal the presence of bound hydrogen in organic material. An analysis of the damage to biological material caused by proton bombardment indicates that the damage should not exceed that caused by an electron probe for comparable amounts of data obtained.

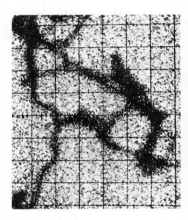

Schematic diagram at right is of the proton scanning microscope described above. The micrograph of myofibrils from rabbit muscle, above, was taken with the instrument. Grid lines correspond to a spacing of 3.6 micrometers.

Courtesy, W. Escovitz, T. Fox, and R. Levi Setti, University of Chicago

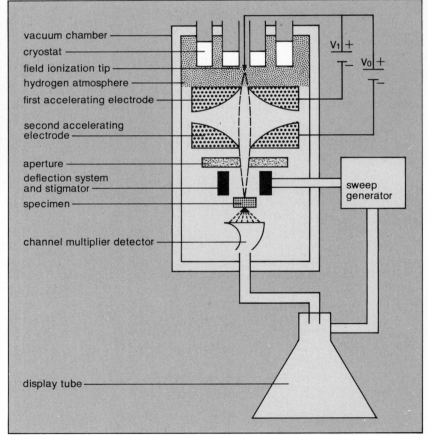

vacuum chamber
cryostat
field ionization tip
hydrogen atmosphere
first accelerating electrode

second accelerating electrode

aperture
deflection system and stigmator
specimen

channel multiplier detector

display tube

V_1 +
V_0 +

sweep generator

throughout the period during which other parts of the same polypeptides evolved to bind different substrates and to associate with each other in different ways.

Very fast kinetics of chlorophyll fluorescence. New and improved lasers continued to provide the basis for spectroscopic measurements that were unthinkable a few years ago. Using a neodymium-glass laser, Michael Seibert (GTE Laboratories Inc.) and R. R. Alfano (City College, New York) measured the kinetics of chlorophyll fluorescent emission in isolated spinach chloroplasts using a picosecond (10^{-12} sec) time scale.

In the chloroplast, light energy is absorbed by light-harvesting pigment systems, which transfer the energy to reaction-center pigment molecules. The latter use the energy for oxidation-reduction reactions that ultimately bring about the splitting of water, the phosphorylation of adenosine diphosphate (ADP), and the reduction of carbon dioxide (CO_2). Plant chloroplasts contain two kinds of reaction-center complexes. One, called photosystem I, produces a strong reductant and a weak oxidant. The other, photosystem II, produces a strong oxidant and a weak reductant. The strong reductant is used to act on CO_2; the strong oxidant

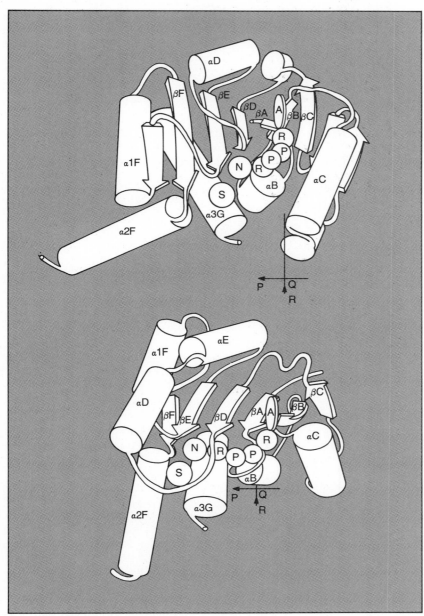

Figure 1. Conformations of the coenzyme-binding portions of the enzymes glyceraldehyde-3-phosphate dehydrogenase (GPDH, top) and lactate dehydrogenase (LDH, bottom) demonstrate similarities that may be of evolutionary significance. Researchers believe that these portions may have evolved very early and been conserved while other parts of the enzymes became differentiated. NRPPRA is the coenzyme NAD and S is the amino acid cysteine. See text for a full explanation of the symbols in the diagrams.

Courtesy, Michael Rossmann

Courtesy, Michael Rossmann

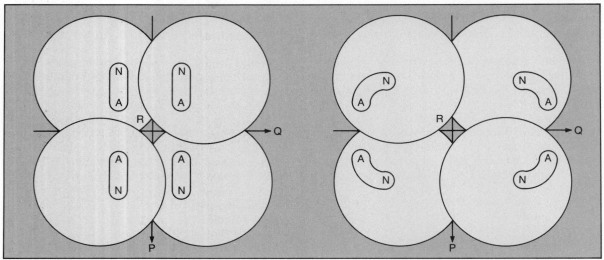

Figure 2. Molecules of GPDH (left) and LDH (right) show similar arrangements of the four subunits on the mutually perpendicular axes P, Q, and R, but different locations for the bound coenzyme sites (NA).

liberates O_2; and the weak oxidant and reductant combine via an electron transport pathway that results in phosphorylation of ADP. All this chemistry and the preceding photochemistry occur relatively slowly with respect to the primary energy transfer reactions, which are completed within a nanosecond (10^{-9} sec).

When the light-harvesting pigment molecules absorb light energy, they are raised to an excited electronic singlet state (having an electronic configuration in which all electrons are paired). That energy can be transferred to another excited singlet state of a different molecular species, or to a triplet state (having an electronic configuration in which two electrons are unpaired; that is, oriented with their spins parallel) or it can be emitted as light accompanying decay to the ground state. The latter process is fluorescent emission. Measurement of the kinetics and the spectrum of fluorescent emission provides information on the identity of the absorbing species and the competing energy transfer processes.

The data of Seibert and Alfano showed two maxima in the time dependence of the fluorescence: one at 15 picoseconds and one at 90 picoseconds after the flash. The fluorescence associated with the first peak is interpreted as arising from photosystem I (lifetime $\leqq 10$ picoseconds) and the second peak from photosystem II (lifetime 200–300 picoseconds in different plants). Earlier measurements using nanosecond flashes showed only one maximum. It therefore seems likely that the instrumentation is now in hand to elucidate the mechanisms of primary energy transfer in photosynthesis.

Proton NMR spectroscopy of transfer-RNA. In general, the proton nuclear magnetic resonance (NMR) signal produced by specific protons on organic molecules cannot be distinguished when those molecules are dissolved in water because the solvent protons overwhelm the specific signal from the solute. Several years ago, R. G. Shulman and his colleagues at Bell Laboratories showed that transfer-RNA (ribonucleic acid) molecules dissolved in water exhibit a number of partially resolved resonances in the low-field region (−11 to −15 parts per million "downfield" from the reference substance, dimethyl silapentane sulfonate) of the proton NMR spectrum. Recently, Shulman's group, together with D. R. Kearns, B. R. Reid, and their associates at the University of California at Riverside showed that these resonances are due to the protons located on the nitrogen in the rings of uracil and guanine. Hydrogen bonding of these protons shifts their resonance peaks sufficiently downfield to be well resolved from water and nonbonded ring protons. Since each of the possible base pairs in the double helix structure of the RNA molecule contains the hydrogen-bonded ring proton of either uracil or guanine, each base pair should contribute equally to the proton NMR spectrum in the −11 to −15 ppm region. The peak positions actually observed depend on the local environment of the base pair. In particular, the resonance peak will be shifted up- or downfield, by an amount corresponding to 0.1–1.3 ppm, by ring current fields from adjacent base pairs. Thus, the extent of shift of an identified resonance peak from its "standard" position is a sensitive indicator of both nucleotide sequence and local helicity.

Examination of the high-resolution proton NMR spectra of six different transfer-RNAs, a number of chemically modified transfer-RNAs, and several specific fragments of yeast phenylalanine transfer-RNA permitted the following conclusions to be drawn: (1) in every case, the observed spectrum agrees with that predicted from the base pairs as they occur in the familiar cloverleaf model for transfer-RNA; (2) the spectrum of yeast phenylalanine transfer-RNA is identical to the sum of the spectra of its fragments, that is, the features responsible for transfer-RNA structure are present in the fragments; and (3) the spectrum of *Escherichia coli* glutamate transfer-RNA agrees with the predicted spectra only if it is assumed that two arms of the cloverleaf (the acceptor stem and the TψC stem) are stacked one upon the other. Clearly, high-resolution proton NMR spectroscopy will provide much structural information about low-molecular-weight nucleic acids in the future.

Sliding filament model of muscle contraction. The 20th anniversary of the sliding filament model of muscle contraction proposed by H. E. Huxley and Jean Hanson and by A. F. Huxley and R. Niedergerke occurred in 1974. According to the model, muscle contraction is accomplished by sliding thick filaments of myosin past thin filaments of actin; each filament remains constant in

length. Both actin and myosin filaments are arranged in an interpenetrating hexagonal array, such that each myosin filament is surrounded by six actin filaments and each actin filament is surrounded by three myosin filaments. The force causing movement is generated by interaction of cross-bridges (which probably consist of the heads of myosin molecules, containing the actin-combining and adenosine triphosphate-splitting sites of myosin) with the actin filaments. The precise structural details of the process remained largely unknown.

During the past year J. C. Haselgrove and H. E. Huxley (Medical Research Council Laboratory of Molecular Biology, Cambridge, Eng.) recorded X-ray diffraction patterns from living muscle fibers at rest, during active contraction, and in rigor. The pattern from contracting muscles was obtained by stroboscopic technique: the muscle was stimulated electrically to give a three-second contraction every two minutes. The circuit controlling the contractions also operated a lead shutter in the X-ray beam that could be opened only when an electrical contact dependent upon the muscle pulling a pivoted lever was closed. Thus, the X-ray exposure was made only while the muscle was generating tension and while it was being stimulated. A usable pattern was usually built up in 10–15 hours.

Muscle fiber demonstrates striated pattern, caused by the arrangement of thick (myosin) and thin (actin) filaments. Contraction of a sarcomere, the unit of muscle between two Z lines, is believed caused by cross-bridges (heads of myosin molecules) moving radially away from myosin toward actin filaments.

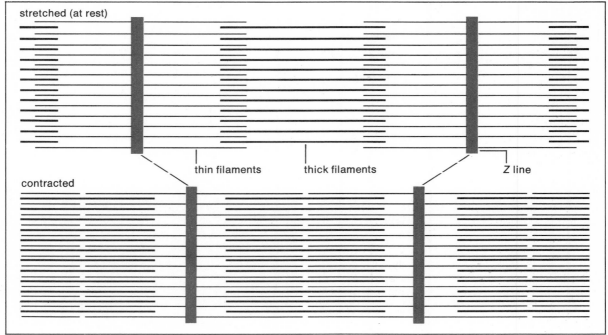

From these experiments Haselgrove and Huxley concluded that the cross-bridges move radially away from the myosin filaments and toward the actin filaments during contraction. Given a constant sarcomere length, the average movement of the cross-bridges during contraction is about 40% of that in rigor. (Sarcomeres are the serial portions into which a muscle fibril is divided by flat protein structures, called Z lines.)

Muscles were also studied generating tension after they had shortened actively against a load. The findings were consistent with the sliding filament theory of contraction. This was the first demonstration that an increase in filament overlap takes place during active shortening exactly to the same extent as occurs during passive changes in sarcomere length.

—Robert Haselkorn

Genetics

Somatic cell genetics, the study of the genetic and epigenetic processes of the somatic (nonreproductive) cells of higher organisms, has developed in recent years into a major field of biological research. The growth of the field has depended on the development and improvement of techniques for the genetic analysis of cultured cells, such as techniques for cell hybridization and for the isola-

tion of mutants. Dealing primarily with somatic mammalian cells growing in vitro (outside the living body) the research has focused on the problems of gene regulation, gene mapping, mutation, and gene transfer.

Gene regulation in somatic cell hybrids. Discovered little more than a decade ago, somatic cell hybridization has become one of the main techniques for the genetic analysis of cultured mammalian cells. Hybridization consists of the fusion of two cells of different types followed by the formation of a single nucleus that contains the genomes of the two parental cells. (A genome is the total chromosome complement of a cell.) Even genomes of cells of different mammalian species can be combined in hybrids that have an apparently unlimited proliferative capacity. Hybrids between cells that differ in the expression of differentiated functions have been used extensively in studies on the mechanisms of gene regulation.

Recently, the quantitative aspects of gene regulation have been investigated. Richard Davidson (Harvard Medical School) and Boris Ephrussi (Center of Molecular Genetics, France) studied the control of melanin synthesis in hybrids between pigmented Syrian hamster melanoma cells and unpigmented mouse fibroblast cells. (Melanin is a dark brown or black pigment in hair and skin.)

Human-mouse hybrid cell culture (left) has been infected with the common cold virus, while that at right has not. Such hybrid cells are susceptible to human viruses that do not affect mouse cells.

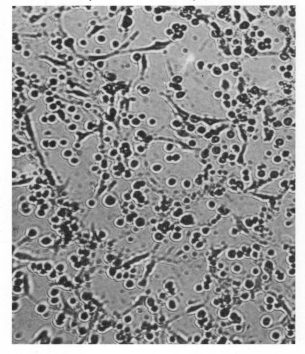
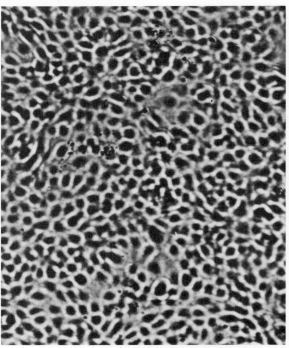

While hybrid cells that contained one fibroblast genome and one melanoma genome were always unpigmented, many (but not all) hybrid cells that contained one fibroblast genome and two melanoma genomes were pigmented (synthesized melanin). These results are consistent with the hypothesis that mammalian cells use diffusible regulator molecules to control the expression of at least some differentiated functions, such as pigmentation. The results also suggested that these regulator substances are produced in rather limited quantities, since a two-fold difference in the genome ratio seems sufficient to determine whether a hybrid will be differentiated or undifferentiated.

Experiments on the regulation of liver-specific proteins in hybrids between rat hepatoma cells that produce albumin and mouse lymphoblasts that do not do so were carried out by Mary Weiss and her colleagues (Center of Molecular Genetics, France). In the hybrids albumin synthesis continued, suggesting that albumin synthesis and melanin synthesis are controlled by different types of regulatory systems. The nature of the albumin in the hybrids was determined through the use of species-specific antibodies, and it was found that the albumin of the mouse (lymphoblast) type was being produced as well as the albumin of the rat (hepatoma) type. These results demonstrated that the genes for a specialized function are maintained in mouse lymphoblast cells which, as a result of embryonic differentiation, do not express the function, and, furthermore, that the genes for the function are maintained in such a state that they can be activated.

Hybrids between cells of different rodent species, such as those discussed above, generally lose relatively few chromosomes of either species. In contrast, hybrids between rodent cells and human cells rapidly and selectively lose human chromosomes during the first several cell generations after the formation of the hybrids. Hybrids of this type have been used in studies on gene regulation. Hilary Koprowski and his colleagues (Wistar Institute of Anatomy and Biology) studied the control of the hormone-mediated inducibility of tyrosine amino transferase (TAT) in hybrids between rat hepatoma cells, in which TAT is inducible, and human fibroblasts, in which it is not. In hybrid cells that retained the human X chromosome, there was no inducibility of TAT. In contrast, in hybrids that had lost the X chromosome, TAT inducibility reappeared. It was suggested that a regulatory gene located on the human X chromosome produces a soluble factor that acts to block the induction of TAT. The results further suggested that this soluble factor has a short-term effect, since enzyme inducibility returned following the loss of the X chromosome.

In all of the experiments described above, fusion occurred between two cells that each contained a nucleus. Nils Ringertz and his colleagues (Karolinska Institute, Sweden) recently demonstrated that it is possible to fuse isolated nuclei with the cytoplasm of cells from which the nuclei have been removed. Studies of such "reconstituted" cells should provide information on the types of nucleocytoplasmic interactions that affect cell growth and differentiation.

The above-mentioned studies and others on the regulation of differentiated functions in hybrid cells have provided some information on the mechanisms that may be functioning to control gene expression in mammalian cells. Further hybridization studies should provide additional information on such questions as whether structural and regulatory genes are linked in mammalian cells. However, the elucidation of specific mechanisms of control will probably require biochemical studies with isolated genes and purified regulator substances, and a wide variety of technical problems must be solved before such studies become feasible.

Human gene mapping in somatic cell hybrids. As mentioned above, hybrids between human cells and rodent cells (of mouse, rat, Syrian hamster, and Chinese hamster origin) preferentially lose human chromosomes. This pattern of segregation has made it possible to use human-rodent hybrids for the assignment of genes to specific human chromosomes. By 1974 studies with human-rodent hybrids had led to the assignment of at least one gene to almost every human chromosome, and some chromosomes had been assigned several genes. For example, hybridization studies indicated that the genes for adenylate kinase-2, fumarate hydratase, guanylate kinase, peptidase-C, phosphoglucomutase-1, 6-phosphogluconate dehydrogenase, phosphopyruvate hydratase, and uridyl diphosphate glucose pyrophosphorylase are located on human chromosome number 1.

Gene assignments have been based on either the selective or random retention of specific chromosomes in hybrids. In the case of selective retention, mutant rodent cells that lack a given enzyme activity are hybridized with human cells that produce the enzyme, and the hybrids are grown in a selective medium which allows for the survival of only those cells that have the enzyme activity. The selective retention method is preferable for gene-mapping studies. However, because of the relatively small number of selective systems available most of the gene assignments in cell hybrids have been based on correlations between human en-

Molecular biology

zymes and chromosomes under nonselective conditions.

Researchers recently demonstrated that cell hybrids can be used to determine the part of a chromosome on which a given gene is located. Frank Ruddle and his co-workers at Yale University hybridized mouse cells with human cells in which there had been a translocation between the X chromosome and one of the autosomes (chromosomes other than sex chromosomes). The results suggested that the genes for hypoxanthine-guanine phosphoribosyl transferase (HGPRT), glucose 6-phosphate dehydrogenase (G6PD), and phosphoglycerate kinase (PGK) are located on the long arm of the X chromosome and are arranged in the order (from the end of the chromosome toward the middle): G6PD—HGPRT—PGK.

In addition to the mapping of structural and regulatory loci, human-rodent hybrids have been used to determine the position of genes that deter-mine viral receptors. Howard Green and his colleagues at the Massachusetts Institute of Technology studied the sensitivity of human-mouse hybrids to polio virus. Human cells have a polio virus receptor and can be infected by the virus, while mouse cells lack the receptor and are therefore resistant to the virus. Recent experiments showed that the gene which determines the polio virus receptor is located on human chromosome number 19.

Mutation in cultured cells. Mutant cell lines exhibiting recognizably altered heritable characteristics were used during the year in a wide variety of experiments on the expression and organization of the genetic material in mammalian cells. Among those isolated were drug-resistant and auxotrophic mutants, temperature-sensitive mutants, mutants that affect enzyme activity, and mutants that affect properties of cell membranes. One question underlying all of the studies with the mutant cells concerns the nature of the event leading

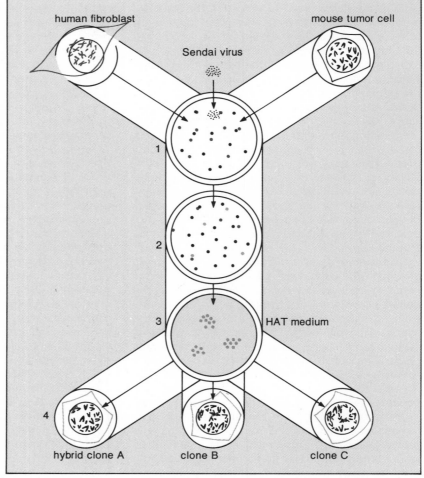

To form hybrid somatic cells human fibroblasts are mixed with mouse tumor cells that are deficient in either the TK or HGPRT enzymes, and a fusion-enhancing agent such as the Sendai virus is added (1). Some of the cells fuse (2), first forming heterokaryons with two nuclei and then new hybrid cells (gray). Plating the cells on HAT, a "selective" medium (3), causes the fused cells to proliferate and form colonies; the mouse parental cells cannot do this because they lack an enzyme, and the fibroblasts do not multiply rapidly. The fused cells retain all the mouse chromosomes but only a few of the 46 human ones; because the human chromosome loss is random each hybrid clone contains a different set of human chromosomes (4).

Adapted from "Hybrid Cells and Human Genes" by Frank H. Ruddle and Raju S. Kucherlapati

to the altered phenotype (observable characteristics of the cells): is the altered phenotype the result of a change in the DNA, and, if so, does the change occur in a structural gene or a regulatory gene? In spite of a broad effort in many laboratories, the genetic basis for most of the "mutant" cells being studied remained unclear.

An example of the complexity that was encountered in studies with mutant mammalian cells is provided by the case of azaguanine resistance. In general, the ability to survive in culture medium containing azaguanine is associated with the loss of HGPRT activity. Some HGPRT$^-$ mutants seem to be "nonreverting," that is, no revertants (to a normal level of HGPRT) can be detected in a population of 10^8 mutant cells. The absence of revertants raised the possibility that these HGPRT$^-$ cells had undergone deletions in the HGPRT structural gene (a nonreversible genetic change). However, Harry Harris and his colleagues at the Medical Research Council laboratories, Eng., observed the apparent reactivation of the mutant HGPRT gene. In attempts to isolate hybrids between HGPRT$^+$ human cells and "nonreverting" HGPRT$^-$ mouse cells, HGPRT$^+$ cells were selected. Some of the HGPRT$^+$ cells contained no human chromosomes, and the HGPRT activity in those cells was found to be of the mouse, not the human, type.

Such reactivation of HGPRT was also observed in other "nonreverting" rodent cell lines. These results suggest that the alteration in the HGPRT$^-$ cells may not be a structural gene mutation. However, the results of some other recent experimental research do not agree with this suggestion. Thomas Caskey and his colleagues at the Baylor College of Medicine, Houston, Tex., showed that some cells that survive in azaguanine produce a protein that has no HGPRT activity yet reacts with an antibody against HGPRT. This protein is presumably the product of a mutated HGPRT structural gene.

Transfer of genetic information. During the past decade, there were many attempts to develop systems of genetic transfer for mammalian cells. The most straightforward approach, allowing one cell to incorporate the isolated DNA of another cell, was relatively unsuccessful. However, other approaches yielded positive results. O. Wesley McBride and Harvey Ozer of the U.S. National Institutes of Health used purified metaphase chromosomes rather than isolated DNA to transfer genetic information. (Condensed chromosomes can be most easily isolated at the metaphase stage of cell division.) HGPRT-deficient mouse cells were incubated with metaphase chromosomes isolated from Chinese hamster cells that had HGPRT activity. Cells with HGPRT activity were selected, and it was shown that the enzyme in those cells was of the donor (Chinese hamster) type. As of 1974 it was still not known whether the selected HGPRT$^+$ cells had incorporated chromosomes that were intact or only small pieces of chromosomes.

Another recent development involved the use of bacteriophages that carry specific bacterial genes as vectors for genetic transfer. C. H. Doy and his colleagues of Australian National University showed that bacterial genes can be introduced into plant cells by means of bacteriophages. Tomato cells are not able to grow in vitro in a medium in which lactose is the carbon source. Cultures of tomato cells were treated with bacteriophage carrying the lac operon (group of genes) of *Escherichia coli* and were then transferred to lactose medium. Some of these cultures were then found to be able to grow in lactose medium for up to nine months. Biochemical and immunological experiments suggested that the tomato cells were producing the *E. coli* form of the enzyme β-galactosidase. Although these results appeared promising, further experiments are necessary to demonstrate that bacterial genes can be transferred into, and expressed in, the cells of higher organisms.

—Richard L. Davidson

Obituaries

The following persons, all of whom died between May 1, 1973, and June 30, 1974, were noted for distinguished accomplishments in one or more scientific endeavors.

Bose, Satyendra Nath (Jan. 1, 1894—Feb. 4, 1974), Indian theoretical physicist, was an expert on quantum mechanics, a branch of mathematical physics that deals with the motion of electrons, protons, neutrons, and other subatomic particles. In 1924 Albert Einstein received a copy of "Planck's Law and Light Quantum Hypothesis," Bose's analysis of particles in connection with photons, or light quanta. Einstein, impressed with the notion that radiation could be considered a form of gas made up of photons, translated the short monograph into German for publication. He also believed that the statistical methods worked out by Bose could be extended to ordinary atoms under an assumption being developed by Louis de Broglie; namely, that material particles have both wave and particulate properties. Though the two scientists never met, they collaborated by mail and gave their names to what is known as Bose-Einstein statistics. Bose graduated from Calcutta University and taught at the universities of Dacca

(1921–46) and Calcutta (1916; 1946–56). He was elected a fellow of the Royal Society in 1958.

Bulman, Oliver Meredith Boone (May 20, 1902 –Feb. 18, 1974), British paleontologist, was a leading authority on lower Paleozoic rocks and fossils and, in particular, on graptolites, extinct colonial marine animals of the Ordovician and Silurian periods. During his lifelong study of graptolites, he was associated with the Imperial College of Science and Technology in London and with Cambridge University. He was elected a fellow of the Royal Society in 1940 and served as president of the Geological Society (1962–64).

Bush, Vannevar (March 11, 1890–June 28, 1974), U.S. electrical engineer, developed the differential analyzer, the first electronic analog computer. In the 1930s, with colleagues from the Massachusetts Institute of Technology, he built an analog computer capable of analyzing differential equations containing up to 18 independent variables. Among Bush's other accomplishments were a network analyzer that simulated the performance of large electrical networks, and the Rapid Selector, a device using a code and microfilm to facilitate information retrieval. Though not widely used, the Rapid Selector was an important step in developing efficient methods for coordinating masses of information.

In 1940 Bush was appointed chairman of the National Defense Research Committee. The following year he became director of the newly established Office of Scientific Research and Development, which coordinated research for the war effort and advised the government on scientific matters. Bush was later named chairman of the Joint Research and Development Board (1946–47) and served on the Research and Development Board of the National Military Establishment (1947–48). He was also president of the Carnegie Institution (1939–55).

Carmichael, Leonard (Nov. 9, 1898–Sept. 16, 1973), U.S. psychologist, educator, and author, was president of Tufts College, Medford, Mass., for 14 years before being named head of the Smithsonian Institution in 1953. He retired in 1964 to join the National Geographic Society as vice-president for research and exploration and remained there until the time of his death. Carmichael brought new vitality and prestige to the Smithsonian by building the Museum of History and Technology and adding two wings to the Museum of Natural History. The lively exhibitions that he initiated eventually attracted more than ten million visitors a year. He earlier edited and compiled *Manual of Child Psychology,* a classic in its field. In recognition of his lifelong scholarship, Macrobius A, a moon crater, was renamed for him.

Condon, Edward U(hler) (March 2, 1902–March 26, 1974), U.S. physicist, was a recognized expert in the fields of quantum mechanics, atomic and molecular spectra, nuclear physics, microwave radio, solid-state physics, and glass manufacturing. He also produced for the U.S. Air Force a widely discussed report on unidentified flying objects. Condon received his Ph.D. in physics (1926) from the University of California at Berkeley.

During his diversified career Condon was associate director of Westinghouse Research Laboratories (1937–45), director of the National Bureau of Standards (1945–51), director of research and development at the Corning Glass Works (1951–54), professor of physics at Washington University in St. Louis (1956–63), and professor of astrophysics at the University of Colorado at Boulder (1965–70). In 1969 Condon became a science adviser to *Britannica Yearbook of Science and the Future.*

Davis, Adelle (Feb. 25, 1904–May 31, 1974), a widely read author and, in recent years, an outspoken guest on television talk shows, gained national fame for her criticism of American eating habits and for her advocacy of organic fruits and vegetables, milk, eggs, liver, fish, vitamin pills, and brewer's yeast. She obtained a B.A. degree (1927) in dietetics at the University of California, Berkeley, and later earned a master's degree (1939) in biochemistry at the University of Southern California. During her many years as a consulting nutritionist in New York and California, she prepared diets for countless clients, many of them referred to her by physicians. Although quite aware of modern research on nutrition, Davis was frequently taken to task for "unscientifically" attributing all manner of diseases to improper diets. Her four books, which together sold more than nine million copies, were *Let's Cook It Right* (1947), *Let's Have Healthy Children* (1951), *Let's Eat Right to Keep Fit* (1954), and *Let's Get Well* (1965).

Ewing, (William) Maurice (May 12, 1906–May 4, 1974), U.S. geophysicist, opened new avenues of oceanographic research by successfully employing shock-wave techniques to probe marine sediments and ocean floors. His seismic refraction measurements of the Atlantic Ocean basins, the Mid-Atlantic Ridge, and the Mediterranean and Norwegian seas became the basis of later computer analyses. Ewing joined other scientists in suggesting that earthquakes are related to the central oceanic rifts that encircle the globe, and he theorized that sea-floor spreading is episodic in nature and probably worldwide. During World War II he developed SOFAR (sound finding and ranging) for the U.S. Navy to provide, through explosion-induced sound waves, a means of underwater communications especially useful as a distress signal. Ewing was educated at Rice Institute, Houston, Tex., and directed the Lamont-Doherty Geological Observatory at Columbia University from 1949 to 1972.

Gast, Paul Werner (Sept. 11, 1930—May 16, 1973), noted U.S. geologist, was chief of the NASA Division of Planetary and Earth Sciences at the Lyndon B. Johnson Space Center at Houston. He developed the rubidium-strontium and uranium-lead isotope methods for dating moon rocks and directed the study of lunar geology after the first moon rocks were brought back to earth by Apollo 11 in 1969. Gast received the NASA Medal for Exceptional Service in 1970 and its Distinguished Service Medal in 1972.

Hess, Walter Rudolf (March 17, 1881—Aug. 12, 1973), Swiss physiologist, was named co-winner in 1949 (with António Egas Moniz of Portugal) of the Nobel Prize for Physiology or Medicine "for his discovery of the functional organization of the interbrain as a coordinator of the activities of the internal organs." He mapped the brain's control centers to such a degree that he could induce the behavior pattern of a cat confronted by a dog simply by stimulating the proper points on the animal's hypothalamus. After abandoning a successful practice in ophthalmology in 1912, he became director (1917–51) of the Physiological Institute at the University of Zurich, where he spent nearly 35 years in research and teaching. His writings include *The Biology of Mind* (1964).

Kahn, Louis Isadore (Feb. 20, 1901—March 17, 1974), undoubtedly one of the most potent forces among U.S. architects of the past two decades. He redefined architecture for himself and developed a style that set him far apart from such other great American architects as Frank Lloyd Wright. His influence on other architects began with the completion (1951) of his first major project, the Yale University Art Gallery, which expressed in brick, glass, and harsh unfinished concrete Kahn's insights into the use of natural light and strong, stark, geometrical forms. More startling still were the bare pipes, uncovered ducts, and open storage

Walter Hess (above) and Louis Kahn (left).

spaces, which Kahn viewed as an integral part of his architectural design. His other well-known works include the Kimbell Art Museum in Fort Worth, Tex., the capitol buildings in Dacca, Bangladesh, the Salk Institute in La Jolla, Calif., and the Richards Medical Research Laboratories at the University of Pennsylvania (where Kahn taught architecture for many years).

Krusen, Frank Hammond (June 26, 1898 — Sept. 16, 1973), regarded by the medical profession as the father of physical medicine and rehabilitation, Krusen turned from surgery to problems of physical disabilities while recovering from tuberculosis. His early work was undertaken at Temple University, Philadelphia (1929–35), where he established and directed the department of physical medicine. He returned to Temple in 1963 to head the Krusen Rehabilitation Center. From 1935 to 1963 he was associated with Mayo Clinic in Rochester, Minn. As a result of his successes with World War II amputees, paraplegics, and other seriously wounded persons, the American Medical Association recognized physical medicine as a specialty in 1947.

Kuiper, Gerard Peter (Dec. 7, 1905 — Dec. 23, 1973), U.S. astronomer known especially for his discoveries and theories concerning the solar system. Kuiper completed his education in The Netherlands before joining (1936) the Yerkes Observatory in Wisconsin, a facility operated by the University of Chicago. His discoveries include carbon dioxide in the atmosphere of Mars (1948), the fifth moon of Uranus (Miranda; 1948), and the second moon of Neptune (Nereid; 1949). He also measured the visual diameter of Pluto (1943–44)

Alfred Romer

Courtesy, Harvard University News Office

and put forward (1949) a theory of planetary origin by condensation. Kuiper was twice director of the combined Yerkes and McDonald observatories (1947–49; 1957–60) and from 1960 headed the Lunar and Planetary Laboratory at the University of Arizona.

Meyer, Karl (May 19, 1884 — April 27, 1974), U.S. virologist, was responsible for important viral discoveries applicable to public health and veterinary medicine. After graduating from the University of Zurich as a veterinarian, he went to South Africa where his success in protecting cattle from a deadly tick-borne infection attracted wide attention. He moved to the U.S. in 1910 and successively isolated the virus of Eastern and then Western equine encephalitis. During his many years (1915–54) with the Hooper Foundation at the University of California in San Francisco, Meyer solved crucial problems related to typhoid, coccidioidomycosis (valley fever), psittacosis (parrot fever), and leptospirosis (an infectious jaundice). He is also credited with saving the canned foods industry by developing a method of flash sterilization that prevents botulism.

Nord, Friedrich Franz (Aug. 9, 1889 — July 12, 1973), obtained his doctoral degree from the University of Berlin in 1914 and worked for a time at the Max Planck Institute for Biochemistry and at the Physiological Institute of the University of Berlin. He joined Fordham University, New York City, in 1938 and remained there until his retirement in 1960. Nord established the concept of cryobiology (1927), the use of low-temperature environments in biological research. He was the first to isolate lignin (1955), the chief component of woody tissue, which is used in binders, adhesives, and briquetting agents and in making the synthetic flavoring agent vanillin. Nord was the founder (1941) and editor of *Archives of Biochemistry,* and editor (1939–71) of *Advances in Enzymology and Related Areas of Molecular Biology.*

Romer, Alfred Sherwood (Dec. 28, 1894 — Nov. 5, 1973), U.S. vertebrate paleontologist and comparative anatomist, served as director (1946–61) of the Museum of Comparative Zoology at Harvard University and president (1966) of the American Association for the Advancement of Science. In such books as *Vertebrate Paleontology* (1933) and *The Vertebrate Body* (1949) he explained the nature and order of the evolution of vertebrate animals in such a way as to revolutionize subsequent thinking on evolutionary biology. He studied under William K. Gregory at Columbia University, where he obtained his doctoral degree. After teaching anatomy at Bellevue Hospital Medical College in New York City for two years, Romer moved to the University of Chicago, where he remained for 11 years. In 1934 he joined the Harvard University faculty, retiring in 1965.

Sutherland, Earl Wilbur, Jr. (Nov. 19, 1915 – March 9, 1974), U.S. pharmacologist, was awarded the 1971 Nobel Prize for Physiology or Medicine for isolating cyclic AMP, the chemical that enables hormones to perform their vital function of carrying messages via the bloodstream. In 1956, while serving as director of the department of medicine at Western Reserve University in Cleveland, O., Sutherland discovered cyclic AMP (cyclic adenosine 3', 5'-monophosphate) and later demonstrated its vital function as a regulatory substance in the chemistry of living organisms. His breakthrough stimulated a vast range of subsequent research.

Sutherland received his M.D. degree (1942) from Washington University Medical School in St. Louis, Mo., where, after serving in World War II, he worked in the laboratory of Carl Ferdinand Cori, himself a Nobel laureate. Sutherland moved to Vanderbilt University, Nashville, Tenn., in 1963 to devote full time to research. From July 1973 he was on the faculty of the University of Miami Medical School.

Virtanen, Artturi Ilmari (Jan. 15, 1895 – Nov. 11, 1973), Finnish biochemist, was awarded the 1945 Nobel Prize for Chemistry "for his researches and inventions in agricultural and nutritive chemistry, especially for his method of fodder preservation." His special concern was fermentation processes which spoil stored silage. By developing a procedure (now known as AIV, Virtanen's initials) that increased the acidity of the fodder through the addition of dilute hydrochloric and sulfuric acids, he was able to prevent destructive fermentation. Fur-

ther experiments showed that the acid treatment had no adverse effects on the nutritive value or edibility of the fodder or on products derived from animals that fed on it. His *AIV System as the Basis of Cattle Feeding* appeared in 1943. Virtanen was also responsible for valuable research on the nitrogen-fixing bacteria found in the root nodules of leguminous plants, on improved methods for preserving butter, and on partially synthetic cattle feeds.

Waksman, Selman Abraham (July 22, 1888 – Aug. 16, 1973), U.S. microbiologist, was awarded the 1952 Nobel Prize for Physiology or Medicine for his discovery of streptomycin, the first specific agent effective in the treatment of tuberculosis. He also isolated and developed several other antibiotics (a term he coined in 1941), including neomycin, that are used in treating infectious diseases in humans, domestic animals, and plants. During his years at Rutgers University, New Brunswick, N.J., Waksman was professor of soil microbiology (1930–40), professor of microbiology and chairman of the department (1940–58), and director of the Rutgers Institute of Microbiology (1949–58). Among his books are *Principles of Soil Microbiology* (1927), one of the most exhaustive works on the subject in any language, and *My Life with the Microbes* (1954), an autobiography.

White, Paul Dudley (June 6, 1886 – Oct. 31, 1973), U.S. physician, obtained his M.D. degree from Harvard Medical School (1911), where he later taught while maintaining a lifelong association with Massachusetts General Hospital. To a great extent, he was responsible for modern advances in cardiology and used every opportunity to acquaint the general public with the causes of

Reportagebild

Earl Sutherland (right) receiving the 1971 Nobel Prize for Physiology or Medicine from Gustav VI Adolf, king of Sweden.

heart disorders and with practical means to prevent them. White, who gained national prominence as cardiologist to Pres. Dwight D. Eisenhower, was among the first to employ electrocardiograms for diagnosing heart trouble and used some 21,000 cardiograms as the basis of *Heart Disease* (1931), one of his many books.

Zwicky, Fritz (Feb. 14, 1898—Feb. 8, 1974), Swiss astronomer, physicist, and jet propulsion expert, was especially well known for valuable contributions to the understanding of supernovae, stars that for a short time are far brighter than normal. He believed that the explosions of supernovae are totally different from those of ordinary novae, occurring only two or three times every 1,000 years in our galaxy. He contended that in supernovae most of the matter of the star is dissipated, leaving little or nothing behind. To confirm his theories, Zwicky studied neighboring galaxies and between 1937 and 1941, while working at the California Institute of Technology, discovered 18 supernovae; only 12 others had been previously reported. In the early 1930s Zwicky contributed substantially to the physics of solid-state matter, gaseous ionization, and thermodynamics. In 1943 he joined Aerojet Engineering Corp. in Azusa, Calif., where he developed some of the earliest jet engines, including the JATO (jet-assisted takeoff) units used to launch heavy-laden aircraft from short runways.

Photography

Photography continued to grow during the year at a rate greater than the average of all other industries. The estimated annual worldwide sales volume for all photographic products for 1973–74 was in the $12 billion to $15 billion range, with about 49% consumed in the U.S., 24% in Europe, and 27% in the rest of the world. This order of growth was expected to continue well into the 1980s.

Prints and print processing. Practically all color-print materials manufactured in the U.S. were available on resin-coated papers only, and both Eastman Kodak and GAF introduced their variable-contrast and some other black-and-white papers on a resin-coated (RC) base. The paper base used for prints had long been a problem, since processing chemicals were absorbed by the paper fibers, necessitating long wash times. In RC papers the paper base is coated on both sides with a water-impermeable polyethylene layer that prevents chemicals from coming into contact with the paper fibers. Wash times for black-and-white papers were reduced from one hour to four minutes, and

drying times were shorter. Furthermore, RC papers remain flat during processing, making it possible to introduce sheet-paper processors in addition to the well-established roll-paper processors for photofinishing laboratories. They had proved difficult to dry mount with existing methods, but Seal, Inc., introduced a two-step color-mount process that was able to overcome this disadvantage.

Many manufacturers of color paper for the ever increasing photofinishing market had by 1974 adopted Ektaprint 3, a three-step process that replaced the Ektaprint C five-step process. Besides taking less time (8 versus 30 minutes) and 80% less water, it reduced effluent chemicals by 40% and eliminated cyanides, zinc, and phosphates.

Films and film processing. Kodacolor II, introduced in 1972 for use with the Pocket Instamatic, was further improved in 1973. Based on new findings in emulsion technology, it provided better color reproduction with decreased graininess and increased sharpness, and hence gave satisfying results despite the higher enlargements needed to get normal print sizes out of the 110 format. This film, which was being made available in 135 and other formats as well, was replacing the Kodacolor-X products. At the same time, the new Kodak Flexicolor C-41 process was replacing the C-22 process. It required only one-third the time, with proportionate savings in water, and offered environmental advantages similar to the three-step color-paper process.

The emulsion improvements in Kodacolor II were also being utilized in other new films such as Vericolor II, which replaced Ektacolor and was also processed in C-41. The process for the improved Ektachrome films, called E-6, also was shorter, incorporated a bleach that was regenerated, and used a reversal bath that eliminated the need for reexposure by light. The new technologies were being introduced to the 16-mm and 35-mm motion-picture field with Eastman Color Negative II film.

The Polaroid Corp. introduced Type 105 Positive/Negative pack film, which incorporated several significant improvements over Type 55, the earlier black-and-white material that made it possible to obtain a permanent negative together with a positive print. The new material was rated at ASA 75, identical to Type 58 and Type 108 Polaroid color film and so close to Kodacolor that it could be used with no exposure correction as a convenient test material to ensure that the more expensive color film was exposed correctly. The negative was coated on 4-mil polyester support, resulting in higher dimensional stability and shorter drying time.

Cibachrome and Cibachrome Print, well known in Europe for several years, found widespread use in the U.S. only recently. Both are direct-positive, multilayer color materials, coated on clear 7-mil polyester and white opaque cellulose triacetate base, respectively. In the first applications of these materials in the United States color transparencies or prints of exceptional sharpness and resolution were obtained. Because of their high resistance to fading, which surpassed that of conventional color materials by five to ten times, Cibachrome transparencies and prints were especially useful for display purposes.

Cameras and lenses. With few exceptions, the small and medium-sized camera field can be divided in two groups, single-lens reflex (SLR) cameras and viewfinder cameras. The single-lens reflex viewing system lends itself easily to exchangeable lenses and to through-the-lens light-metering systems. Viewfinder cameras, on the other hand, need adjustable or exchangeable viewfinders if lenses of different focal lengths are to be used. Hence, most viewfinder cameras currently being offered had a noninterchangeable lens, and emphasis was placed on size and convenience of operation. These cameras, which in-

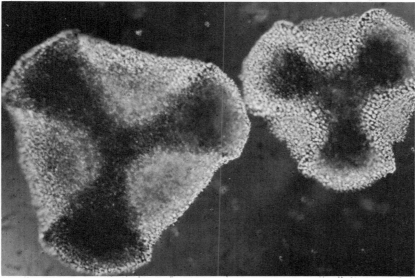

Photos, F. W. Goro

Photomicrographs made on Polaroid's new Type 105 Positive/Negative pack film show skeletons of radiolarians, tiny marine protozoans recovered in cores from the ocean floor by the Deep-Sea Drilling Project.

Sidney Harris

*"Oh, that's not Dr. Zakheim.
That's a hologram."*

cluded the Kodak Pocket Instamatics, Konica C35, Minolta Hi-Matic F, Olympus 35RC, Rollei A26 and 35, and Yashica Electro-35MC, were known as automated compacts.

E. Leitz, Inc., maker of the Leica, had been one of the few exceptions in the viewfinder field. By offering exchangeable lenses with parallax compensation and through-the-lens metering, it had remained the choice of photographers who preferred the faster action of the viewfinder camera. In 1974, however, Leitz was entering the automated-compact market, teaming with Minolta to produce the Leica CL, a sophisticated compact with exchangeable lenses, parallax compensation, and a through-the-lens metering system similar to the Leica M-5.

Among single-lens reflex cameras, the Olympus OM system was the most noteworthy new entry. The Olympus OM-1 was significantly smaller and lighter than the average SLR. Apparently produced with a considerable degree of precision and designed with operational simplicity in mind, it was offered with a complete system of accessories.

More lenses using multiple coating, new glasses, and floating elements were introduced. Among the variable focal length lenses, several now allowed very close focusing and some covered the moderate wide-angle to moderate tele-

photo range. New entries included the 38–100-mm f/3.5 Auto Tamron Zoom, the 70–210-mm Vivitar Series 1 Macro Zoom, and the 35–100-mm f/2.8 Konica Varifocal Hexanon AR.

To bridge the gap between the fixed design of the small camera and the swing/tilt and shift adjustments of the view camera, Schneider had introduced its 35-mm PA-Curtagon lens, which permitted a shift of up to 7 mm in four directions. Nikon followed with its 35-mm PC-Nikkor and Olympus with the 35-mm Zuiko Shift, both of which shifted up to 11 mm. Now Canon introduced the 35-mm f/2.8 TS Canon S.S.C. lens, which not only allowed shifting of up to 11 mm to correct converging lines but also tilted to ± 8° to increase depth of field.

The Sinar-p was the first view camera to move the swing and tilt axes away from the center or the base of the standards. Rather, Sinar was swinging or tilting the standards around axes located about one-tenth from the side or about one-fifth from the bottom of the image. This created a "two-point image focus" that permitted an image to be kept sharp while tilting or swinging the standards.

Lighting. Small electronic flash units had become increasingly popular over the past few years. With the introduction of monitoring devices in the form of an electric eye and a dumping tube

306

(an additional flashtube concealed from view), it was possible to make automatic flashlights that metered the amount of light according to distance. The surplus charge was cut off and dumped, thus making *f*-stop adjustments with changing distance unnecessary. The flash duration in automatic flashes typically had a range of 1/500 to 1/50,000 sec.

In order not to lose the dumped energy and to increase the number of flashes per battery charge, thyristor circuitry was introduced during the year. The Braun F-022 and F-027 automatic thyristor flashes had a flashhead that swiveled around a horizontal axis independently of the electric eye, thus permitting the automatic feature to be retained when light from the flash was bounced from the ceiling. Not to be outdone, Metz subsequently introduced the model 218TR, called Quadrolight since it permitted automatic flash action while swiveling the flashhead up, down, or to both sides.

Motion pictures. Eastman Kodak introduced two Super 8 motion-picture cameras, the Ektasound 130 and 140, with sound-recording capabilities. Simultaneously, two films, Kodachrome II and Ektachrome 160, were offered with magnetic sound striping. The latter, together with high-speed lenses and 230° shutters, gave the Ektasound cameras the same available-light features as the Kodak XL cameras introduced earlier. Two sound projectors, Ektasound 235 and 245, completed the line. The Ektasound 245 projector allowed recording as well as playback.

Kodak also extended the Super 8 sound system for use in industrial and educational applications as well as in television. For the latter, a Supermatic 200 sound camera, which would accept 50-ft and 200-ft silent- or sound-film cartridges, was available. A Supermatic 8 processor was automated to the point where it could virtually be operated in a lighted office by untrained personnel. It could process the newly introduced Ektachrome SM 7244 film directly from the 50-ft cartridge in 13 minutes or from the 200-ft cartridge in 28 minutes. Supermatic 60 and 70 sound projectors could project on a built-in screen or an external screen, silent or sound. Perhaps the most advanced element of the Supermatic series was the Kodak Supermatic film videoplayer VP-1, which made it possible to display Super 8 film on a television receiver simply by attaching it to the antenna terminal or to generate a signal that could be distributed to any number of television receivers for local or remote display.

Editing film without losing picture-to-sound synchronization is difficult with sound-movie systems like Ektasound or Supermatic because the sound is recorded some 18 frames ahead of the picture. In 16-mm and 35-mm motion-picture

work, a magnetic-coated sound film of the same dimensions and with sprocket holes of the same size as the picture film has been used for years to record sound parallel with the picture film. Because its dimensions are the same, this magnetic film can be run through the same editing machines as the picture film and the two can be cut and spliced on a length-for-length basis.

An adaptation of this system for Super 8 was devised by Robert Doyle, an astrophysicist at Harvard University, and equipment was being marketed. This included Super 8 magnetic film with sprocket holes and a modified Sony quarter-inch reel-to-reel tape recorder, capable of being linked to several makes of Super 8 cameras, that could transmit the camera frame rate to the magnetic film in such a way that the sprocket holes on the magnetic film and the picture film were matched precisely.

—Lothar K. Engelmann

Physics

With concerns ranging from superconductivity and the energy crisis to the constituents and configurations of atomic nuclei, physicists continued to perform research on a broad front during the past year. Among the highlights, the search to determine the internal structure of elementary particles led to modifications in the previously conceived quark-parton model.

High-energy physics

The goal of research in high-energy physics is to understand the structure of matter at its most fundamental level and the nature of the forces that determine this structure. In order to probe more deeply into this structure, it is necessary to study interactions between elementary particles at ever-higher energies.

Over the past few years, two powerful new high-energy research facilities have been completed and brought into operation. One is the Intersecting Storage Rings (ISR) at the European Organization for Nuclear Research (CERN) in Geneva. It provides an energy equivalent to a proton of 2,000 GeV (billion electron volts) striking a proton at rest, by means of head-on collisions between two protons of 31 GeV each. The other is the synchrotron of the Fermi National Accelerator Laboratory (NAL) near Chicago. It accelerates protons to an energy of 400 GeV, and, with an intensity (energy flux of the particles per unit area normal to the direction of propagation) very much greater than the ISR, is also able to produce intense beams of secondary particles. Both of these facilities have

1675-1975

Among the many achievements credited to Dutch scientist Christiaan Huygens is the design in 1675 of the balance spring. This mechanism was later applied to watches, greatly improving their accuracy.

passed the first blush of exciting novelty and have settled into a systematic program of large-scale research productivity.

Scaling and "asymptopia"; what is a proton? The particles of the atomic nucleus, protons and neutrons, interact primarily through a powerful, short-range force referred to by physicists as the strong interaction. Since the early 1950s a large number of other "elementary" particles that share this strong interaction have been discovered. These hadrons, as they are called, include many kinds of mesons (such as pions and kaons) as well as baryons (nucleon isobars, hyperons, and the more familiar protons and neutrons). All of the mesons as well as all baryons except the proton and neutron undergo spontaneous radioactive decay in less than a millionth of a second. Therefore, they can be observed and studied only during high-energy collisions, either produced in cosmic rays or with particle accelerators.

Physicists have wondered whether these elementary particles have an internal structure, and, if so, what its nature might be. One hint seemed to come in the organization of these particles into families, with certain properties common to the members of each family. In order to explain this organization, Murray Gell-Mann and George Zweig proposed that each particle may consist of two or three fundamental entities called "quarks." Assuming three kinds of quarks and three corresponding antiquarks, the physicists suggested that baryons contained three quarks while mesons were made of a quark and an antiquark. These quarks would have an electric charge of one-third or two-thirds that of an electron, making them readily detectable should they indeed exist.

Continued sensitive searches, most recently at NAL and at the ISR, failed to find any direct evidence for the existence of such quarks. Theoretical physicists, nonetheless, believe that the quark model has an essential validity and that some properties of the forces conspire to prevent the appearance of quarks as free, separate particles. Many experiments did, in fact, support the predictions of the quark model as it pertains to properties of the known elementary particles. These include values of the various particles' magnetic properties, mass differences, and angular momenta.

Another set of experiments explored the inelastic collisions of high-energy electrons with protons that result in the production of pions. (In inelastic collisions the total kinetic energy of the colliding particles increases or decreases.) These experiments produced data indicating that the pattern of the angular distribution of the scattered electrons is independent of the energy. This property of the data, referred to as "scaling," was interpreted as resulting from scattering on tiny internal constituents of the proton. These constituents were dubbed "partons" by Richard Feynman and J. D. Bjorken. It is natural that physicists have sought to marry the two concepts, and so the "quark-parton" model for internal structure of the proton and other elementary particles has come to the fore.

During the past year several experiments contributed evidence that the interaction of protons, and thus presumably the internal structure of these particles, is somewhat more complex and subtle than the still-embryonic quark-parton model suggested. First, the size of a proton, as measured by the probability of one proton striking another (the cross section) was found to increase at very high energies. Experiments at both NAL and the ISR showed an increase of almost 15% in effective proton cross-sectional area between 50 GeV and 2,000 GeV. Second, the production of mesons with a high component of momentum per-

pendicular to the initial direction of motion of a proton colliding with another proton was found to be more probable at high energies than at low energies, other things being equal. While some aspects of these results might find compatible interpretation within the context of the quark-parton model, the results were unexpected and generated a theoretical scramble.

A third, still preliminary, result concerned the process whereby an electron and a positron collide head on and annihilate each other, producing in the process a number of strongly interacting particles, mostly pions. The quark-parton model had predicted that this process would proceed at a rate that would decline with the inverse square of the energy, just as the corresponding process of electron-positron pairs annihilating into a pair of mu mesons (muons) is known to behave. During the past year, however, experiments at a new facility at Stanford University, in which electrons and positrons of up to 2.5 GeV energy apiece were made to collide head on, suggested strong disa-

Fermi National Accelerator
Laboratory near Chicago,
as seen from the air.
In the foreground
are the booster accelerator
and cooling pond.
The auditorium and 16-story
Central Laboratory building
rise behind them,
and at the right is a portion
of the main accelerator ring,
4 mi in circumference.

Courtesy, Fermi National Accelerator Laboratory, photo, Tony Frelo

greement with the predictions. Instead of declining with increasing energy, the data show that this meson production process takes place at a rate essentially independent of the energy. While preliminary, the extensive Stanford data are in substantial agreement with earlier results obtained at Harvard University with the Cambridge Electron Accelerator and with a similar electron-positron facility of lower energy at the Frascati Laboratory in Italy.

At one time physicists expected that the energy region above about 50 GeV would reveal a somewhat monotonous asymptotic behavior of strongly interacting particles. For example, one manifestation of asymptotic behavior would be the approach of all cross sections to constant values at high energies. On the basis of recent research it appeared that this high-energy realm, dubbed "asymptopia," must either lie at a yet higher energy than particle accelerators now achieve or indeed may not exist at all.

Viewed another way, there was a hope that high-energy interactions and the production of various particles could be predicted from a universal set of equations that do not involve the energy. This formalism possesses a property referred to as scaling, and indeed scaling does hold with varying degrees of precision for many processes involving the collision of two hadrons. Scaling as described here is closely related to the scaling discussed above for electron-proton inelastic scattering. The yields of energetic pions close to the initial direction of motion of protons in proton-proton collisions scale very well, and even cosmic-ray data on muon spectra, indirectly the products of energetic pions, agree with scaling. However, a recent experiment at NAL on the inelastic scattering (during which particles are produced) of muons on protons demonstrated preliminary evidence of scaling violation. Some other data from the very-high-energy interactions of cosmic rays also has suggested significant departures from scaling. Furthermore, the electron-positron results described above appeared to contradict the predictions of scaling not only in degree but in a rather fundamental way.

What conclusion, then, can be reached? Most experimental evidence is compatible with the concept that a proton is made up of three quarks (and there may be several additional and virtual quark-antiquark pairs) and that high-energy electrons may probe the interior of the proton and interact with the quarks independently. (A virtual particle is one that is formed and then reabsorbed too quickly to provide direct evidence for its existence.) Perhaps scaling violations are due to the finite size or to the internal structure of the quarks; it is also conceivable that there is a breakdown in man's concept of space and time at such high energies and small distances.

Equipment at the Fermi National Accelerator Laboratory used to study muon–proton interactions includes a Cerenkov counter, the long tube at the right. It identifies high-speed particles by means of the radiation they emit when moving faster than the speed of light in a medium. At the left are scintillation counters. These are made of special plastic that scintillates when a charged particle at high energy passes through them, the light then being detected by photoelectric tubes.

Courtesy, Fermi National Accelerator Laboratory, photo, Tim Fielding

Neutral currents and the weak interaction.
Besides the strong interaction responsible for nuclear forces, another major topic of study in high-energy physics was the "weak" interaction. The radioactive beta-decays (spontaneous emission of electrons or positrons) of nuclei and of some elementary particles are governed by this weak interaction, but as its name implies it is incredibly weak compared with the strong interaction or with the better-understood electromagnetic interaction. However, it may be explored at high energies, where its strength is somewhat greater, by isolating reactions that cannot proceed in any other way. The particles that participate in the weak interaction, besides hadrons, include electrons, muons, and two kinds of neutrinos. These four particles, as well as their corresponding anti-particles, are referred to collectively as "leptons." Neutrinos in particular are unique in that they interact with other matter only through the weak interaction, and consequently are able to penetrate through miles of solid material with a virtually negligible probability of interaction.

Until 1974 the only class of reactions observed involving the weak interaction was that in which the two leptons involved in a reaction were of different electric charges; for example, a neutrino incident on a proton resulted in an emerging positive muon, leaving the proton transformed to a neutron. During the past two years, however, theoretical physicists considered that they had made a breakthrough in their theoretical understanding of this mysterious force, allowing it to be related directly to the electromagnetic interaction so that explicit calculations and predictions can be made.

One prediction of this new theory has been that weak interactions should be seen at high energies involving no lepton charge change. Thus, a neutrino at high energy might strike a proton, producing several pions (hadrons) from the original proton, and then recoil as a neutrino of lower energy rather than being transformed into a muon or electron. This process may be thought of as involving the exchange of a virtual particle, which propagates the weak interaction. In the charge-changing case, this particle would be electrically charged; in the charge-conserving case, it would be neutral. Such a hypothetical particle has been referred to as an intermediate vector boson, or W boson (for "weak"). As in the case of the quark or parton, there has never been any evidence for the real, physical existence of the W boson, although experimental searches are continuing.

The new theories of weak interaction prompted experimentalists to look for examples of charge-conserving weak processes, and during the year positive evidence seemed to have been found. An experimental team at CERN used a very large bubble chamber filled with a heavy liquid (Freon) in

Courtesy, Fermi National Accelerator Laboratory, photo, Tim Fielding

Eight proportional chambers and their associated electronic equipment are used at the Fermi National Accelerator Laboratory to find the paths of charged particles. Each chamber consists of wire grids maintained at high voltages and surrounded by special gases. When a charged particle passes through them, it ionizes the gas and causes a breakdown between the grids in that particular area. In this way the particle's path can be traced.

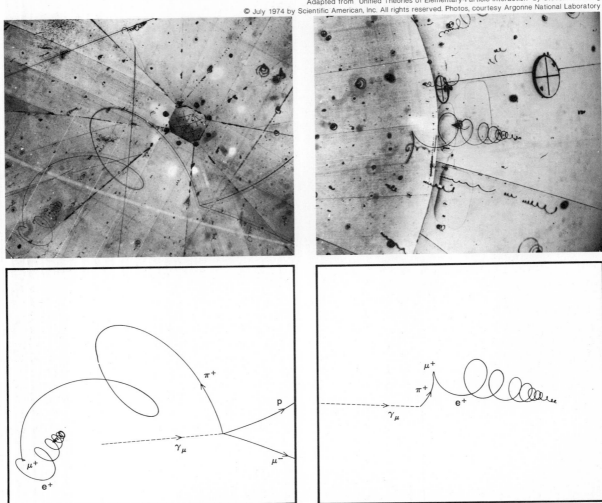

*Evidence supporting the existence of neutral currents was obtained in experiments using a neutrino
beam and a 12-ft bubble chamber filled with liquid hydrogen at Argonne National Laboratory.
In the bubble chamber photograph at the top left and in the map below it can be seen a familiar
charged current process ($v_\mu + p \to \mu^- + p + \pi^+$), in which a unit of electric charge is exchanged
between the leptons (v_μ, μ^-) and the other particles. In the photograph and map at the right
there are no outgoing negative muon (μ^-) or proton (p) tracks, a distinguishing characteristic
of the neutral current process ($v_\mu + p \to v_\mu + n + \pi^+$). In such photographs tracks are left only by charged
particles, thus causing the incoming neutrino (v_μ) and outgoing neutrino and neutron (n) to be invisible.
In both photographs the positive pion (π^+) is seen to decay into a positive muon (μ^+)
and then into a positive electron (e^+), visible as a tightly wound spiral.*

pure neutrino or antineutrino beams (all other particles produced by the 28 GeV synchrotron were
shielded out by many tons of steel). Along with
many hundreds of neutrino and antineutrino reactions of the familiar type involving a lepton charge
change, the team discovered 166 examples of
reactions that showed no evidence for an emerging charged lepton. These reactions were interpreted as a neutrino or antineutrino colliding with
a nucleus in the chamber liquid, causing the emergence of a number of tracks of pions, protons, and
other hadrons, but revealing no trace of charged
muons or electrons. Physicists believed that such
results provided evidence for neutral W boson exchange or neutral currents. Even more convincing
as evidence was a demonstration of the elastic
scattering (scattering during which no energy is
surrendered) of a neutrino on an atomic electron.
While such events are predicted if neutral currents
exist, they seemed virtually impossible to explain
by any other process.

—Lawrence W. Jones

Nuclear physics

A substantial number of the world's nuclear physicists and chemists met in Munich, W.Ger., in August–September 1973 for the first general international conference in their field in four years. An enormous amount of new work was reported. Most striking was a major change in the precision attained. Experimental measurements typically were more precise than they had been four years earlier by factors of 10 to 100.

Fundamental nuclear science can be classified into three broad but interrelated and interdependent areas. The first represents the wide range of detailed measurements made using many techniques, on many nuclei, and in laboratories throughout the world; this is the continuing base of nuclear study on which all else rests. The second is the area of heavy-ion research, perhaps the most rapidly growing field in international nuclear science, which takes advantage of the recently gained ability to produce intense ion beams of essentially all nuclei in nature and use these in nuclear reaction studies. The third is the area of so-called medium-energy nuclear science. There, not only higher energies than any previously used but also the new pionic and kaonic meson beams, which become available in this energy regime, are used to probe deeply into nuclei in regions inaccessible to earlier studies. New discoveries were made in all of these areas.

Nuclear lifetimes. One of the most fundamental measurements that can be made on a nuclear state is its lifetime against decay. Over the years, increasingly sophisticated electronic instrumentation has made it possible to measure shorter and shorter lifetimes, and at the beginning of 1973 nuclear lifetimes as short as 10^{-13} sec (0.1 picoseconds, or one ten thousand billionth of a second) were measured. But the greatest interest for physicists has been in the range of 10^{-18} sec (1 attosecond, or one billion billionth of a second; it requires 30 million attoseconds for a beam of light to move one centimeter) because this marked the rough dividing line between two important nuclear processes: direct ones, in which the projectile simply chips a fragment off the target or induces a tidal wave in its surface, and compound ones, in which the target "swallows" the projectile and approaches statistical equilibrium before emitting a reaction product.

During the year Georges Temmer and his colleagues at Rutgers University, New Brunswick, N. J., succeeded in measuring nuclear lifetimes in the attosecond range by using the very regular atomic lattice planes in high-quality crystals to provide time markers. This afforded an improvement of almost a million in the range of lifetimes open to measurement.

Giant vibrations. One of the simplest modes of nuclear motion is the so-called giant dipole resonance, in which all the neutrons in a nucleus oscillate in a simple dipole fashion against all the protons present. (In the dipole oscillation, when all the neutrons move right all the protons move left, and vice versa, so that the total shape changes regularly from prolate to spherical to oblate and back again.) This mode has been located and studied in all nuclei. Physicists also have long recognized that other simple vibrations are possible; among these are the monopole, wherein all neutrons and protons oscillate radially so that their density and size vary periodically, and the quadrupole, which can be considered most simply as two dipoles back-to-back. The puzzle was why these other vibrations had not been observed.

During the past year impressive evidence for the existence of the giant quadrupole resonance, in nuclei throughout the periodic table, was obtained by Fred Bertrand and Monty Lewis and their collaborators at the Oak Ridge (Tenn.) National Laboratory in systematic studies on the inelastic scattering of protons, helions, and alpha particles. (In inelastic scattering the total kinetic energy of the colliding particles increases or decreases.) Whereas the giant dipole resonance occurs at an excitation of about $78A^{-1/3}$ million electron volts (MeV), the giant quadrupole resonance was located at $63A^{-1/3}$ MeV. (A is the total number of neutrons and protons present.)

But major puzzles remained concerning the giant quadrupoles. Measurements on inelastic electron scattering by Y. Torizuka and his collaborators in Tokyo and by F. R. Buskirk and his collaborators in Darmstadt, W.Ger., suggested that only part of this resonance had been located. A similar conclusion followed from pioneering work by Stanley Hanna and his collaborators at Stanford University on the radiative capture of polarized protons.

Bubble and toroidal nuclei. Like all other isolated quantum systems, nuclei assume equilibrium configurations that minimize their total energy. Many years ago John Wheeler at Princeton University speculated about the possibility that some nuclei might lower their electrostatic energy by assuming bubble configurations so that the protons could get farther from one another.

During the past few years C. Y. Wong and his collaborators at Oak Ridge undertook detailed microscopic nuclear model calculations in order to investigate this suggestion; they predicted that nuclei such as argon-36, selenium-84, cerium-138, and mercury-200 should indeed be bubbles in their ground (lowest energy) states. Moreover, they predicted that toroidal (doughnut-shaped) nuclei, once formed, might be expected to be stable.

During the year intense interest in these predictions was generated by a series of measurements of Ernst-Wilhelm Otten and his collaborators at the laboratories of the European Organization for Nuclear Research (CERN). In systematic measurements of the isotope shift in mercury isotopes they found a striking increase in radius as they approached ^{200}Hg to indicate the possible occurrence of the predicted bubble.

Various alternative explanations involving the sudden onset of strong deformations were advanced, but the situation remained open and extensive efforts were being devoted to searches for other possible bubble nuclei. Also a new look was being taken at the toroidal nuclear predictions in the light of the possible production of such configurations in a reaction wherein a very-high-energy heavy ion such as nitrogen might create the doughnut hole in a heavy nuclear target such as lead or uranium.

Whatever the situation on stable nuclei, Dietmar Kolb and Ronald Cusson at Yale University and Oak Ridge demonstrated, in detailed calculations concerning the distribution of mass in heavy nuclei and the changes in this distribution during fission, that in many nuclei fissioning involves a bubble stage that thickens at its outer ends and tears apart around its equator as fission proceeds. Such calculations would have been completely impossible even a few years ago.

Prehistoric fission. The first demonstration of a nuclear-fission chain reaction is universally attributed to Enrico Fermi and his collaborators at the University of Chicago on Dec. 2, 1942. Reports from the French Atomic Energy Commission appear to suggest that nature beat Fermi to the chain reaction by about 2,000,000,000 years. All the evidence suggests that roughly 2,000,000,000 years ago the uranium deposits now being mined from the Oklo deposits in Gabon in Africa had been laid down by the normal geological processes. Although the fissionable isotope of uranium, ^{235}U, is now only 0.7% of natural uranium (the remainder essentially all ^{238}U), its lifetime against spontaneous radioactive decay is roughly ten times shorter than that for ^{238}U. This means that 2,000,000,000 years ago the fraction of ^{235}U was substantially higher.

This situation did not change until a geological cataclysm added water to the formerly dry uranium deposits; this provided the missing moderator to slow down neutrons in the uranium to the optimum speed for inducing fission—and the whole deposit went critical. Unmistakable fission products provide the evidence for this. Thus, more than 1,000,000,000 watt-years of nuclear energy were released to a prehistoric region.

Pion and kaon capture. Preliminary studies of the interaction of negative pions and kaons with nuclei were reported by Peter Barnes and his associates from Carnegie-Mellon University, Pittsburgh, Pa., and Argonne National Laboratory and by Robert T. Siegel and his colleagues from the Space Radiation Effects Laboratory of the National

Proton synchrotron at the Joint Nuclear Research Institute at Dubna in the Soviet Union.

Sovfoto

Aeronautics and Space Administration (NASA). A totally unexpected result was obtained. Because the pions and kaons give up the energy equivalent of their mass when captured, the target nucleus receives hundreds of MeV of energy in each capture; however, it gains essentially none of the momentum it would have received had energy been delivered by, say, a proton. Researchers found that the residual nuclei which are detected after the capture are those that would be obtained by systematically removing one, two, three, and even more alpha particles (complexes of two neutrons and two protons) from the target. Nuclei corresponding to the removal of even one neutron or proton are much more rare.

From these experiments physicists postulated that the neutrons and protons in all nuclei may spend a significant fraction of their time clustered into alpha particles so that they can be released easily when the meson delivers its energy. This, again, remained an open question but a fascinating one in terms of nuclear structure.

Abnormal nuclear states. The availability of heavy-ion beams at ever higher energies led T. D. Lee and G. C. Wick at Columbia University to speculate about totally new types of nuclear matter that might become accessible in ultra-high-energy (about 2,000,000,000 electron volts per nucleon) heavy-ion collisions. They noted that, in the past, physicists doing work on elementary particles had concentrated on putting more and more energy into smaller and smaller volumes. With heavy-ion beams, however, the intent would be to examine what happens when more and more energy is delivered to a larger volume, such as a heavy nucleus.

Fascinating predictions emerge from a consideration of such an experiment. Among them is the possible existence of totally new nuclear states containing from 350 to 100,000 nucleons, each with an effective mass very much less than in normal nuclei. Even more striking is the prediction that in such collisions it may be possible to create an excited vacuum state, a limited region of vacuum with properties different from the normal vacuum. This is clearly a fundamental question—does the vacuum have structure—and one that is providing major impetus for the large super-high-energy heavy-ion facilities now under construction.

Facilities. Among the major facilities now in use for high-energy heavy-ion work are the super-HILAC (heavy-ion linear accelerator) at Berkeley, Calif., and the coupled cyclotrons at the Joint Nuclear Research Institute at Dubna in the U.S.S.R. Major facilities under construction or in final planning stages are the UNILAC at Darmstadt, the 30

million-volt tandem electrostatic accelerator at Daresbury in the U.K., and the GANIL coupled linear accelerator and cyclotron in France. After a period of many years when no new facilities were initiated in the U.S., approval was granted by the U.S. Atomic Energy Commission for installation of a 25 million-volt tandem electrostatic accelerator at the Oak Ridge National Laboratory, where it was to be coupled with Oak Ridge's existing isochronous cyclotron. Approval was also given for the coupling of the Berkeley super-HILAC to the Bevatron to create a super-high-energy facility that will be used both in nuclear physics and in biomedicine. Preliminary design studies were recently reported for even larger electrostatic accelerators having terminal voltages in the 50–60 million volt range; these, too, would be enormously powerful heavy-ion facilities.

In the area of medium-energy nuclear physics, the Los Alamos Meson Physics Facility became operational during recent months; it was designed to be the world's most powerful facility in the medium range. Large cyclotron facilities at Vancouver, B.C., and at Zurich, Switz., that were somewhat more restricted in capability were also nearing completion, as was the cyclotron project at Indiana University. With their proton beams, these machines focus on experiments involving the strong nuclear interaction; complementing them in the U.S. in its focus on the electromagnetic interaction is the newly operational Bates 400 MeV electron linear accelerator at the Massachusetts Institute of Technology.

After a long period of dwindling support, the U.S. federal budgets for the coming year provide grounds for cautious optimism regarding the future of U.S. nuclear science. Fiscal 1975 marked the first year since 1968 in which an increase in operating support (in dollars of constant value) over the level of the previous year was programmed and the first since 1968 in which construction of a major new facility was initiated.

—D. Allan Bromley

Solid-state physics

In solid-state physics during the past year considerable activity took place in research on superconductivity and in attempting to meet challenges posed by the energy crisis. The discipline was honored when three of its outstanding practitioners shared the 1973 Nobel Prize for Physics.

Superconductivity. In 1973 the discovery of intriguing results from an organic material, (TTF) (TCNQ), was reported. TTF is tetrathiofulvalene and TCNQ is tetracyano-*p*-quinodimethan. When combined chemically, these two compounds form needlelike crystals that have interesting conductivity properties. For example, the conductivity is much higher along the axis of such a crystal than it is perpendicular to it. But much more striking is the temperature dependence of the conductivity found in a few crystals out of the many studied to date. Near 58° K (−215° C) the conductivity was reported to rise very sharply, reaching a value comparable to the best metals (even though the density of conducting electrons is much smaller than in metals). This rise is followed, at slightly lower temperatures, by a very large drop in conductivity.

The excitement over these results was due to the possibility that the conductivity peak near 58° K was caused by the onset of superconductivity (a condition in which electrical resistance disappears) and the hope that improvements in materials technology by using (TTF) (TCNQ) or closely related materials could lead to practical high-temperature superconductors. For example, if the superconducting transition temperature could be raised well above 20° K, then liquid hydrogen could be used as a refrigerant rather than the very expensive liquid helium now required. This would make practical applications of superconductivity much more attractive.

A group from the University of Pennsylvania first reported the striking results from a few (TTF) (TCNQ) samples at the annual American Physical Society solid-state meeting at San Diego, Calif., in March 1973. Other groups, however, were unable to reproduce the Pennsylvania work, and lively discussions arose as to whether the results were due to the growing of more perfect samples by the Pennsylvania group or whether the Pennsylvanians' results were simply artificial products of the peculiarities of their measurements.

To appreciate the intensity of the arguments, one must realize that a 50° K superconductor would have an immense economic impact. For example, much energy and money could be saved by using superconducting motors and generators and by transmitting electricity on superconducting power lines. On the personal level, if the Pennsylvania work led to high-temperature superconductors, that group and perhaps others who made important contributions would be obvious candidates for Nobel prizes.

All of this excitement was generated by a single paper from the Pennsylvania group at the 1973 American Physical Society meeting. At the 1974 meeting, more than 30 papers from ten different laboratories were presented. However, there was still no resolution of the question as to the source of the conductivity peak near 58° K. The Pennsylvania group reported that it had reproduced its results with new samples and more refined measurements; however, these results were not reproduced elsewhere. Clearly, considerable activity in this area was expected to continue through 1974 and 1975.

With the more conventional superconductors, such as Nb and Nb_3Ge, steady progress was being made both in learning how to incorporate the superconductors in machinery and electrical transmission lines and in slowly but steadily developing higher temperature superconductors. In the latter area, however, progress was measured in fractions of a degree, the present upper limit being about 23° K.

Nobel Prize. Workers in solid-state physics were encouraged by the award of the 1973 Nobel Prize for Physics to three of their colleagues: Leo Esaki of IBM's Watson Research Center, Ivar Giaever of the General Electric Research and Development Center, and Brian D. Josephson of Cambridge University. The prizes were for pioneer work in electron tunneling in solids. Tunneling is a pure quantum mechanical effect in which a particle penetrates a barrier that according to classical theory should be impenetrable. As Giaever, originally trained as a mechanical engineer, stated in his Nobel lecture: "To an engineer, it sounds rather strange that if you throw a tennis ball against a wall enough times, it will eventually go through without damaging either the wall or itself. That must be the hard way to a Nobel Prize! The trick, of course, is to use very tiny balls (electrons) and lots of them." The wall also must be very thin, usually 100 Å (about 50 layers of atoms) or less.

Esaki concerned himself with tunneling across the forbidden gap in semiconductors. By making use of this effect, he developed the Esaki, or tunnel, diode, which exhibits negative resistance; that is, when the voltage rises above some critical value, current will decrease rather than increase. By virtue of this characteristic, an Esaki diode can act as an amplifier of electrical signals.

The experiments for which Giaever received the Nobel Prize concerned the energies that electrons could have in superconductors. A few years before he did his experiments, John Bardeen, Leon Cooper, and John Robert Schrieffer had published their (BCS) theory of superconductivity, for which they themselves received the Nobel Prize in 1972. A striking characteristic of this theory was the prediction that when a metal becomes superconducting there will be a range of energy in which no electrons can exist; that is, a forbidden gap. Giaever found that the tunneling at certain applied voltages decreased or disappeared when he cooled his samples through the superconducting transition temperature. This disappearance of current was explained by the forbidden gap of the BCS theory, which prevents electrons from tunneling into the superconductor. Thus, Giaever provided support for the BCS theory and was able to study the superconducting gap of many materials in detail.

Josephson's prize came for research that demonstrated tunneling between two superconducting metals, work that again was closely related to the BCS theory. He showed, theoretically, that tunneling between two superconductors could have very special characteristics. For example, a direct current could flow with no applied voltage across the insulating layer. (This is known as the DC Josephson effect.) Experiments confirmed his predictions.

Energy shortage and physics. As with most areas of science, solid-state physics has languished recently because of reduced research support from the federal government. However, a selective increase of funding as well as new challenges were received in 1974 as a result of the energy crisis. The funding and challenge are closely coupled. If solid-state physics can make worthwhile contributions to the solution of the energy crisis, the discipline will prosper; if not, hard times probably lie ahead.

The supreme practical contribution of solid-state physics to date has been the transistor. The invention of this device at Bell Telephone Laboratories just after World War II led to a major technological revolution. The transistor made possible a huge range of developments, from large computers to space travel and communication satellites. Solid-state physics was central to all of these developments and grew with them, spawning huge new laboratories such as those of IBM.

The challenge from the energy crisis is quite different. The question being asked is whether solid-state physics can be coupled to other fields

Research scientist Svante Prochazka subjects ceramic turbine vane made from silicon carbide to a flame test. Prochazka invented a simple and inexpensive method of fabricating ceramic parts from silicon carbide, one of the most heat-resistant materials known. The application of this technique may increase the efficiency of gas turbines used for generating electric power and also remove a major barrier to the development of an economical gas turbine engine for automobiles.

already highly developed in their own right and produce improvements that will be worth the investment. A further constraint is that these improvements are being requested within a few years rather than over several decades.

One example, touched on earlier in this article, is the practical application of superconductivity. Can motors, generators, and electrical transmission lines be made more economical by incorporating superconductors into their design? Efforts were under way in this direction in many countries. Although some progress was being made, it was too soon as of 1974 to draw any conclusion as to the ultimate success of these efforts.

Another area showing increased activity is the application of techniques and knowledge developed in the study of solid surfaces and in electron spectroscopy to catalytic processes. Catalytic processes play a central role in the production of energy because they provide the key for transforming one form of fuel into another. For example, through the use of catalytic processes, coal can be transformed into gasoline. (The U.S. has vast coal resources compared to its oil reserves.) The efficiency of the catalytic processes used in such a transformation is instrumental in determining its cost and, therefore, its practicability. If improved processes can be developed, they can be instrumental in meeting the energy crisis.

The catalytic processes in use in 1974 were principally the result of trial-and-error research. However, methods developed recently in solid-state physics permit an examination of surfaces and molecules adsorbed on surfaces at an atomic level. This is important because in catalysis one set of molecules is adsorbed on a surface and the subsequent interaction between the surface and these molecules causes a rearrangement of the atoms in the molecules; this, in turn, causes the surface to give off another set of molecules more desirable for accomplishing a catalytic reaction.

Many questions remain to be asked about solid-state research in catalysis. For example, why do two metals, nickel and platinum, produce different end products when the same molecules are placed on their surfaces? The major question that must be answered concerns the relationship between the electronic structure of the solid surface and its catalytic behavior. If the nature of this relationship can be established, then the development of new catalytic processes can be accomplished in a more efficient manner than the old trial-and-error method. The attempts to determine the relationship, using solid-state physics, were just starting in 1974, and researchers expected that it would be a number of years before they could tell whether success was being achieved.

One area in which a major breakthrough may be developing is the transmission of information. For cases where this is now done using electrical wires, there is a growing possibility that the wires will be replaced by small, flexible fibers of glass, silica, or plastic. Information would be sent down these fibers using light from small semiconductor lasers or light-emitting diodes. The U.S. Navy as of 1974 was replacing the electrical communication systems of both a destroyer and a jet plane with such an optical system. Among the advantages of such a system are reduced weight and higher security of information. For example, it is very easy to tap a conventional telephone line, but it will be almost impossible to tap a telephone using optical transmission lines. (*See* Feature Article: FIBER OPTICS: COMMUNICATIONS SYSTEM OF THE FUTURE?)

—William E. Spicer

Science, General

One of the great ironies in the intellectual development of mankind is that the scientific enterprise, which embodies some of the noblest aspirations of the mind, has received its greatest economic stimuli from acts of war. This was true with the ancient Greeks and the Romans, and even Leonardo da Vinci found it desirable to invent new engines of war. Every major power today is constantly reminded of the extent to which its continued sovereignty depends on the genius and productivity of its scientists and engineers.

Not least among these powers is the United States. In the mid-19th century, when its fledgling scientific enterprise was barely visible, the U.S. brought the National Academy of Sciences into being and assigned it to work on military problems. World War I saw the creation of the National Research Council and the beginning of independence in areas of science and technology that previously had been exclusively European. The great leap forward for U.S. science, however, came during and after World War II, when the development of the atomic weapon by U.S. scientists and their collaboration in the development of radar and jet aircraft made it clear that henceforth military supremacy would be derived as much from the laboratory as from courage and generalship.

It was also clear that the level of effort needed to maintain that supremacy could be sustained only by national governments. In the U.S. this meant the development of a vast military program in support of research and development (R and D) and, later, the application of a nearly equivalent level of support for biomedical research in a war against disease and premature death. In parallel with these two developments was a third pattern of research support, based partly on public gratitude and partly on the recognition that there is a definite if not always discernible link between basic research and its useful applications. It was this that led to the establishment of the National Science Foundation (NSF) and to even larger programs in support of basic research under the Department of Defense and the National Institutes of Health.

The steady growth of federal support for science and technology produced plain evidence of success in the increased number of Nobel prizes won by U.S. scientists and in the country's worldwide supremacy in many fields of high-technology industry. But any complacency that might have arisen as a result vanished in October 1957, when a Soviet artificial satellite orbited the earth in solitary triumph.

Thus it was that, in 1957, the U.S. moved into a new relationship with its scientific community—one in which both scientists and engineers would share in the formulation of national policies for the support of R and D designed to maximize the benefits to the nation. To provide a broad base for this relationship, the President's Science Advisory Committee and, later, the Office of Science and Technology were created in the Executive Office of the President.

By mid-1973, however, both agencies were gone, obliterated by an executive order of Pres. Richard Nixon. It was against this background that almost all science-policy discussion among the national leadership during the year took place, focused on one question: U.S. expenditures for R and D were running close to $30 billion per year. This immense national enterprise was almost entirely guided by government decisions. Who was now responsible for managing this enterprise and how well was it being done?

The search for science policy. During the year it was not possible to single out one locus of government concern for the health of the national R and D effort. In 1973 H. Guyford Stever, director of the NSF, had been designated by President Nixon as, simply, "science adviser." Later, the president specified that Stever was to be *his* science adviser, but added that the advice would be channeled through then Secretary of the Treasury George Shultz.

As head of the only U.S. agency responsible for the orderly growth of basic research, Stever was authorized to draw upon the intellectual resources of the NSF for a continuing examination of the federal R and D program. He chose, however, to establish within the foundation a Science and Technology Policy Office under the direction of Russell Drew, who had been a staff officer of the

Sidney Harris

"But we just don't have the technology to carry it out."

now-defunct White House Office of Science and Technology. By mid-1974 Drew had completed the difficult task of simultaneously planning a program, selecting a professional staff of about 20 people, and preparing a budget. Admittedly, his task had been made easier by the fact that the White House had conspicuously excluded the science adviser from questions involving military R and D—one of the central tasks of the old OST.

In addition, since the new office was separated from the White House both geographically and hierarchically, Drew could select problems on the basis of whether or not they were feasible rather than whether or not they were on that day's agenda at the top policy-making levels. Among the areas selected for priority study were energy, international food shortages, the abundance of nonrenewable material resources, and the application of social science research to national issues.

How well Stever, Drew, and the professional staff of the Science and Technology Policy Office would provide scientific guidance to Nixon's policy-makers would depend on the instrumentalities at their disposal. One of these was the Federal Council for Science and Technology, composed of representatives of all major federal agencies involved with science. In commenting on this arrangement, Philip Handler, president of the nongovernmental National Academy of Sciences, told an interviewer for the *Washingtonian*: "The idea is that Stever, with his clout as Science Adviser and his NSF range over the whole of science, can use the Council as a vehicle for inter-agency transfer of knowledge, for sharing of projects and facilities, for cross-fertilization of concepts and insights, and so on. But the Council has never lived up to its potential, and Stever is handicapped in turning things around because the Adviser's clout diminished the moment he left the White House and because NSF is a small agency in a league of big agencies. Those are simply the realities of bureaucratic life. . . . Any way you look at it, it is inescapable that the Science Adviser has been pulled down to a lower level. When he was in the White House, he was the President's in-house problem solver. The President needs him close at hand, needs him as an expert who can serve almost as an adversary to the cabinet departments submitting science proposals. Somebody is going to have to re-invent the Science Adviser at the White House level."

A few months after this interview, however, Handler had some cheering words for Stever. Delivering his annual report to the members of the Academy, with Stever in the audience, Handler observed that the science adviser "seems to have been more successful in that role than I—perhaps than he—anticipated, in no small measure because . . . of the impact of the fuel shortage and the changing political character of the post-Watergate White House."

A similar observation appearing in the *Washingtonian* was made by William D. Carey, who, before becoming a vice-president of the management-consulting firm of Arthur D. Little, had spent more than 25 years on the staff of the Bureau of the Budget, specializing in science. Said Carey: "I've seen evidence that Guy Stever is operating with considerable confidence, developing channels to industry and the academic institutions, and putting together a fine group of people in NSF, but he has an uphill struggle to infuse his influence into a Presidency in great disarray."

That disarray was heightened by the resignation of Secretary Shultz, who was nominally Stever's point of contact with the White House. William Simon was appointed to the Treasury post a short time later, but it was made clear that his assignment was considerably narrower than that of his predecessor, and there was no mention of any responsibilities involving the transmission of scientific advice to the White House. Drew, however, pointed out that he customarily sat with the Domestic Council of the White House when it was in session and frequently was invited to attend reviews of science budgets within the Office of Management and Budget.

That the entire situation had given rise to a feeling of almost desperate concern within the Washington scientific community was amply illustrated by an article in the February 1974 issue of the *Washingtonian* magazine, "Who Unplugged America's Science Machine?" Its author, Vernon Pizer, found no shortage of quotable authorities:

Handler: ". . . take the fiasco of the President's War on Cancer where a 'disease of the month' was picked and people and resources were pulled from other health programs to attack it. Focusing unduly large efforts on finding a quick payoff on cancer unbalances the total quest for medical knowledge, pinches NIH's [National Institutes of Health] ability to perform research over the spectrum of biomedicine, research from which could come the answers to a host of medical riddles including—ironically—cancer. This kind of thinking permeates the whole fabric of science because the fabric is woven largely on looms controlled and funded by inept, myopic federal decision-making."

Glenn Seaborg, Nobel-laureate, physicist, and former chairman of the Atomic Energy Commission: "We need to fill our policy vacuum. We need to announce a strong program of support for basic research, and a workable mechanism for establishing priorities in the various fields of science/

technology consistent with our national requirements. It is long past time that we recognized that science has a potent capacity to determine the welfare of the nation and so must be accorded a central and continuing role in the decision-making process."

Emilio Q. Daddario, director of the Office of Technology Assessment set up to advise the U.S. Congress: "Because there is no definitive science policy we are forced to fall back on short-range responses jerrybuilt to meet each crisis at its apex. We simply have to fashion a national policy on a rational anticipatory basis with the executive and legislative branches, the public sector, the academicians, and industry all influencing its ultimate shape so that it is a national consensus. It has to look ahead at our needs and goals and provide the scientific-technological vehicle to get us there, and it has to have enough flexibility so that it does not stifle initiative."

Philip Abelson, editor of *Science,* the weekly magazine of the American Association for the Advancement of Science: "Every time a field of science generates a wave of popular enthusiasm every government agency tries to get on the gravy train; as soon as popular enthusiasm switches to something else they immediately change trains. What we need desperately is a sound, coherent government way of handling science, one that cuts out the train changing."

It was only fair to point out that the absence of a national science policy was not the result of dereliction on the part of any one national administration. Except for a national commitment to support a certain amount of basic research with government funds and to permit the mission agencies to support rather large programs in both basic and applied research generally related to their missions, there never had been a U.S. science policy. Its absence seemed far less critical, however, when national support was growing and—even as that support leveled off—when an implicit embassy of science existed in the White House.

The apparent lack of concern over this situation at the highest levels of the Nixon administration impelled a new and higher level of political action on the part of the national scientific societies. Traditionally, they had abjured direct intervention in the science-policy process, except to offer occasional reminders of the long-range benefits of R and D. Now they were beginning to seek friends in high places and to propose specific new approaches to the management of the scientific enterprise.

One of the most active leaders in this effort was Alan C. Nixon, who had already defied tradition when he was elected to the presidency of the American Chemical Society for 1973 without having been placed before the ACS membership by the formal nominating committee. He was, in effect, a write-in candidate, representing the anxieties of the membership over the shaky economic situation for many professional chemists. In the spring of 1973 he called together a Committee of Scientific Society Presidents (CSSP), and in October the committee unanimously recommended to the other President Nixon that, in order to provide the government with the best scientific and technological information, there be established either a Cabinet-level Department of Science and Technology or a Council of Science and Technology in the White House, or—alternatively—that the science adviser be elevated to Cabinet rank.

At a later meeting, in January 1974, the committee refined its ideas and concentrated on the establishment of a Council of Science and Technology Advisers, similar to the Council of Economic Advisers. The new idea indicated an increasing political sophistication on the part of the society presidents. It was noted that the council could be created by the Congress and, in the words of one member of CSSP, could not "be abolished or starved by executive fiat."

Even the American Association for the Advancement of Science, the largest and one of the oldest scientific organizations in the U.S., began to shed some of its corporate reticence. It provided a home for a campus-based Science and Public Policy Studies Group and converted it into a Committee on Science and Public Policy. One of the first tasks undertaken by the committee was to assist the Committee on Science and Astronautics of the U.S. Congress in planning an extensive series of hearings on national science policy. And when the AAAS executive officer announced plans to return to academic life in early 1974, a high-ranking officer of the association let it be known that a search was under way for a "politically effective scientist of national distinction" to replace him.

The public attitude. The struggle of the scientific leadership to rekindle enthusiasm for federal funding of basic research appeared to have public support behind it. Concerned lest the environmentalist movement of the early 1970s had led the public to equate scientific research with a mindlessly growing technology that eroded the quality of life, the National Science Board of the NSF commissioned a study of public attitudes toward science that indicated the reverse was true. Most Americans felt science was a positive factor in the human condition and conferred high social status on scientists in general.

On the other hand, it was also apparent that public interest in science in the abstract had diminished. There was noticeably less coverage of scientific advances in the daily press, and fewer reporters were assigned to cover science exclusively. Even the annual meetings of the AAAS, regarded nationally as the World Series of science, were featuring fewer sessions on science itself and an increasing number on national policy issues in which science was a central factor. Congressional interest in science qua science dwindled markedly.

At the same time, it was apparent to those who cared that science was beginning to permeate almost all aspects of national life. A study of the major concerns of the U.S. public, published in 1973 by Potomac Associates, revealed that of the ten leading concerns, designated by 80% or more of those surveyed, six depended for solutions on advances in science and technology. Even though members of Congress were not crowding each other for berths on the science committees, the pages of the *Congressional Record* were full of references to technical questions relating to pollution abatement, technological productivity, health care, and similar matters.

It was as though science itself was less visible than it had once been simply because it had been diffused throughout the entire fabric of government. This process, while presumably beneficial in the long run, created a source of anxiety for those concerned about the lack of a national science policy and—more specifically—the impoverished R and D budget. In fact—according to the Office of Management and Budget —there really was no R and D budget. Making a rare public appearance, Hugh Loweth, deputy associate director for energy and science at the OMB, said: "We often hear the phrase, the R and D budget, as if there were such a thing. Most of you realize, of course, that what is called the R and D budget is actually a collection of funds invested in research that derive from a number of programmatic decisions. It is the decisions made outside of R and D that determine the amount of money that goes into R and D and I think that's right. You don't do R and D for R and D's sake."

And that might well be the heart of the matter. The government did not do R and D for R and D's sake, but many scientists do.

Science and the budget. Loweth spoke in late February 1974. A few weeks earlier President Nixon had delivered his state of the union message and the budget for fiscal 1975 had been announced. It was plain from both documents that the circumstances in which the country found itself had forced a number of Loweth's "programmatic decisions" favoring R and D. Thus, according to the president, "One of the great

strengths of this Nation has been its preeminence in science and technology. In times of national peril, we have turned to the men and women in the laboratories in universities, Government, and private industry to apply their knowledge to new challenges."

He then proceeded to list the challenges facing the nation. The need to deal with the energy crisis had resulted in the allocation of $1.8 billion for direct R and D on energy and an additional $216 million for supporting research—an increase of 80% over fiscal 1974. Scientific research was to be used to fight drug abuse, prevent infant mortality, combat venereal disease, and to aid in treating mental patients. The services of the scientific community were called upon to design better forms of transportation, to increase agricultural productivity, and to lower pollution. It was necessary to develop better and safer methods for the extraction of minerals, and science was summoned to find ways to protect the population from natural disasters. Even the social sciences were promised funds to study the social effects of such programs as social security, welfare, and health insurance, as well as experimental educational methods.

When the fiscal 1975 budget was released on February 4, it included $19.6 billion for federally sponsored R and D, an increase of $1.7 billion that was hailed by the science adviser as the largest dollar increase in ten years and a 10% rise over fiscal 1974. The R and D total included an increase of $1 billion for civilian-oriented programs such as energy, health, education, environment, urban problems, and transportation, representing a gain of 17%, while the request for defense R and D, including the military aspects of the Atomic Energy Commission, was $900 million or about 10% above the previous year. Space program activities were budgeted to decrease by 7%, reflecting the completion of the Apollo and Skylab programs and the transition to the space shuttle system. The overall budget for the National Aeronautics and Space Administration rose by almost $100 million, however, representing a new interest in aeronautics and applications programs.

The increase in budgetary requests for civilian R and D programs was indeed remarkable. In the budget for fiscal 1966, defense and space activities absorbed 76% of total R and D, leaving only 24% for civilian programs. The portion alloted to the civilian sector in fiscal 1975 was 35%. Since 1969 civilian R and D had increased by almost 120%, from $3.2 billion to $6.8 billion. In addition to the increase in energy R and D, there were notable gains for biomedical research, with almost all the increase going to the National Cancer Program; environmental protection, with appropriations for the Environmental Protection Agency more than doubling; and educational research.

Although the NSF had been the only federal agency specifically charged with maintaining a balanced, across-the-board program of support for fundamental science, its budget also showed the effect of current governmental anxieties. Of the $694.7 million budgeted for research support in fiscal 1975, the fraction devoted to the acceleration of energy research and development amounted to $252.6 million.

Despite the many indications of relief with which the scientific community greeted the new budget, there were a few caveats. One of them came from Harvey Brooks, then dean of engineering and applied physics at Harvard University. Speaking to a group of science-policy scholars in San Francisco, he noted:

"There is considerable stress in the President's budget message on the fact that R and D outlays will increase by 10% over FY74. This is indeed true, and a welcome sign to the scientific and engineering community, but a comparison of recent trends with the consumer price index gives a less optimistic picture. In terms of 1967 dollars, total federal R and D obligations were $16 billion in FY67, $13.5 in FY73, $13 in FY74 and $13.2 in FY75. Thus in real terms total federal R and D sup-

Sidney Harris

"These days everything *is higher."*

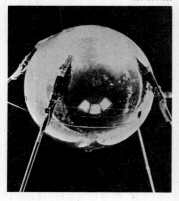

"First it was space, then environment, and now energy" describes how the major targets of U.S. scientific funding have changed over the last decade. Above is Sputnik 1, the Soviet Union's first orbiting satellite, which stimulated the U.S. effort in space. Right is a poster on endangered animals, a major focus of environmentalists, and on the opposite page is an oil derrick in Saudi Arabia.

port will still be 17.5% below its peak level of FY67, perhaps more."

Another came from the biomedical community as it watched the shape of research support within the National Institutes of Health change from support of untargeted research through the traditional granting process to the support of targeted research through the contract mechanism. It was apparent in the fiscal 1975 budget that the administration viewed the prime targets as cancer and heart disease. The Department of Health, Education, and Welfare summary on the budgets for the several National Institutes of Health listed the National Cancer Institute ($600 million) and the Heart and Lung Institute ($309 million) separately, then lumped all the rest under the heading "Other Research Institutes" ($794 million; much less than the other two and only $1 million more than in fiscal 1974).

Targeted research. The traditional pattern of research support within the NIH had emphasized diversity and peer review. Biomedical scientists with promising projects would submit a proposal for additional research to the NIH, and study groups of distinguished biomedical researchers would select those proposals that appeared most deserving of support. The focus of this system was on contributions to knowledge as opposed

to achieving a designated mission assigned by the agency.

Now the pattern was changing to one in which NIH management teams identified a specific goal and contracted with a research team to achieve it. It was this system that Handler had criticized. In fiscal 1974 NIH spent $291 million in support of 4,266 new, competitively selected research grants. In fiscal 1975 it would spend only $195 million to support 2,875 such grants. In contrast, money available for research contracts had risen significantly, from $261 million in fiscal 1973 to $335 million in fiscal 1974 to $362 million, or 20% of the total NIH budget, in fiscal 1975. Furthermore, all the increased contract funds in the fiscal 1975 budget were for programs in the cancer and heart institutes.

The transition to targeted research in the NIH did not originate with the Nixon administration. It could be traced back to several speeches by Pres. Lyndon Johnson, in which he made it clear that the purpose of federal support of biomedical research was to find cures. The developing interest in quick payoffs from research programs was heightened in time by a growing concern over the size of the overall federal budget. The pressure to place a lid on expenditures was especially severe on the research sector because—unlike such segments as

interest on the national debt—it was considered to be entirely discretionary. In 1967 the discretionary portion of the federal budget was only 40% of the total, but in fiscal 1975 it had fallen even further, to less than 30%. It was not a happy state of affairs, but those who understood the situation were reluctant to cry havoc over the fact that the Nixon R and D budgets were barely keeping up with the effects of inflation.

Meanwhile, the mounting insistence on quick payoffs that had marked both the Johnson and the Nixon White House was beginning to appear in Congress. Within the House of Representatives, the Committee on Science and Astronautics had always been most protective of the support program for academic research, but in the spring of 1974 John D. Holmfeld, science policy consultant to the committee, made some sobering observations to a session at the 1974 meeting of the AAAS.

"During the past three or four years, there has been a focusing of interest in the Congress on a number of science-policy issues, chiefly based on the reluctance of the Office of Management and Budget to support certain academic activities of a sustaining nature, such as graduate training and summer institutes. During the past two years, Congress has put the money back into the budget for these purposes. Whether the new demands on the

Congress to devote their attention to the energy problem will diminish the amount of attention that can be paid to these other issues is yet to be determined.

"Another question that will emerge this year or in the next two is whether we can still use the rationale of supporting basic research by 'riding the wave', as Dr. Stever called it. First it was space, then environment, and now energy. Much of the incremental funding has been justified by these concerns. I think it is a good guess that if not this year, then in future years, that kind of rationale will come under question. . . . We have on the House Science and Astronautics Committee one or two members who are very skeptical about basic research and as the high-seniority members begin to leave, there will be a notable switch there."

Holmfeld questioned the wisdom of the NSF in embracing the prevailing concept of quick payoffs through its burgeoning RANN program—Research Applied to National Needs—in partial justification of an increase in its budget for the support of fundamental research. The arguments offered by the foundation were that, first, the nation must continue to generate new information so the information will be there when it is needed for application to national problems, and, second, that with so much money going into the RANN program, it is

325

necessary to supplement the basic-research program in order to preserve the historic function of the agency. Holmfeld continued:

"These justifications have been used to support large increases in funding, but whether or not the research supported by these large increases will actually yield results that constitute a useful pool of knowledge is something that NSF, or the congressional staffs, or—even better—the scientific community will have to look at. It's going to be necessary in our committee and within the Office of Management and Budget to find better methods to ask the right questions on the basic-research area.

"I must also point out that there are substantial research programs where we are beginning to look failure in the face. The most significant one and one that has not yet surfaced publicly is the nuclear-power program. That's been on the way for 15 or 20 years now and with the changing of the guard in the Joint Committee on Atomic Energy, I think that there's no doubt that unless the Liquid Metal Fast Breeder Reactor, which has been severely criticized in the scientific community, begins to show that it has the capability of operating not only as a pilot plant but also making a substantial impact, there's a real possibility that the kind of disenchantment that we've already begun to see in the military field could very well hit us and then the reaction would be an over-reaction. Looking down the road, not two years perhaps, but five years, that is the real danger—that the pendulum will swing back—not to the middle, but all the way to the other side."

Congress and the NAS. These somber words, unreported in the press, sounded very much like the end of what had been called the blank-check era of U.S. science. For it had been Congress that had uncritically accepted the thesis that the more money one put into one end of the scientific enterprise, the more benefits came out the other. It was Congress that appropriated more money for the NIH in the late 1950s than that institution could sensibly use. It was Congress that rejected a logical candidate for the post of NASA administrator because he truthfully testified that he could not make proper use of the millions the country wanted to spend to catch up with the Soviet Sputnik. More recently, as Holmfeld noted, it was Congress that restored administration cuts in the budget for the training of scientists.

Thus it was all the more ironic that as Congress showed signs of losing its enthusiasm for the support of academic research, it was turning more and more to the science community for advice in dealing with some of the nation's most intractable problems. Nowhere was this more evident than in its relationship with the National Academy of Sciences (NAS).

The Academy is an organization of distinguished scientists with a unique charter. Passed by the Congress and signed by Abraham Lincoln in 1863, the Act of Incorporation establishing the Academy designated it as officially responsible for advising the federal government on matters of science and technology but specified that it was to be completely independent of the government in both finances and procedures. It receives no appropriations from the government and its employees are not under Civil Service. Although the Academy performed a small number of services for Congress in the 19th century, the congresses of the 20th century had neither the funds nor the inclination to make use of it. Instead, the Academy worked primarily for the Executive Branch, through individual contracts with federal agencies.

In the 91st Congress, however, the act authorizing funds for the Department of Defense ordered the department to contract with the Academy to conduct a study of the ecological and physiological dangers inherent in the use of herbicides and the effects of the defoliation program in South Vietnam. That same Congress was also wrestling with the question of whether or not it was reasonable to demand that the U.S. automobile industry substantially reduce the emission of carbon monoxide, unburned hydrocarbons, and nitrogen oxides from engines in the 1975 and 1976 models. When the automobile industry protested that the dramatic reductions being contemplated—in some cases as much as 90%—were not technologically feasible in so short a time, Congress turned the problem over to the Academy.

The Academy recognized that although both sets of problems could be expressed in scientific and technological terms, it would be impossible to extract from them all their emotional components and possibilities of professional bias to which even Nobel laureates are subject. The study on the use of herbicides in Vietnam led to divisions within the Academy, traceable in large measure to the bitterness generated by the Indochina war itself. Much of the controversy centered on the estimate of damage to the inland forests of Vietnam, and those who were close to the study were astonished to find reputable scientists, studying the same sets of aerial reconnaissance photographs through the same optical instruments, emerging with estimates of damage that differed by a factor of approximately 200.

After almost a year of negotiation between the Academy, its study committee, and its review panels, the report as issued in February 1974 had narrowed the area of disagreement to a factor of four.

It also made clear that more than a century would be required to restore the mangrove forests of Vietnam unless extraordinary measures were taken, and in no case could restoration be accomplished in less than 20 years. How long it would take to repair the damage that had been done to the image of scientific objectivity was another question. As one of the scientists on the study told a reporter, with more than a little bitterness, "The fact that I was a dove on Vietnam doesn't change the way I count trees!"

The Academy fared better with its report on automobile emissions. By pointing to recent technological advances that offered more efficient methods of reducing noxious emissions, the Academy managed to gain respectful attention from both Congress and the automobile industry. It also learned that if it wished to collaborate successfully with Congress, its reports must be not only scientifically sound but also politically viable.

The aftermath of this successful effort was quite remarkable. The 92nd Congress passed new legislation calling on the Academy to assist the National Study Commission on Water Quality in the investigation of all technological, economic, social, and environmental effects of achieving or not achieving the very stringent effluent limitations imposed by the Federal Water Pollution Control Act Amendments of 1972; to conduct reviews of scientific questions raised by enforcement of the Federal Environmental Pesticide Control Act of 1972; and to study methods of equitably reimbursing physicians for their services in teaching hospitals. The 93rd Congress mandated a three-year review and study by the Academy of Veterans Administration medical facilities; assistance to the Environmental Protection Agency in strengthening the scientific basis for environmental regulation programs; and a review of the scientific basis for an army program aimed at developing the water resources of the Washington, D.C., area.

Because the Academy is not a government agency and cannot receive appropriations directly from Congress, it was necessary in each piece of legislation to direct the appropriate federal agency to contract individually with the Academy. In one case, however, the Senate Committee on Public Works received permission and funds, under a Senate resolution, to contract with the Academy directly. Partly as a result of the earlier Academy study, it had become clear to the Senate committee that the very low levels of engine emissions imposed by the Clean Air Amendments of 1970 had been set rather arbitrarily and could impose severe economic costs on the public. Accordingly, the committee contracted with the Academy to study the health effects of ambient air quality stan-

dards and the social and economic costs *and* benefits of compliance with the new standards.

By the end of April 1974, the second session of the 93rd Congress was considering an additional half dozen or more major pieces of legislation calling for similar studies by the Academy. While it may have been true that the Congress had lost much of its earlier enthusiasm for nourishing the national science effort, there was little doubt that by mid-1974 it had a far greater respect for the value of scientific judgment in questions of high policy.

—Howard J. Lewis

Transportation

Technological advances in transportation, already slowed somewhat by inflationary trends, received another setback as a result of the energy crisis. Since transportation is such a heavy consumer of energy—*i.e.*, about 24% of total consumption in the U.S.—and since about 96% of the fuel consumed by transportation in the U.S. is petroleum, the changing oil-supply situation caused major changes in thinking about where future research and development efforts in the field should be directed.

Some transport technology programs were adversely affected, especially those seeking to advance innovations that would consume large quantities of fuel. Examples were the supersonic air transport (SST), vertical-lift aircraft, and the rotary engine for use in automobiles. On the other hand, electrification of railroads, the trans-Alaska pipeline, and nuclear-powered merchant ships gained support. The trucking industry, long prevented from utilizing a number of technological advances because of limits on truck size and weight, obtained considerable backing for easing the restrictions.

Aviation. The fuel shortage forced many airlines to ground their B-747 jumbo jets—16 at one period—and further orders for the 350–400 passenger versions were stopped. Boeing Aircraft, the builder, offered the airlines a modified special-performance version, called the B-747SP, which was shorter and carried only 280 passengers but had a longer range. It was designed to compete with long-range versions of two other wide-body jets, the L-1011 and DC-10. Pan American World Airways ordered 7 and took options on 18 more, with deliveries to start in the first half of 1976. Japan Air Lines also announced plans to place orders.

The cargo-carrying capabilities of the B-747, hitherto largely ignored by the airlines, finally

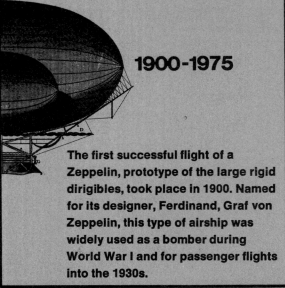

1900-1975

The first successful flight of a Zeppelin, prototype of the large rigid dirigibles, took place in 1900. Named for its designer, Ferdinand, Graf von Zeppelin, this type of airship was widely used as a bomber during World War I and for passenger flights into the 1930s.

became sufficiently attractive to stimulate orders, and ten all-cargo or convertible (passenger or all-cargo) versions were expected to be in service by 1975. They would handle up to 100 tons of cargo, usually in containers.

Lockheed Aircraft received its first orders for the new, long-range L-1011 wide-body trijet, with Cathay Pacific Airways ordering two and taking options on two more. The extended range (up to 4,500 mi) of this model was designed to compete with the long-range version of the DC-10. Deliveries were to be made in late 1975. Orders for two long-range versions (up to 6,500 mi) of the DC-10 were placed by Japan Air Lines. The planes would seat 277 passengers and were capable of flying nonstop on such long routes as Tokyo–New York. JAL also ordered four 327-passenger DC-10s for its domestic routes. Deliveries were to start in 1976.

Airbus Industrie, the European consortium building the Airbus A-300, a twin-engine, wide-body aircraft designed largely for short-to-medium range markets, won certification for its long-awaited aircraft. The first models began commercial service with Air France between London and Paris in late May. The initial models would hold from 250 to 280 passengers and have a range of 1,400 nautical miles. Later models probably would include a smaller, shortened version carrying about 200 passengers and designed specifically for the short-haul market, where commercial air travel had failed to capture any sizable percentage of the total. The builders of the Airbus claimed it would be especially attractive in this market, since it would use considerably less fuel, in terms of capacity, than its three- and four-engine competitors, would be much quieter, and would cause less pollution. They also claimed the aircraft would cost less to operate and could break even with a load factor (percentage of seats filled) as low as 36% on a 500-mi flight.

The other European air transport development combine, the British and French builders of the Concorde SST, continued plans to build 16 such aircraft, even though total development costs had risen to $2.9 billion. The only firm orders were with the nationalized Air France and British Airways; virtually every U.S. air carrier had canceled its options to buy, although commercial service was scheduled to begin in 1976. A historic transatlantic flight from Washington, D.C., to Paris in 3 hr 33 min by the Concorde highlighted an eight-day demonstration trip of over 10,000 mi in South and North America. While it clearly proved the technical feasibility of the aircraft, the high fuel consumption, small carrying capacity (about 140 passengers), and very high unit cost (about $50

B-747 jumbo jets lie grounded in the New Mexico desert, victims of declining passenger traffic and fuel shortages. Two airlines offered 11 of the giant planes, worth $250 million each, for sale.

million) obviously had discouraged buyers. The new Labour government in Great Britain was taking a hard look at the program and might well pull out after the initial 16 aircraft had been built.

The government-owned Airtransit Canada, called "the world's first true STOL [short takeoff and landing] airline," began commercial service between Ottawa and Montreal. Initially, it was using small DHC-6 Twin Otters, specially modified by de Havilland Aircraft, with area navigation gear and a new, sophisticated air data acquisition system (ADAS) that permitted on-board programming of the entire flight. Advanced microwave landing systems, which could succeed the instrument landing systems in general use, were being used at each landing site. The twin-engine, high-wing aircraft could seat 11 and would operate between exclusively STOL ports over a reserved low-altitude route. The next phase of the STOL development program in Canada would use the DHC-7, a 48-passenger, four-engine turboprop that was expected to be ready for U.S. scheduled commuter air service by early 1976.

Two new, huge, fully modernized airports were opened to commercial traffic. Built for the future, the partially completed Dallas-Fort Worth (Tex.)

Airport had 17-in.-thick concrete runways designed to handle million-pound aircraft, heavier than any yet built. Two cargo terminals at each end of the complex would be capable of handling 200 aircraft the size of the Air Force's C-5A, the largest all-cargo transport in current use. The airport featured large half-loop terminals split by a ten-lane intercity freeway, with access roads leading directly to parking spaces close to the aircraft. A 13-mi, automated tracked system, called Airtrans, utilized rubber-tired vehicles to move passengers between any two points within the four completed terminals in about ten minutes. Airtrans was also designed to move employees, luggage, mail, supplies, and trash. The next two phases of growth were scheduled for completion in 1985 and 2001, when 13 terminals should be in use.

While not quite as large, L'Aéroport de Charles de Gaulle outside Paris had similar technological advances in lighting, automated passenger flow, baggage handling, and communications. In a unique system, incoming passengers checked in at the drive-in entrance to the terminal, where their baggage was routed directly to the outgoing aircraft and they were computer-directed to the parking space closest to their aircraft. The central

329

terminal was large and cylindrical in shape and was connected with satellite loading/unloading stations by underground moving walkways. Cargo operations were also being stressed, with 750 ac devoted to freight facilities. Since the airport was located in a relatively underpopulated area, it could operate jet flights around the clock.

The fuel shortage revived interest in the possible development of lighter-than-air ships, and the world's first international airship conference was held in London. Most observers believed that large airships were at least ten years away, but the coordinator of the conference claimed one could be built more cheaply than a B-747 with a cargo-carrying capacity five times as large (500 tons) and the ability to fly at speeds up to 100 mph on a fraction of the jumbo jet's fuel consumption. According to *Jane's Freight Containers*, the Soviet Union was seriously considering development of a nuclear-powered airship capable of hauling 180 tons of containerized freight or 1,800 passengers at a cruising speed of 190 mph.

Another futuristic aviation innovation by the U.S.S.R. was a giant hybrid aircraft that could fly 25 to 50 ft above water at speeds up to 300 knots, utilizing both aerodynamic lift and ground effect. A prototype was being flight tested on the Caspian Sea. It had a main wingspan of 125 ft and a forward 40-ft stub wing housing eight turbojets topside. Thrust from these turbojets is deflected downward to create an air-bubble lift on the main wing for takeoff, and it is then redeflected over the main wing to create additional aerodynamic lift. Two additional jets were positioned on the craft's V-tail.

Highway transport. To help truck lines conserve fuel while handling the same or more traffic, at least seven Western states in the U.S. raised truck size and weight limits on their major highways. Nationally, the administration urged congressional approval, during the fuel shortage, of similar increases on the nation's 42,500-mi Interstate Highway System. Such changes opened the way for carriers to increase payloads, especially when twin-trailer operations were permitted, and to make better use of improved engines.

The fuel shortage also forced delays in deadlines for very strict emission standards for motor vehicles in the U.S., with auto makers given another year to conduct research and development and to make evaluation tests. The Environmental Protection Agency (EPA) reported that the 1975 model autos should get significantly better mileage as compared with 1974 models. According to the EPA, preliminary tests of 1975 model prototypes indicated gains of 20% or more for larger cars and somewhat less for smaller cars, improvements made possible through the introduction of catalytic converters to supplement the engine changes used to control pollutants.

Courtesy, Dallas-Fort Worth Airport

High fuel consumption was one of the major reasons cited by the Ford Motor Co. for ending its research and development program on the Wankel rotary engine. About 100 engineers and researchers were shifted to other projects after two years of effort. Ford said it would concentrate on other engines with a greater potential for improved fuel economy and low emission levels. General Motors was also taking another look at the Wankel as it delayed the introduction of Wankel-powered autos from the fall of 1974 until at least late 1975.

The U.S. Department of Transportation (DOT) began a $1,250,000 program with three truck manufacturers (White, Freightliner, and International Harvester), under which nine heavy-duty diesel tractors were turned over to three interstate truck lines (Overnite, Mid-American, and Ryder) for a two-year evaluation of ways to reduce noise without loss of performance. The new tractors reportedly could operate at half the noise level of conventional diesels without losing efficiency.

DOT also announced delivery of the first units of three competing Transbus vehicles to Phoenix, Ariz., for testing. This represented another step in a $26 million program to develop basic changes in urban transit buses. Improvements included wider doors, safety bumpers, extra strong body construction, lower floors and better lighting for easier and faster boarding and discharging, and less noise and engine emissions. The three new buses would operate in Miami, Fla., New York City, Kansas City, Mo., and Seattle, Wash., for in-service evaluation. The program was scheduled for conclusion in 1975, when the winning prototype would be named.

Another DOT program sought to stimulate low-capital-intensive improvements in urban mass transport—which meant greater and more efficient use of buses and car pools. This was being done in a number of ways, including exclusive lanes, priority use of lanes, and priority control of traffic signals. The highly successful exclusive busway into Washington, D.C., was opened to car pools with four or more occupants, and a $53 million exclusive busway project in Los Angeles was nearing completion. Priority treatment for buses was being provided in Seattle through exclusive use of on-off ramps servicing an express highway; in Minneapolis through easier access to a metered (controlled traffic volume) freeway; and in New York City through bus lanes running counter to the peak traffic flow on major commuter highways. Miami was testing a system giving city buses traffic signal preemption, while a similar system under way in Washington, D.C., would provide 450 buses with special signal preemption equipment.

Pipeline and tunnel projects. After several years of effort, the builders of the highly controversial 790-mi, 48-in.-diameter trans-Alaska crude oil pipeline from the North Slope fields to

Courtesy, Dallas-Fort Worth Airport

Rubber-tired vehicles on an automated track system move passengers between any two points in the four completed terminals of the Dallas-Fort Worth Airport (opposite page). The terminals, shaped like half-loops (left), are bisected by a ten-lane expressway.

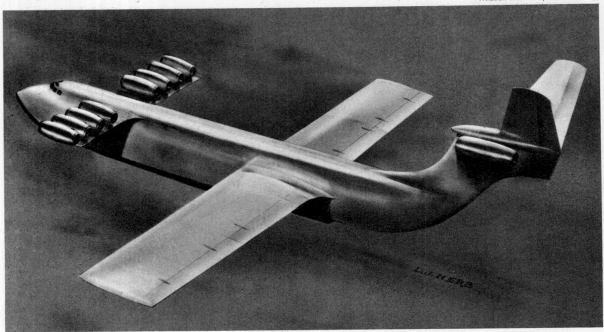

Experimental Soviet aircraft, at 500 tons the world's heaviest, is tested over the Caspian Sea.
The jet blasts from the eight engines hit the water and then bounce back under the main wing to create
a lifting bubble of air on which the craft will ride, some 25–50 ft above the water's surface.

the ice-free port of Valdez finally won both congressional and administrative approval for start of construction. The delays and repeated changes required to meet environmental objections had increased the cost from $900 million to an estimated $4.5 billion, and this could use up $5 billion if year-round construction was needed to meet the mid-1977 completion goal. The builders agreed to double the starting capacity to 1.2 million bbl a day to help meet long-range U.S. fuel needs.

The long-proposed and hotly contested submarine tunnel connecting Great Britain and France finally won the approval of the British government. If actually built, the English Channel tunnel (or "chunnel") would extend 32 mi across the bottom of the Strait of Dover and would consist of three tubes, two enclosing single-line rail tracks and the third a service tube. Current plans called for construction of the more than $2 billion project to start around mid-1975, with completion by 1980. The trains, which would include auto-carrying cars, were expected to make the London-to-Paris run in less than four hours, a saving of about two hours as compared with current rail-ferry runs.

Railroads. The list of commodities moving by unit train (shuttle-type freight trains hauling a single commodity) on U.S. railroads was expanded to include petroleum. In what was seen as the start of an effort by the railroads to raise their share of U.S. petroleum traffic from the current 2%, Southern Pacific began operating a 70-tank car unit train, capable of hauling 1.6 million gal of crude oil per trip, between Salt Lake City, Utah, and a refinery at Richmond, Calif. Four similar trains were expected to begin service in the near future.

Following several years of research, General American Transportation Corp. unveiled its new Tank Train. The concept called for operating 40-car, million-gallon unit trains equipped with a unique interconnect system that would permit loading and unloading of all cars at a single point, at a rate of 3,000 gal per minute. Two three-car, full-size Tank Trains were undergoing extensive evaluation tests on runs between Anchorage and Fairbanks, Alaska, and between Great Falls and Miles City, Mont., both areas where severe climatic and operating conditions are encountered. After six months the Association of American Railroads would examine the results and, if the tests proved successful, production could begin in 1975. The innovation could make it feasible to operate marginal oil wells situated close to rail lines but without enough flow to justify pipeline construction.

Unlike many European railroads, U.S. roads had little mileage under electrification. The high performance and relatively low cost of the diesel engine made the initial costs of electrification too high, particularly for an industry that had found it increasingly difficult to attract capital for right-of-way facilities. The rapidly rising cost of diesel fuel, however, resulted in a serious reevaluation by the U.S. government and the railroad industry, and a joint task force concluded that electrification offered the only feasible means whereby the railroads could utilize coal or nuclear power for high-volume, long-distance freight movements and intercity movements of passengers. The task force recommended that the U.S. government play a sizable role in stimulating the switchover to rail electrification. A step in this direction was taken when the Black Mesa and Lake Powell Railroad, called "the world's first railroad electrified with 50,000-v alternating current," started coal unit-train operations between mines in Arizona and a utility 80 mi away near the Utah border. The new railroad was designed for full-scale automated operations.

The railroads continued their efforts to satisfy the automobile manufacturing industry's desire that all of its rail shipments of new autos be in fully enclosed cars or containers, thus avoiding vandalism and pilferage en route. The Penn Central, Chessie System, and Grand Trunk Western railroads joined in testing triple-decked, enclosed RailPac cars able to carry up to 18 automobiles. The superstructure was designed so that it could be used on specially built flatcars or on flatcars currently in operation. The Southern Railway introduced a new auto carrier called Autoguard. The enclosed 124-ft car was articulated into three sec-

tions to permit easy rounding of curves, which Southern contended was one of the problems with the 89-ft standard open-rack cars still in use. Autoguard could carry 18 standard automobiles on its three decks. The two prototype cars being tested had easy-opening doors at either end and were equipped with adjustable ramps to expedite handling.

The quasi-governmental organization Amtrak continued to experience both rising passenger volumes and bigger deficits. Its tests of a turbine-powered passenger train along the high-density Chicago–St. Louis corridor proved sufficiently successful to justify announced plans to purchase 20 additional trains of this type. The French train, leased from ANF-Frangeco, was considered to be an especially good performer from the standpoint of Amtrak officials.

The Bay Area Rapid Transit (BART) urban rail system opened another ten-mile segment of its 75-mi network in the San Francisco-Oakland area and made successful test runs through the long underwater tube connecting San Francisco to the bulk of the network. However, major troubles with BART's automatic control system forced manual operations, and by the spring of 1974 approval to use the tube—an obvious vital link—still had not been granted. In an effort to remedy the problem a new backup computer system was designed to monitor and correct the false electronic train-loca-

Drawing by Stevenson; © 1973, The New Yorker Magazine, Inc.

"It seems only yesterday they were giving away crystal stemware."

Amtrak's turbine-powered passenger train, made in France and designed to operate at speeds of up to 125 mph, prepares to leave Newark, N. J., for Chicago on its first long-distance test run, on Aug. 10, 1973. Amtrak officials expressed satisfaction at the train's performance.

tion signals being given by the basic system. BART also encountered serious mechanical problems with many of its specially developed cars, and after receiving 280 it imposed an embargo on further shipments. Rohr Industries, the builder, later reported that it had corrected the problems and resumed deliveries.

Water transport. Both the U.S. Navy and U.S. merchant marine stood to gain when the Federal Communications Commission authorized RCA, ITT, and Western Union to participate in and share the estimated $100-million-plus cost of Marisat, the Communications Satellite Corporation's interim two-ocean, ship-to-shore satellite communications system. The three private companies have already agreed to provide about 20%, or nearly $15 million, of the estimated $73.3 million initial cost.

Plans for construction of several offshore, deepwater ports in the Atlantic Ocean and the Gulf of Mexico progressed, though legislation to clarify which U.S. agency would have authority to approve them—and how much authority it would have—had not yet been passed. Without such offshore ports, the giant tankers being used to haul petroleum could not unload at Atlantic and Gulf coast ports.

The U.S. Maritime Administration reported that American President Lines' newest containership would be equipped with an electronic monitoring system, developed by MarAd through a university research team, that would evaluate engine and machine functioning and wear and tear. By maximizing the ship's efficiency, MarAd claimed, it would be possible to save up to 6% in fuel consumption as well as to reduce maintenance costs.

Litton Industries' new integrated tug-barge "Presque Isle" began its first ore-carrying assignment on the Great Lakes. The 50,000-ton-capacity, 975-ft barge contained a notch in its stern into which the 152-ft tug fitted for integrated operations. Locked together, the $35 million tug-barge unit could operate at 16 mph, and through the use of special on-board gear, it was able to off-load its entire cargo in about five hours. Unlocked, the tug could be used for other purposes.

The goal of year-round navigation on the Great Lakes drew a step closer as a result of a trip through the frozen Straits of Mackinac in late February by the tugboat "James A. Hannah." Helped by a Coast Guard icebreaker, the 4,000-hp diesel tugboat, strengthened with heavy three-quarter-inch, steel-reinforced plating, pushed a barge loaded with a million gallons of critically needed gasoline and heating oil through solid ice to Mackinac Island. The ice was reported to be over six feet thick, and the vessel had to operate in subzero temperatures with winds of more than 50 mph and heavy rolling seas. A unique new bow design on

U.S. Coast Guard icebreaker "Southwind" (foreground) breaks path for the ore carrier "John G. Munson" in the Straits of Mackinac, part of an effort to keep the upper Great Lakes navigable throughout the year.

the barge helped make the hazardous trip possible.

The rapidly rising costs of fuel narrowed the cost gap between conventional ships and nuclear-powered merchant ships sufficiently for the Maritime Administration to ask again whether any operators were interested in acquiring nuclear ships for the movement of bulk cargoes. While considerable interest in such ships had been expressed by U.S. operators in the past, the government had been reluctant to help finance the sizable initial construction costs. One prominent British shipping spokesman said the necessary technology was already known, and that the initial construction-cost gap had been narrowed to less than $5 million per ship. He also noted that sizable savings in time and fuel were possible with nuclear ships, but that many nations were reluctant to allow such ships in their waters and ports.

The U.S. Navy's program to develop a 2,000-ton prototype surface effect ship (SES; using air-cushion lift but with rigid sidehulls for water movements only) for use on the open seas at speeds up to 80 knots was delayed following frequent failures by two 100-ton SES prototypes being tested

in saltwater operations. Nevertheless, a Navy advisory council selected Bell Aerospace and Lockheed Aircraft to continue the next phase of the program: development of detailed designs for such a vessel. The Navy subsequently approved a three-month extension of rough-water testing of a 100-ton SES model by Aerojet-General. Bell Aerospace of Canada was operating an air-cushion vehicle (with flexible skirts) over water and level ground, but operations had been limited because of the vehicle's sluggishness when changing direction at high speeds and because of its yawing tendencies.

—Frank A. Smith

Veterinary medicine

A major concern of the veterinary profession during the year was how best to meet the increasing demand for its services. A study by the National Academy of Sciences-National Research Council, reported two years earlier as "New Horizons for Veterinary Medicine," identified the areas in which additional personnel would be required and projected a need, by 1980, for a work force of 42,000 veterinarians. Since this was about 4,000 more than the existing educational system could provide, added impetus was given to finding more efficient means for delivering veterinary services, as well as to expansion of the schools. Although the NAS-NRC report dealt only with the U.S., the problems were much the same in Canada.

Changing concepts. Over the preceding two decades, an increasing proportion of new graduates had elected to work with companion animals (dogs, cats, pleasure horses), thus creating a shortage of veterinarians with an interest in food-animal practice. At the same time, several factors conspired to make this shortage more acute. During the 1971–72 outbreak of Newcastle disease of poultry in California, for example, it was necessary to place reserve military veterinarians on extended duty in order to implement control and eradication measures. Requirements of the U.S. Environmental Protection Agency increased the need for veterinarians able to advise on such matters as feedlot waste disposal, and recently strengthened Food and Drug Administration (FDA) regulations concerning use of potent drugs and feed additives placed added responsibilities on the profession.

An "outreach" program established by the veterinary college at the University of Georgia in early 1974 was designed to provide more direct service to livestock growers in that state, especially in areas lacking adequate veterinary attention. Included in the comprehensive planning were

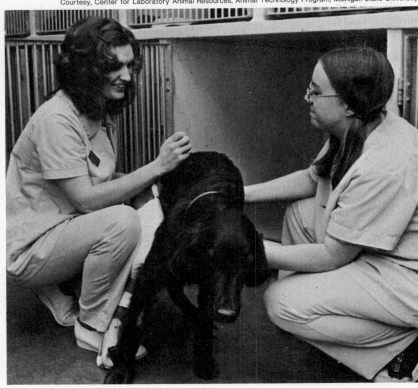

Students undergo training in the animal technology program at Michigan State University, the only program of its type in the U.S. Animal technologists aid veterinarians by working in such capacities as laboratory technicians and surgical nurses.

loan and scholarship incentives for students expressing a willingness to practice in rural areas and an externship program whereby senior students would be required to work for two to four months in an established, preferably rural, practice. The latter provision, it was hoped, would also benefit the practitioner-preceptor, who would absorb first-hand much of the newer information being taught to students.

In early 1974 the U.S. Department of Health, Education, and Welfare announced a loan repayment plan to encourage veterinarians and other persons in the health professions to settle in shortage areas. Persons licensed to practice in a state where such areas existed could qualify for repayment, by the government, of 60% of unpaid educational loans in return for agreeing to practice in a specified area for two years and for repayment of 85% for agreeing for three years. All but a few of the states had listed counties they considered to be shortage areas.

To attract veterinarians into long-term rural practice, the province of Manitoba agreed to provide local agencies with animal hospital facilities and annual operating grants of up to $5,000, which would be matched by locally raised funds. This extended to the veterinary profession the same support already offered hospitals for humans, and up to 35 veterinary service districts were expected to

be formed. By mutual agreement the veterinarian could either be salaried or practice on a regular fee basis, the primary objective of the program being to provide the facilities and atmosphere that would permit veterinarians to utilize their expertise fully and most efficiently. The program was designed to be completely voluntary, and any veterinarians who preferred to remain in strictly private practice were free to do so.

New schools. Although admissions to the veterinary schools in the U.S. and Canada rose 30% from 1968–73, the number of qualified students seeking to become veterinarians was increasing even more rapidly. Most schools anticipated having eight to ten applicants for each available place in the class to be admitted in 1974, and thus the chances of acceptance were less than for applicants to schools of medicine. One reason for this was the recent upsurge in applications from women students, enrollment of whom had increased nearly 250% (from 367 to 888) over a five-year period. To accommodate this rising demand, several schools were planning to increase their enrollments by as much as 50% as soon as additional facilities could be provided. The 19th veterinary school in the United States, at Louisiana State University, started its first class in January 1974, and a school in Florida was scheduled to open in 1976.

With only 18 states and three provinces having veterinary schools, a major problem had been the difficulty encountered by non-residents in gaining admission. For this reason, several states, including New Jersey, Wisconsin, and Oregon, were actively considering establishing schools or entering into regional compacts with existing schools, and federal training of meat inspectors and regulatory veterinarians was proposed. In late 1973 the governors of the six New England states approved plans for a regional veterinary college and agreed to provide planning money for a teaching and research facility to be located at one of the existing universities. When fully implemented, this would be the first truly interstate institution of higher education in the U.S.

Even with expanded facilities, the school at the University of California (Davis) would graduate only 125 students annually, less than half the number required by that state. To help alleviate the shortage, the dean of agriculture at California State Polytechnic University (Pomona) suggested a "nontraditional" curriculum utilizing other schools offering training in animal science, together with extensive fieldwork supervised by practicing veterinarians.

Another approach had been the establishment of programs for veterinary technicians, ranging from correspondence courses to four years of university-level study, at some 30 institutions. The diversity of these programs had made it difficult for veterinarians to evaluate the credentials of the graduates, and professional licensing bodies were loath to recognize the value of such training. In late 1973, however, the American Veterinary Medical Association accredited the programs offered by the University of Nebraska (Curtis), where this type of training had been pioneered, and by Michigan State University.

New swine disease. Several outbreaks of a previously unreported swine vesicular disease (SVD) in Great Britain involved 132 herds, and 71,-000 pigs were destroyed at a cost of some $4 million. Seven new outbreaks occurred in October 1973, and because the virus was highly resistant outside the host, fear was expressed that importation of the disease into the U.S. was inevitable. There were no vaccines or drugs that were effective in treatment of the disease. SVD resembles foot-and-mouth disease but is more difficult to diagnose, and the virus can be spread in raw or improperly cooked garbage containing pork scraps. Likely sources of the SVD virus would be imported frozen pork or garbage that is discharged from intercontinental aircraft.

Diethylstilbestrol feeding. In January 1974 the year-old FDA ban on the use of diethylstilbestrol (DES) in beef cattle feeds was declared illegal, together with the later ban on subcutaneous implantation of the growth-promoting hormone in pellets. The ban had been instituted because of the suspected carcinogenic potential of DES, minute traces of which had been found in cattle after the drug had been improperly used. In no case, however, had cancer in humans been associated with the eating of meat containing such residues, and it was later decided that no real hazard to human health existed. In numerous studies made prior to the ban, DES used at approved levels had lowered the cost of beef production by several cents a pound.

—J. F. Smithcors

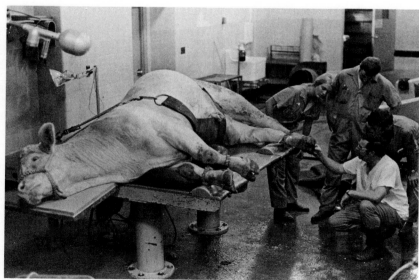

Charolais bull is examined at the Large Animal Clinic of the University of Georgia to determine the presence of interdigital fibroma. Such examinations are part of the university's "Outreach" program to aid cattlemen in the state.

Courtesy, College of Veterinary Medicine, University of Georgia

Zoology

Few fields offer the variety and depth of discovery available in zoology—or as many surprises. In an age of computers, applied technology, and a "war on cancer," who would expect the 1973 Nobel Prize for Medicine or Physiology to be awarded to three zoologists for their work in the study of animal behavior? The award was given to Konrad Lorenz, Nikolaas Tinbergen, and Karl von Frisch, not for the solution of a particular problem but for the creation of a new zoological field study—ethology, the biological study of behavior. (*See* Year in Review: HONORS.)

The three ethologists worked along classical lines of observation, insight, and analysis, but many of their contemporaries were carrying out their investigations in laboratories equipped with the most sophisticated devices. In many cases, the chief obstacle to computer applications was the absence of a large, reliable base of data from which the computer could calculate, and the compilation of such data was expected to occupy zoologists for many years to come. In this connection, millions of seemingly useless bits of information gathered by zoologists, physicians, and other scientists over the years were beginning to pay off.

Essentially, computer services are performed by translating a logical question or process into a complex arithmetic problem for the computer to solve. Computers are in no way capable of thought. If set up correctly, they can use probability statistics based on the information in their data banks in place of a deduction. Thus, by feeding into a computer measurements from forearm bones of a number of species of fossil and recent monkeys, chimpanzees, gibbons, apes, and other primates, investigators were able to decide how to classify a certain fossil primate for which only the forearm had been recovered.

Other workers were using the computer for the stuffy business of classification. Some classical taxonomists reacted negatively to computerized classification, identification, and nomenclature of specimens in their field, but many believed that computerization could make this most subjective of biological endeavors into a "hard" science. In the second edition of their work on numerical taxonomy, Peter Sneath and Robert Sokal stated: "Numerical taxonomy is a revolutionary approach to biological classification. . . . Instead of qualitatively appraising the resemblance of organisms on the basis of certain favored characters, a taxonomist using this new methodology will attempt to amass as many distinguishing characters as possible, giving equal weight to each. . . . The aim of the new system is to rid taxonomy of its traditionally subjective nature, so that any two scientists, given the same set of characters but working entirely independently, will always arrive at identical estimates of the resemblance between two organisms."

One large group of zoologists, the cellular and molecular biologists, were entirely dependent on

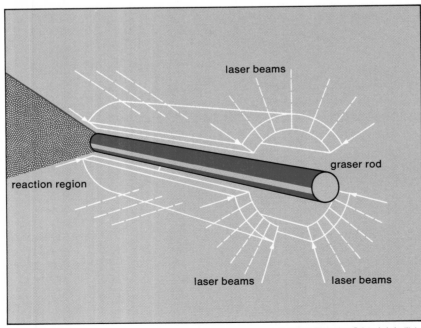

Gamma-ray laser, or graser, emits beam of gamma rays, which scientists believe might be able to produce three-dimensional photographs of molecules. Pulses from an optical laser act on a rod of fissionable material to produce intense neutron densities and thereby stimulate gamma-ray emission from nuclei in the rod.

laser beams

graser rod

reaction region

laser beams

laser beams

the most sophisticated kinds of instruments. The challenge in this field was to find a microscope or other instrument that would enable the investigator to observe molecules. Nobel prizes had been awarded to the scientists who determined the molecular structure of insulin, DNA, and gamma globulin by indirect methods, but the molecules described were never seen by the investigators. During the year a joint U.S.-Soviet effort was being discussed to produce three-dimensional photographs of molecules with a laserlike beam, and the implications for the study of cells, organelles and components of cells, genes, viruses, and structures of proteins could be incalculable. The instrument, called a graser, used gamma rays, which lie toward the short end of the electromagnetic wave spectrum. Their significance for zoology is that they are shorter than the spaces within molecules, thus enabling them to resolve distances as small as a fraction of an angstrom. Theory indicates that a beam of parallel, single-wavelength gamma rays could produce holographs of molecules.

Another joint U.S.-Soviet development was a process called Kirlian photography (after its Soviet inventors, Semyon D. Kirlian and his wife and co-worker, Valentina). The Kirlian process resembles the X-ray process and is simple and relatively inexpensive. It is claimed that Kirlian pictures capture on color film previously invisible emanations from the human body, plants, and animals. The emanations change in color and size depending on the physiological states of the subject. According to the *Wall Street Journal*, William A. Tiller, professor of materials science at Stanford University, believed that Kirlian photographs might make it possible to "monitor energy states" in living things. While he cautioned that scientists were still in the early stages of understanding the process, he believed "it will turn out to be a most efficacious device with great value in biology, medicine and agriculture."

Cellular investigations. "The casual reader will be forgiven for believing that many cell biologists nowadays spend a considerable amount of their time moving proteins around a sea of membrane lipid in a delightful molecular version of childhood pursuits." The writer of this statement caught the exuberance of the men and women carrying out investigations at submicroscopic levels. One would think that, as small as cells are, the work would have been completed long ago, but each study set off further investigations, and the time might never come when everything about the cell is known. For instance, several studies of the surface of normal and transformed cells with the scanning electron microscope were reported, and some of the most beautiful and informative pic-

tures published to date appeared in the May 1974 issue of the *Journal of Cell Biology.* The investigators followed changes in the shape of the cell's surface in a culture of Chinese hamster ovary cells during different phases of the cell cycle, and found that, after mitosis, the cells begin to spread out, and there is a high incidence of "blebs" or bubble-like protuberances of the cytoplasm, intermingled with numerous fingerlike microvilli.

At an international meeting on the structure and function of the microtubules and microfilaments found in cells, a Canadian scientist presented evidence that actin-like microfilaments interact with microtubules and speculated that these two components act like "muscle" and "bone" in causing the cell or its parts to move. Another investigator described the structure of some microtubules found in the cilia (fine, hairlike extensions from the surface of some cells) as being built up from a thirteen-fold, left-handed helix. It was also demonstrated that filaments from a particular marine worm are protein α helices supercoiled into several higher-order helical forms. Only when one remembers that these filaments and tubules are measured in angstroms does the elegance of such descriptions become apparent (one angstrom equals one hundred-millionth of a centimeter).

Cellular investigators at the University College of North Wales developed a chimera using very early mouse and rat embryos. (A chimera is a living structure or organism in which the tissues are of different or varied genetic origin, sometimes as the result of grafting.) These investigators suggested that rat and mouse embryos, in which the patterns of differentiation are similar but proceed on a different time scale, might develop successfully when aggregated together. The chimera developed through an apparently normal blastocyst stage, an early embryonic stage in which the embryo is a hollow ball composed of millions of cells. The remarkably flexible developmental potential of the rodent egg, thus demonstrated, offered an opportunity to follow the interaction of genotypically different cell populations during further stages. It could also elucidate aspects of grafts and transplants, as well as fetal-maternal interactions.

General zoology. Since Léon Dufour published his observations in 1833 on the anatomy of the gut in the Homoptera, it has been known that most members of this group of plant-sucking bugs have a more or less complicated association between the beginning of the midgut and its termination, in conjunction with the Malpighian tubules, in the first segment of the hindgut. Dufour named this arrangement the "filter chamber" in which, he supposed, the watery contents of the plant juices were filtered off directly to the hindgut, while the nutrient material was carried on to the midgut. A physiological study made by W. W. T. Cheung and A. T. Marshall on a variety of cicadas in Hong Kong and Australia strongly suggested that Dufour was right; the chamber really is a "filter chamber" and the ionic concentrations needed to sustain its operation are the result of secretory activity elsewhere in the Malpighian tubules and the regions of the gut wall.

The wealth of chemical studies on ants was illustrated by M. S. Blum (University of Georgia), who noted that at least 47 hydrocarbons had so far been identified in the Dufour's gland of ants. Many of the secretions of ants contain compounds unrecorded elsewhere in the animal kingdom. The Dufour gland secretion is commonly used as part of a defense alarm and recognition system, and investigators at the Free University of Brussels reported that the ant *Myrmica rubra* uses the secretion of this gland to attract nest mates to a source of disturbance. A mandibular gland secretion that incites the ants to sting any intruder may also be produced and, if the danger persists, additional nest mates are recruited with the aid of a trail pheromone (chemical signal) produced by the poison gland.

Although numerous theories of smelling and tasting had been published, there was little experimental evidence on the energy-transfer mechanisms involved in the interactions of chemical messengers with receptors of sensory neurones. In an investigation of one such mechanism, various chemicals that inhibit feeding in the American cockroach, *Periplaneta americana*, and the smaller European elm bark beetle, *Scolytus multistriatus*, were shown to react with a particular large molecule that is physically accessible to these chemicals through pores and tubules in the walls of chemoreceptor nerve membranes of the insect's antennae. In response to specific stimuli, such nerve membranes can bring about alterations of inorganic ion flows. The study was of particular interest because scientists believe learning and memory in higher animals probably involve this fundamental mechanism.

Because fisheries biologists have been unable to see and count fish in the sea directly, they have used indirect methods involving tedious sampling and elaborate mathematical models. During the year new acoustic echo techniques were reported that could not only find fish but could quantify the received echoes. Electric counting, integrating, and computing systems were being developed in

David Hughes from Bruce Coleman Inc.

Leatherback turtles provide anatomical evidence that they undergo a countercurrent heat exchange. This was the first evidence of such a heat-retention mechanism to be found in a reptile, and suggested that warm-bloodedness first appeared in reptiles.

many countries for assessing fish populations by high-speed surveys. W. Acker of the University of Washington overcame the problem of counting salmon in surface waters by submerging echometers that transmitted upward. Through an ingenious use of the Doppler effect, D. Holliday of TRACOR Inc., San Diego, Calif., estimated the mean size of fish by using the number of body lengths traveled per second at average cruising and burst speeds.

The life cycles of various species of fish include long-range migrations for feeding and spawning purposes. Long stretches of these migratory routes are often crossed without food intake, so that efficient use of the store of internal energy is essential to survival. D. Weihs reported that swimming speed affects performance and efficiency of motion appreciably, suggesting that theoretical predictions of optimal cruising speeds can be of assistance in tracking and locating fish during migrations. In terms of energy of movement as a proportion of a given total energy store, Weihs indicated that it is theoretically advantageous if the speed of fish through the water is maintained at about one body length per second.

Recent studies of four-legged mammals have shown that the energetic cost of their locomotion varies in a regular manner with size, and the oxygen consumption of a running mammal can be predicted quite closely if its speed and weight are known. A group at Harvard University reported that below 18 km per hr the hopping kangaroo expends more energy than a four-legged animal running at a comparable speed, but above 18 km per hr it appears to use less. Kangaroos are known to travel at speeds of up to 40 km per hr for several kilometers at a time, and the energy saving involved in hopping may give some insight into why hopping herbivores in Australia survived while their quadrupedal counterparts became extinct 20,000 to 30,000 years ago.

One group of investigators was trying to determine whether dinosaurs were warm-blooded or cold-blooded. The evidence was inconclusive so far, but another group at Harvard presented anatomical evidence for a countercurrent heat exchange in the leatherback turtle, the first evidence for this kind of heat-retention mechanism in a reptile. Phylogenetically, reptiles stand between amphibians, which are cold-blooded, and birds and mammals, which are warm-blooded, suggesting the possibility that warm-bloodedness first appeared as a reptilian feature.

Meanwhile, another group was trying to determine what the blood pressure of sauropod dinosaurs would have been. These 80-ton animals are among the largest quadrupeds known. Some were as long as 30 m, and so maintaining an adequate blood supply to the brain of such an animal would require enormous blood pressures and physiological adaptation. The adult giraffe stands five to six meters, has a vertical profile similar to dinosaurs, and when standing quietly has a systolic blood pressure of 125 mm measured below the jaw. Calculations for sauropods such as Apatosaurus (Brontosaurus) and Brachiosaurus suggest that their comparable average arterial pressures would have been 216 mm and 568 mm, respectively. Obviously, the exact measurements would never be known.

It is one thing for an animal to be able to take up a consistent orientation and find its way home by using a simple compass (sun, star, and magnetic compasses are found throughout the animal kingdom). It is, however, a feat of quite a different order for the animal to be able to home consistently from wherever it is released. Pigeons are able to do this over long distances, but nobody is quite sure how. W. T. Keeton reported that the birds set off in a direction that consistently deviates from home and correct themselves later on in the flight. What is more, this initial deviation is often characteristic of a given release site. If these "release-site biases" are due to some local distortion of geophysical cues that the birds are using, then finding the nature of this distortion might explain what navigational system the birds are using. Keeton believed that his birds were not making an error but were probably taking readings correctly from a distorted "map."

For a century naturalists have argued over whether the giant panda is a kind of bear or a member of the raccoon family. During the year Vincent M. Sarich (University of California at Berkeley) reported findings indicating that giant pandas are definitely bears. His opinion, published in Nature, was based on an analysis of certain characteristics of albumin and transferrin from the blood of Chi-Chi, the giant panda at the London Zoo that died in 1972. It is thought that serum proteins mutate at a known rate, thereby enabling researchers to trace the evolution of various species. As measured in Berkeley, panda blood is no further removed from bear blood than dog blood is from that of foxes, but it is quite far from the blood of raccoons.

Behavior. The synchronous flashing behavior of Asian fireflies has interested Western naturalists for many years. From his observations of Pteroptyx, a Melanesian firefly, and species in other firefly genera, J. E. Lloyd of the University of Florida was able to construct a model for the basic mating protocol. Males and females are attracted toward flashes with the appropriate species-specific

Tsetse fly is a carrier of trypanosomes, parasites that infect cattle and thereby deny cattle-growing to much of Africa. Tsetse-fly tissue cultures were being used to grow trypanosomes for laboratory studies.

rhythms and congregate in swarm trees that already harbor numbers of their species. Males land and flash synchronously with nearby males, thus enhancing the brightness of the tree and the likelihood of attracting females.

Differences in the modulation pattern and frequency of signals made by different species are common among other fireflies and in the acoustic signals of insects, and probably are involved in maintaining separation of the species. Ronald Hoy and Robert Paul in New York showed that females from two species of Australian cricket respond to males of their own species because they recognize the rhythm of the male's song. However, the two species can be induced to hybridize, and the hybrid males had a new chirp that primarily attracted hybrid females.

Blue jays were included among the tool-using animals by T. B. Jones and A. C. Kamil of the University of Massachusetts. They reported that blue jays would rip paper from the newspaper below their cage, crumple it, and use it to get food that could not be reached by beak alone. They would also use other objects, such as plastic bag ties and straws. The investigators suggested that this is a behavior one jay learns from another. Another group of investigators reported that one species of prairie dog (*Cynomys ludovicianus*), which lives in long narrow burrows (about 12 cm wide and 30 m long), constructs them so that the air in them is completely changed every ten hours or less.

Ecology. New avenues for pest control were being opened up as a result of the deeper understanding of insect behavior gained from studies of pheromones. F. J. Ritter, A. M. van Osten, and C. J. Persoons of The Netherlands reported the discovery in fungus-infected wood of terpenoid substances that may be identical with the trail pheromone of the termite *Reticulitermes santonensis*. Related compounds from sandalwood oil are also very effective in inducing trail following. Such substances could be used to divert members of a colony to an area where they could be contaminated with slow-acting insecticide. J. M. Cherrett and his colleagues at the University College of North Wales and Rothamsted Experimental Station in Great Britain found that the trail pheromone of the leaf-cutting ant *Atta texana* markedly increases the effectiveness of baits used for controlling several species of leaf cutters.

California scientists reported that they had found a way to control pesticide-resistant mosquitoes without drenching the countryside with harmful chemicals. Synthetic hormones are sprayed on wet areas where mosquitoes breed, disrupting the normal growth pattern of mosquito larvae. Some grow so rapidly that they split their shells and die, while others grow slowly and die before maturity. The growth hormones are environmentally safe and dissipate quickly. Unfortunately, the procedure was much more expensive than conventional methods. (*See* Feature Article: GROWTH REGULATORS: A NEW APPROACH TO INSECT CONTROL.)

Twenty workers from several countries met to discuss progress in controlling cattle trypanosomiasis, a disease that denies large areas of Africa to cattle. Immunization of calves against trypanosomes would be an ideal solution but, for reasons that were imperfectly understood, long-lasting immunity cannot be guaranteed. Much work was needed, but this would be possible only if large quantities of fly-adapted trypanosomes were available; in this connection, development of tsetse fly tissue culture techniques for growing the infective invertebrate forms of trypanosomes could become of great significance in the future. Another possible method of biological control consisted of blocking the anticoagulant of tsetse fly saliva, which would cause blood to coagulate in the fly's food canal and so prevent further feeding.

The most advanced biological control method was based on the release of sterilized male tsetse flies, which then mate with the wild female flies. This technique had been used with varying success on several other types of flies and other insects. One of the problems that often arose when radiation was used to sterilize male insects was that the sterilized males could not compete with untreated native males because radiation impaired their vigor. In an effort to overcome this difficulty, a group at the University of Notre Dame irradiated mosquitoes (*Aedes aegypti*) in air and in nitrogen and tested their competitive mating ability against untreated males. They found that males irradiated with a dose of 7,000–10,000 roentgens in the presence of nitrogen were as competitive as normal males, whereas air-irradiated males were not.

Evolution. Paleontologists found the world's oldest clam, *Fordilla troyensis,* in a drawer in the Smithsonian National Museum of Natural History; it had been stored there for 75 years after being mistaken for a crustacean. This discovery contributed to a reclassification of the species that added 70 million years to the history of clams and dated their origin at about 565 million years ago.

The current view among zoologists was that, among invertebrate groups, amphioxus is closest to the vertebrates. According to *Nature* magazine, R. P. S. Jefferies reported finding fossil evidence for his view that this position belongs to a hitherto obscure group, previously held to be primitive echinoderms, which he renamed the calcichordates. He considered one group of calcichordates, the mitrates, to be the common ancestors of amphioxus on the one hand and of the urochordates and vertebrates on the other. Unfortunately, since the calcichordates are very different from any group living today, there was no way of knowing whether the soft parts had been accurately identified. This being the case, it was difficult to see how hypotheses concerning the evolutionary relationships of such animals could be tested.

Paleocene fossil primates are known only from the European and North American continents. Fourteen genera and numerous species from North America have been described, and four genera (*Plesiadapis, Chiromyoides, Berruvius, Saxomella*) are recorded from Europe. Only one primate genus, *Plesiadapis,* was previously known from the Paleocene of both continents, but during the year a new species of *Chiromyoides* was described, extending the range of this genus to North America and bringing to seven the number of mammalian genera known from the Paleocene of both Europe and North America. This indicated that the primate faunas of the two continents were probably very similar during late Paleocene time, as was to be expected from evidence of continental drift suggesting that Europe and North America were connected via Greenland throughout the early Eocene.

A great deal of work continued on the evolution of primates and man. A Canadian investigator concluded from his studies of tooth chipping in specimens of *Australopithecus robustus* and *A. africanus* that they did not differ substantially in the amount of grit in their diets. This was of importance because earlier studies by others suggested that the robust forms were vegetarians while the gracile forms were omnivores. Another investigator, using a technique called regression analysis and correlating information available for modern primates and man, suggested that the gestation period for australopithecines was somewhere between 250 and 300 days.

—John G. Lepp

Fordilla troyensis, *the world's oldest clam, was first thought to be a crustacean. Studies of its shell revealed its true nature and helped add 70 million years to the history of clams.*

Courtesy, Smithsonian Institution

Can Venice Be Saved?
by Irving R. Levine

Menaced by high water, by air pollution, even by its beloved pigeons, the Queen of the Adriatic faces enemies far more insidious than the Normans and Byzantines of its fabled past.

Oh Venice! Venice! when thy marble walls
Are level with the waters, there shall be
A cry of nations o'er thy sunken halls,
A loud lament along the sweeping sea!
　　　　　　　　　　　　　—Lord Byron

It was more than a century and a half ago that Byron wrote his "Ode on Venice," a poetic cry of anxiety over the survival of that unique and historic city. There was cause for concern before Byron's era and there is even greater cause now. But, in view of Venice's precarious position astride water and its vulnerability to assaults by both nature and man, it is perhaps less surprising that Venice is in danger of extinction than that it has endured until now.

From the time of its foundation about A.D. 570, Venice has been imperiled by the sea, and its earliest inhabitants took constant precautions to safeguard their community. With the passage of time, not only has the sea grown more menacing but also the city and its art treasures confront perils stemming from subsidence of the city's level, industrialization, air pollution, and alterations of the delicate lagoon in which Venice lives.

Major roads
Other roads
Railroads

City limits
Greenbelts
Built-up areas

Dangers are not new to Venice. The city was founded in response to danger. The degeneration of the Roman Empire and repeated foreign invasions drove refugees from the northeastern part of the Italian peninsula to seek safety on clusters of islands lying two and a half miles from shore in a sheltered lagoon of the Adriatic Sea. The primitive community developed into a wealthy center of empire with distant possessions and widely respected military, political, and commercial power. Ornate palaces were built on foundations consisting of millions of closely grouped larch-tree piles driven down about 25 ft into the clay and mud terrain of the islands to solid ground.

The city was positioned safely out of the sea's grasp; the buildings were generally placed one to four feet above mean sea level, which meant that the highest tides experienced at that time would not reach them. At present, though, almost all of Venice rests no more than 28 in. above mean sea level, and its margin of safety is steadily diminishing.

The city occupies 118 islands of varying sizes, interlaced by a labyrinth of 180 canals, spanned by nearly 400 bridges. It is a fantasy site without peer that has charmed visitors through the ages, moving poets to rhapsodize, as did Edwin Markham who described "storied Venice where the night repeats / The heaven of stars down all her rippling streets." The city boasts elegant and opulent architecture of some eight centuries past and preserves in its public squares, government buildings, churches, and private palaces a great array of art masterpieces.

The primitive founders of the original island community, as well as the sophisticated doges who later ruled Venice, appreciated that the city's endurance depended on the protection of the lagoon in which it lay. One of the chief officials was, and is, the magistrate of the waters, with responsibility for the lagoon's welfare.

The lagoon

To understand how Venice exists and why it is in jeopardy, it is necessary to look to the lagoon that shelters the city. Roughly half-moon in shape, the lagoon has an irregularly curved edge formed by the shoreline of Italy's peninsula. The half-moon's straight edge is a narrow strip of low-lying sandbars that separate the waters and islands of the lagoon from the Adriatic. Through three inlets—openings in the sandbars—the Adriatic's tides enter the lagoon and withdraw twice in each 24-hour period.

From sandbars to shoreline, the average distance is about 8 mi, and the length of the lagoon, from tip to tip of the half-moon, is approximately 35 mi. Shallow water permanently covers about three-quarters of this 212-sq-mi basin. The rest is covered by the islands of Venice and by *barene*—tiny, often-submerged mud flats that form a marshy area fringing the mainland.

The *barene* are thought to play a vital role in stabilizing the lagoon. In the past they have acted as a kind of sponge, soaking up, in times of dangerously high tides, excessive seawater that might other-

IRVING R. LEVINE is a news commentator for NBC News. He served as chief correspondent in Rome during the 1960s and is the author of Main Street, Italy (1963).

346

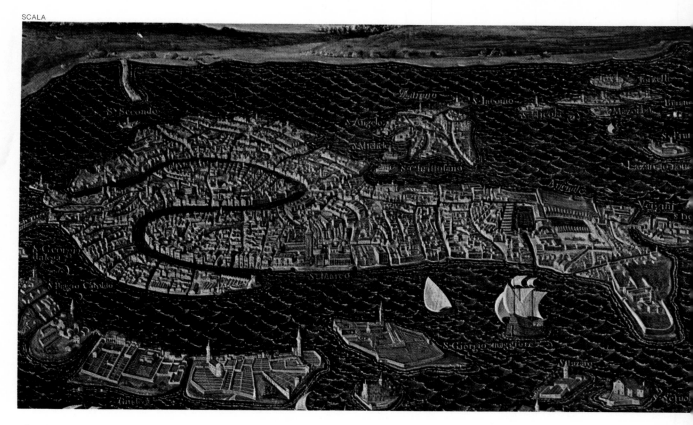

wise have swamped the islands on which Venice resides. But in 1919 expanses of *barene*, which had helped absorb the daily tidal influxes, were transformed into solid land to provide a base for the construction of factories in a newly founded industrial and port area of the mainland bordering the lagoon near the towns of Marghera and Mestre.

Marghera has expanded to become one of Italy's most important industrial areas. Today some 250 industrial plants stand on 5,000 ac of what used to be *barene*. In addition to these acres, comprising Marghera's so-called industrial zones one and two, 10,000 more acres of *barene* have been marked off to be filled in to form industrial zone three. Almost 30,000 ac of *barene* elsewhere in the lagoon have been diked for a commercial fishing area and denied to the searching tides in times of overflow.

Man has altered the lagoon in other ways as well. To enable large freighters and tankers to reach the industrial zones of Marghera, engineers dredged channels through two of the sandbar openings and across the lagoon to the mainland. These channels also provide an open avenue for the tidal advances of the sea into the lagoon, particularly in times of storm, and have rendered the tidal currents through the lagoon swifter and more erosive.

Man's tampering with the lagoon is not exclusively of recent origin. Venice itself was an artificial intrusion. Also, early Venetians diverted the courses of rivers, causing them to empty outside the boundaries of the lagoon in order to forestall silting.

Modern Venice can still be recognized from the fresco (above) depicting the city in the 16th century. The Grand Canal winds through the western portion of the city, while in the foreground the Doges' Palace and St. Mark's Basilica stand guard over the port. The map on the opposite page shows the modern city's metropolitan area.

347

In studying the lagoon, scientists have difficulty in determining which elements in the lagoon's hydrology are the consequences of man's intervention and which were intended by nature. What is known is that the lagoon is a complex and fascinating mechanism. In fact, the lagoon, it has been learned, is not one single, great container of water. It is in a sense partitioned by nature and comprises three distinct basins. Although boats can pass from one basin to the other by following natural channels, the waters do not have the same liberty of movement. The sea that rides the tides through one of the entrances never goes out through another one. The tidal waters that enter through the three entrances spread out in their respective basins until they reach nature's demarcation lines, undetectable to the eye, and briefly merge, only to separate when the time has come for their withdrawal from the estuary, and leave by separate courses. The seawater that enters the lagoon by one sandbar opening leaves, by and large, through the same opening.

Acqua alta

Alterations inflicted on the complex lagoon are thought to be responsible, at least in part, for the phenomenon known as *acqua alta*, or high water. In the normal pattern, the waters of the lagoon ebb and flow with the rhythmic regularity of the tides. For a period of roughly six hours the waves of the rising tide enter the glassy stretches of the lagoon through the three inlets. The waters of the canals rise with the incoming tide, but well within the bounds of safety. Then, for another six hours, as the tide retreats, the lagoon returns to the sea the waters that have visited it. However, when high winds coincide with excesses of tide, low barometric pressure, and other elements, the peacefully repetitive process is broken and water spills out of the canals, flooding sections of the city.

In recent years a constant vigil has been kept. A computer is fed data on meteorological and marine conditions, and warning sirens are sounded whenever it indicates a danger of *acqua alta*. The early warning is followed by additional blasts if the water actually starts to intrude on the city. Although the first floors of most Venetian buildings have long ago been converted into lobbies or storage spaces, the alerts cause merchants to move goods to safety, and ground-level (or rather, sea-level) residents to hasten upstairs.

The seawater rises slowly and almost silently over the stone banks of the *rii*, the canals, and bubbles forth through drains in courtyards and squares. Implacably, the water converts each narrow pedestrian street, or *calle*, into a placid stream and steals into buildings through sills and grates and cracks. As the water begins to rise, booted workmen construct elevated sidewalks with sawhorses and long boards, which are prudently stacked against walls during the autumn and winter months when *acqua alta* is most prevalent. On occasion the water rises above the sawhorses and the planks float away, along with gondolas torn from their moorings and errant terrace furniture.

"As the water begins to rise, booted workmen construct elevated sidewalks with sawhorses and long boards . . . . Water everywhere reflects the sky and the intricate stonework of Venetian architecture." (Above) Pedestrians cross St. Mark's Square on temporary elevated sidewalks set up in front of the clock tower. (Opposite page) High tide at flood level in St. Mark's Square.

348

Gianni Tortoli from Photo Researchers

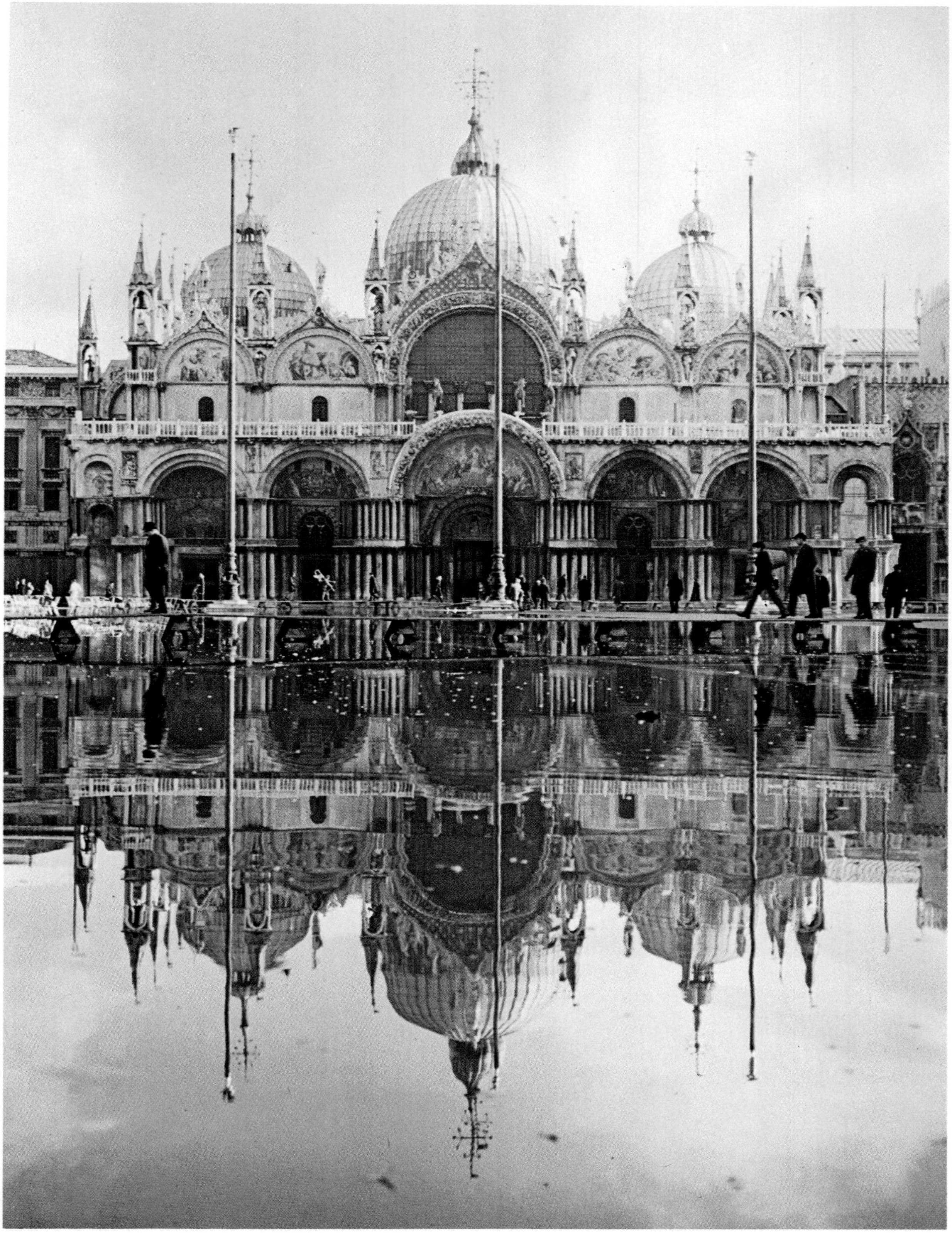

If one ignores its destructive capacity, the high water is a sight of surrealistic beauty, the ultimate blending of Venice with the sea when the borderline between the sea's domain and that claimed by man is erased. Water everywhere reflects the sky and the intricate stonework of Venetian architecture. Venetians make their way to work in hip boots, youngsters float about on rafts of rubber or wood, and tourists assume playful poses of distress to be photographed by friends. St. Mark's Square becomes a lake bounded by stately columns and the Byzantine mosaics and domes of the Basilica of St. Mark. And then, when the tide turns, the waters slip away, leaving behind small pools, drowned rats here and there, a saline residue on walls and floors to be scrubbed off, and—everywhere—imperceptible damage.

The great flood of Nov. 4, 1966

Venice has experienced *acqua alta* from the time of the earliest available records. A document from the year 885 recalls that "the water invaded the whole city, penetrating the houses and churches." The next written reference is dated March 9, 1102, when the city was the victim of "a great inundation." In 1240 "the water was the height of a man in the streets," and in 1430 "the water overflowed and ruined very many bridges and canal-sides." An account of 1600 tells of high water "causing a terrible fear through the whole of Venice."

350

A rare occurrence in times past, high water has become a profound worry in recent years because of its increasing frequency and often swollen dimensions. *Acqua alta* now regularly occurs more than 100 times a year and sometimes 200 times. On most of these occasions the water reaches a depth of only an inch or so in some neighborhoods, but on Nov. 4, 1966, a large part of Venice was under 6½ ft of water for some 20 hours. Never, as far as is known, had the city been inundated for so long a period or to so great a level.

In early November of 1966, heavy rainstorms driven by winds that reached 60 mph swept much of Italy. The wind was the sirocco, which blows northward from the Libyan deserts. The Adriatic ends in a cul-de-sac near Venice, and high winds from the south force the water to rise at that end of the sea. The very low barometric pressure of the storm system provided an additional invitation for the waters to rise; high atmospheric pressure would have acted as a damper on the surging Adriatic.

Another phenomenon came into play. Inequalities in barometric pressure over the length of the Adriatic are known to activate a phenomenon called a seiche. Seiches, which occur in large, landlocked, or almost landlocked, bodies of water such as lakes and seas, are oscillations, varying in duration from a few minutes to many hours. The Adriatic is shaped like a giant bathtub and a seiche sets the water in it sloshing, much as water sloshes in a tub that is gently rocked. The characteristic rhythm for a seiche in the Adriatic is such that the highest phase of the oscillation at any one point recurs every 22 hours, and this may continue for days or even weeks.

On November 4 the forces of nature combined against Venice. The storm hit the sea in step with the seiche's sloshing, a high wind piled additional water onto the lagoon, and low atmospheric pressure offered no resistance to the rising waters. It was in these circumstances that the scheduled high tide crept over Venice before midnight. The tide should have retreated by 5 A.M., but that hour came and passed and the water remained. As noon approached, the waves of a new high tide began to swell the waters still further and Venice lay submerged.

An agonizing wait began for that day's second and last ebb tide, which should have occurred about 6 P.M. Residents of Venice recall how each passing hour intensified fears that laws of nature, considered immutable in the past, had somehow lapsed and Venice would remain drowned, an Atlantis of the Adriatic. When the time for the onset of the ebb tide arrived, the waters, instead of withdrawing, started to creep higher. A Venetian later wrote: "Toward evening everyone realized that an age-old balance had been broken, that the city and the lagoon had lost a link, who knows which, from their delicate interlocking mechanism. Venice was drowning in the lagoon and was waiting in anguish."

At about 9 P.M. the sirocco suddenly abated, releasing pressure at the sandbar openings to the sea, and the waters began to swirl out of the walkways and edifices of Venice as the lagoon tardily voided its sur-

Photos, Angelo Samperi from Sygma

"The condition of each church, palace, statue, painting, fresco, and mosaic was assessed by experts." (Top, center) Restoring the Sansovino Loggetta at the base of St. Mark's bell tower. (Bottom) Laboratory for the restoration of paintings in the former Oratorio of S. Gregorio.

351

plus back into the Adriatic. Behind, the waters left sodden destruction, foundations weakened from the weight and force of the tidal flood, and a frightening realization of how precariously close Venice is to being claimed by the sea. It is said that during the single 24-hour period of Venice's agony, the city aged 100 years.

Actions to save Venice

The events of Nov. 4, 1966, stimulated international concern for the future of Venice. In a number of countries committees were formed, often by private citizens and educational institutions, to provide money and experts to help the city, which has been described as part of the heritage of every country. The United Nations Educational, Scientific, and Cultural Organization (UNESCO) undertook to bring the problems of Venice to the attention of governments and to solicit aid.

With funds donated for the purpose, UNESCO, as a first step, compiled a detailed inventory of the monuments and art works of Venice. The condition of each church, palace, statue, painting, fresco, and mosaic was assessed by experts and an appraisal made of the work needed to restore those in disrepair. This inventory has provided a practical guide for determining priorities in the enormous task of restoration. The survey established that more than half of the artistic monuments in marble and other stone were seriously affected by some form of deterioration and that this figure was increasing by 6% annually.

In April 1973, after several years of intermittent debate, the Italian Parliament passed a law authorizing the government to raise $500 million to be spent over the next five years on the rehabilitation of Venice. The legislation had been delayed by factional disputes; local and national government departments and groups within political parties quarreled over who was to administer the funds and over priorities for their use. Even after passage of the law, there remained the problem of obtaining the actual funds, mainly through a loan from foreign financial institutions and from Italian taxes.

At such time as the funds become available, scientists and administrators, who believe Venice can be saved, stand ready to expand work already under way and to implement drawing-board projects. The problems awaiting solution are many.

Venice is sinking

For years scientists have been aware that the levels of the world's seas, including the Adriatic, are rising because glaciers and the polar ice caps are in one of their periodic melting phases. Quite independently of this, Venice itself is slowly settling. Precise calculations of these phenomena have been undertaken by the Laboratory for the Study of the Dynamics of Large Masses, established after the 1966 onslaught in the ornate Palazzo Papadopoli on the Grand Canal under the direction of Roberto Frassetto. The primary task of the laboratory

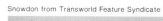

is to explore the interactions of water masses, air masses, and plastic masses (such as the earth) in order to discover the habits of the lagoon and find ways to predict future assaults by the sea. The objective is to find ways to prevent the further subsidence of Venice and its surrender to the sea. The International Business Machines Corp. is cooperating in the project, and an IBM 360-44 computer has been programmed to handle weather reports from distant stations, tidal figures, information on Adriatic seiches, and other data that may eventually provide accurate long-range forecasts of *acqua alta.*

According to Frassetto, the northern Adriatic is rising a tenth of a centimeter (one inch equals 2.54 cm) a year, while Venice itself is sinking at an annual rate of half a centimeter. Intensive surveying of about 1,000 reference points in and near Venice has shown that some areas have dropped more than five centimeters (two inches) in the past 15 years. While the city has been settling very slightly throughout its lifetime, its subsidence has accelerated alarmingly in recent years. Although some other cities, including London, appear to be sinking at a faster rate, Venice has less distance to sink to its doom. Equally alarming is Frassetto's finding that sections of the sandbars separating the Venetian lagoon from the Adriatic are sinking at twice the rate of Venice's islands—or by a full centimeter annually.

One generally accepted explanation for the accelerated subsidence of Venice and its protective sandbars in recent years is that the factories of Marghera have pumped profligate amounts of water from artesian wells deep in the ground. Groundwater levels have dropped 35 to 50 ft throughout the region in the past half century. Frassetto explains that the resultant loss of natural underground pressure enables the weight of the ground and of buildings to cause "a rapid and often irreversible compaction of alluvial sediment forming the subsoil." In short, the removal of the subsurface water causes the ground to sink permanently. There is an additional cause of the subsidence, not only of Venice but of other areas down the Adriatic coast and inland as well. Enormous quantities of methane gas have been pumped out of nearby mainland areas since World War II, creating similar conditions.

A number of solutions have been proposed and some actually put into practice to arrest the sinking of Venice.

● Industry has begun a five-year program to cap the many artesian wells that pump up underground water for the industrial processes of the chemical and aluminum factories and the oil-cracking plants of Marghera. Drilling of new wells has been forbidden by law. Water will be supplied instead by two aqueducts, which will carry water to the Marghera complex from the Sile River, a distance of approximately 12 mi.

● Consideration is being given to suggestions that prodigious amounts of water be pumped back into the underground strata from which it was removed, thereby, it is said, buoying up the depressed land. Some scientists, however, consider this unfeasible.

● Pumping of methane gas in certain areas has been curtailed.

"Venice itself is slowly settling . . . .
The objective is to find ways
to prevent the further subsidence
of Venice and its surrender
to the sea." (Opposite page)
Pedestals on the colonnade of this
building have sunk below the normal
water level, and the rate
of subsidence is believed
to be increasing. (Above) Seen here
in an unusual view across the water,
the church of S. Giorgio Maggiore
and its auxiliary buildings stand
on a separate small island.
The group, one of the architectural
treasures of Venice, is the work
of the great Renaissance architect
Andrea Palladio.

Air pollution

Marghera's industry also has contributed to air pollution, the cause of extensive damage to the art trove of Venice. Industry is not the only, or even the main, culprit in this respect, however. Although the industrial processes emit an estimated 15,000 tons of concentrated sulfuric acid into the atmosphere each year, the prevailing wind directs most of it away from Venice. The homes and hotels of Venice, using low-grade fuel oil for heating, have spewed the greatest quantities of pollutants into the atmosphere.

Another source of pollutants are the *vaporetti* and *motoscafi*, the water-buses that, along with small motorboats, have superseded the traditional graceful gondola (only 400 of which now remain in use) as the city's main mode of transportation. Air pollution is not the only danger from these big boats. As they ply the Grand Canal, which winds through Venice like the letter "S" drawn in mirror image by a child, the waves lap erosively against the stones of buildings, weakening the wooden piles underneath as well.

Combined with humidity, often near 100% in Venice, the pollutants produce a condition in the widely used marble called stone cancer, which defaces statues and causes the surfaces of buildings to crumble and peel away. There are a depressingly large number of cases. For instance, the ornately worked 15th-century doorway of the church of SS. Giovanni e Paolo, a notable example of Gothic-Venetian architecture begun in 1246, is now completely covered by a dark patina. This surface crumbles like salt when touched. On other buildings, stone cancer causes "layers" of stone columns to shed as if made of bark.

A gloomy prognosis is offered by Francesco Valcanover, superintendent of galleries and works of art: "We are still helpless as far as the problems of marble and other stones are concerned. The combination of age, humidity, and lastly the grave insult of air pollution is seriously afflicting our sculptures. It can be said, therefore, that unless science can help us with new methods of restoration, in 15 or 20 years at most, Venice will lose the greater part of her artistic face forever."

Nevertheless, work on the problem is going forward.

● A recent regulation forbids using heavy oil for heating. Homes and hotels are required to convert to other fuels within a specified time. Mains are being laid to carry cleaner-burning natural gas to the city. The conversion is expected to eliminate 50% of the air pollution.

● Smoke-control equipment has been made mandatory for Marghera's factories.

● Venice is experimenting with a process for halting the decay on statues and other small stone works. The statues are wrapped and soaked in deionized water, placed in a vacuum chamber to absorb silicones, and finally baked at controlled temperatures in a furnace. Later, the missing parts of the sculptures can be replaced by skilled artisans. Larger stone works are less radically treated by blasting away the pollution-induced grime, where it is not too profound, by means of minute glass beads shot from an air-gun-type instrument.

357

"Marghera's industry ... has contributed to air pollution, the cause of extensive damage to the art trove of Venice .... Combined with humidity, often near 100% in Venice, the pollutants produce a condition ... called stone cancer." (Opposite page, top) View of the industrial center of Marghera, a borough of Venice built on landfill in the 1920s. (Bottom) The Doges' Palace through winter fog.

"Rain combines with pigeon droppings to form an acid that eats into stone, causing it to flake. . . . Stone cancer . . . defaces statues and causes the surfaces of buildings to crumble and peel away." (Above) Pigeons flock to their midday feeding, at city expense, in St. Mark's Square. (Below) Building facade shows the combined ravages of air pollution and water erosion.

● A protective coating of hard wax is expected to arrest further deterioration of metal art works. Among the city's most famous treasures are four magnificent horses of hand-beaten copper on which traces of the original gilding remain. These Greek works, which date from about the time of Alexander the Great more than 20 centuries ago, perch above the main portal of St. Mark's Basilica. They exhibit the ravages of pollution; the metal is worn thin and is punctured in spots. One by one, they are being transferred to the basilica's museum workshop where a wax coating will be applied to insulate them from sulfur dioxide in the air.

● Extensive restoration work has been completed or is in progress on numerous buildings, including a dozen churches rich in art, and on more than 1,500 frescoes and paintings by such masters as Titian, Tintoretto, and Veronese. In many cases interested committees in other countries have "adopted" specific churches or series of paintings and are paying for their rehabilitation. When the paintings are movable, the work is carried out in art restoration laboratories established in the church of S. Gregorio and in the Scuola di S. Rocco. In some cases the paintings were damaged by air pollution or flooding, but old age and neglect are responsible in many instances.

Pigeon pollution

This form of pollution may sound ludicrous, but it has, nevertheless, inflicted serious damage on stone works. The pigeon population of Venice is variously estimated at from 50,000 to 100,000, compared with about 125,000 human inhabitants of the island community. Pigeons begin to swarm expectantly over the vastness of St. Mark's Square shortly before the piazza clock tower chimes the hour of the midday feeding at city expense. Bags of grain are strewn on the pavement, to the delight of the pigeons and camera-armed tourists alike. The pi-

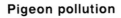

geons also thrive on the tourist practice of purchasing paper cones of corn from vendors in order to pose for photographs with pigeons alighting on fingertips, shoulders, and head.

These delights notwithstanding, the pigeons wreak destruction. Rain combines with pigeon droppings to form an acid that eats into stone, causing it to flake. Pigeon droppings have also been held responsible for the destruction of frescoes, and the sheer accumulated weight of droppings is said to have hastened the collapse of the roof on an old church.

Some Italians regard pigeons as good omens, and since the birds have long been considered one of Venice's traditional charms, it is not easy to arrange for their destruction. However, a recent law provides city officials with the authority, if they choose to exercise it, to reduce indiscriminate feeding, to treat grain with contraceptive drugs, or to thin out the exploding pigeon population by some form of painless execution.

"Among the city's most famous treasures are four magnificent horses of hand-beaten copper ... which ... perch above the main portal of St. Mark's Basilica. They exhibit the ravages of pollution."

359

The floods

The pigeon problem, of course, is minor in contrast to Venice's major preoccupation—the encroachment of the sea. At the present rate, in a few generations or less all Venice could be constantly under water and quite uninhabitable. Some experts give Venice fewer than 50 years to survive unless the trend is reversed.

Not only do the recurrent floods erode the stones of structures and exert tremendous pressures that weaken foundations; they also leave corrosive salts behind. After an onslaught of high water, moisture and salt remain inside the walls of buildings and creep upward. Eventually, this efflorescence of salts causes the walls to decay. In some cases the lower portions of walls have been lined with metal sheeting to stop the saltwater seepage. But far more significant measures are being employed to cope with the critical problem of *acqua alta:*

● The construction of the planned third industrial zone of Marghera has been interrupted for an indefinite period by government order, thus preventing further sacrifice of *barene* into which the lagoon's waters can expand. Unfortunately, the directive came only after landfill had obliterated one-third of the approximately 10,000 ac of *barene* intended for the zone.

● Further dredging of the shipping channels through the lagoon has been stopped, but not until the deepest channel had been dredged to a depth of 50 ft for 17 mi, almost its full distance.

● The *murazzi* or "great walls," which stretch along 3¼ mi on the seaward side of the southernmost sandbars, have been repaired and reinforced after having been breached in several places by the storms of Nov. 4, 1966. The first stone of the original *murazzi* was laid on April 24, 1744, and the stonework barriers against the Adriatic were completed 39 years later. Consisting of great blocks of marble cemented in place and standing more than twice a man's height to form a huge breakwater, the walls were an acknowledgment by the Venetians of more than two centuries ago of their increasing vulnerability. Extension of the barrier and additions to its height may be necessary to buttress Venice's thin sand frontier with the sea.

● Work is going forward on what is expected to be the most important project to prevent the Adriatic from swamping Venice in times of *acqua alta.* The plan is to fashion movable barriers across the openings in the sandbars that screen the lagoon from the Adriatic.

One set of plans calls for construction of inflatable barriers across the sandbar inlets at a cost of about $80 million. It is a method that has been successfully employed on a much smaller scale to protect low-lying regions of The Netherlands. In periods of normal tide, great rubber bags would lie inert and dormant on the seabed. When distant sensors in the Adriatic alerted computers to imminent tidal incursions capable of flooding Venice, the great bags, looking much like giant sausages, would be inflated by compressors on the sea floor, blocking the sea's access to the lagoon. When the danger passed, the bags would be deflated.

If there is to be time to close the cumbersome barriers during the periods of slack water between tides, it is essential that the forecasting techniques (of the IBM computer in the Palazzo Papadopoli) be perfected to give six to eight hours' warning of sea surges.

There are several variations to this barrier concept, including great metal gates that would be swung into place when *acqua alta* threatened. In all its forms, however, the scheme is not without shortcomings. Some estimates indicate that, assuming the frequency of *acqua alta* remains constant, ships would be denied passage to and from the lagoon for about 200 hours a year, disrupting shipping schedules and imposing a substantial financial burden on Marghera's industry. Other researchers, however, believe that with careful organization the closings could be reduced to 40 hours a year, less than the interruptions to shipping caused by fog.

Furthermore, the tides have always provided Venice with a natural sewage-disposal system. Much of Venice's sewage is discharged directly into the canals. If the flushing action of the tides were inhibited for any length of time it would create at least an odoriferous annoyance and possibly a health hazard. However, this objection could be overcome by the planned construction of a sewer system for the city and installation of pollution-control equipment to eliminate the industrial effluents that Marghera has been discharging into the lagoon.

The outlook

There is no guarantee that Venice can be saved, however great the effort. The interactions of the "large masses" that are under study in the Palazzo Papadopoli may prove too large, complex, and unpredictable for the resources of man to master. Yet, unless the steps that have been taken are continued, the evidence indicates that Venice, long known as the Queen of the Adriatic, may perish.

Man came belatedly to the task of trying to salvage his most beautiful artifact, as art critic and scholar Bernard Berenson described the city. The progress made so far has been considerable, but the work remaining is daunting and the treadmill process of deterioration leaves much to be done. Nevertheless, there was, in 1974, reason to expect that Byron's words of anguish would not prove a prophecy:

Oh! agony—that centuries should reap
No mellower harvest! Thirteen hundred years
Of wealth and glory turn'd to dust and tears.

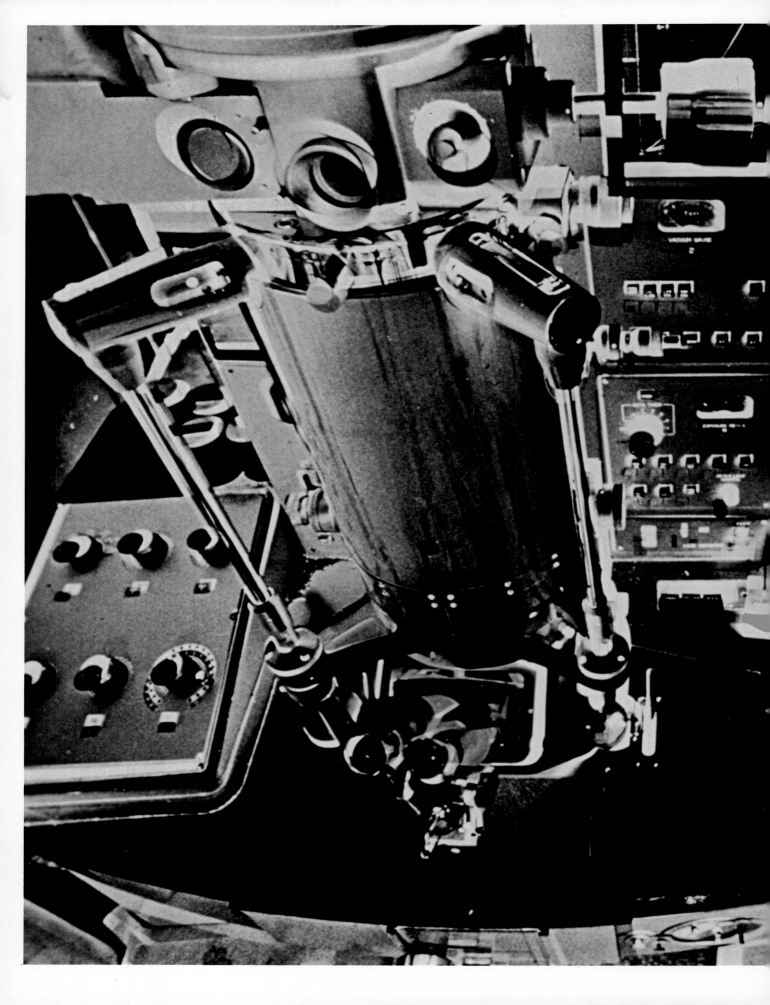

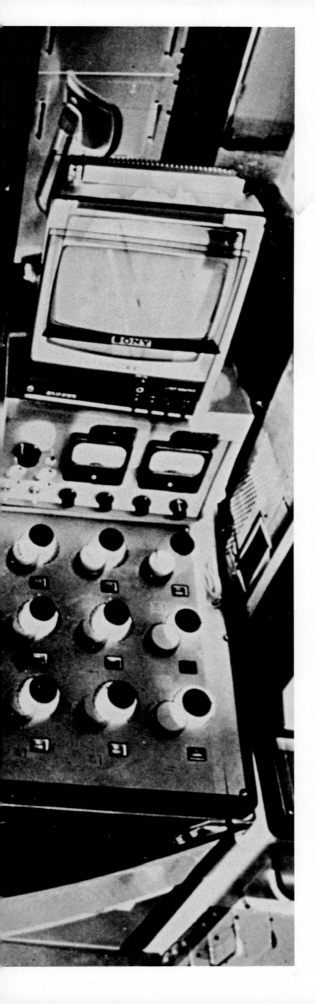

Probing the Invisible World

by Humberto Fernandez-Moran

Electron microscopes of increasing power and sophistication are enabling scientists to understand how life functions at the previously invisible level of atoms and molecules.

Coiled inside every human body is a complicated network of nerve cells, which, laid end to end, would stretch to the moon and back more than three times. Individually these cells cannot be seen with the naked eye, yet within this invisible world lie the keys to understanding every life process and many diseases. In the time that it has taken men to perfect the rocketry and communications systems that have revealed the surface of the moon, scientists from many countries have evolved a powerful tool—the electron microscope—that has opened this new world of "inner space" for the first time.

Very simply, the electron microscope is a hollow vacuum tube fitted with magnetic or, less commonly, electrostatic lenses that are used to focus beams of electrons on a specimen in such a way that minute groups of molecules become visible to the human eye. This was the basic design with which Max Knoll and Ernst Ruska began experimenting in the 1930s. By the 1970s the sophisticated grandchildren of this first model were large enough to fill entire stories of buildings.

Transmission electron microscope can achieve greater magnification than any other instrument —up to 500,000 times—but is restricted to flat, thin specimens. Electrons projected from a filament are accelerated by an anode (a positively charged plate) and then concentrated on the specimen by an electromagnetic condenser lens. The specimen is mounted inside the objective lens. As the electrons pass through the specimen, the magnetic field of the lens deflects them and a magnified but distorted image is formed. One or more intermediate lenses correct the distortion, and finally the projector lens focuses the image and projects it onto a screen.

H. FERNANDEZ-MORAN is the A. N. Pritzker professor of biophysics at the University of Chicago.

Light microscopes

In microscopy, light is the conventional medium of operation, and to understand how a microscope works one should consider a few basic principles of magnification and optics. For example, the print on this page can be magnified by bringing the page gradually closer to one's eyes. However, eventually a point is reached where it becomes nearly impossible for the viewer to focus clearly on the print. Moving the page any closer only makes it increasingly difficult to distinguish the letters and words. This ability to form distinguishable images of objects is called resolving power. In order to make the print appear still larger and yet still readable—to provide the necessary resolving power—one must use a lens external to the eye, such as a magnifying glass. Resolving power is dependent on focal length (distance from the focus to the surface of the lens), which varies with the length of the illuminating electromagnetic wave.

The conventional light microscope works on this principle, comprising a system of lenses and concentrated light that illuminates the object and magnifies it up to 2,000 times. Of course it is not possible to see an entire page at one time in this way, but in the section under the microscope one can begin to distinguish between points on the object that are about 2,000–8,000 atoms apart. Microscopists use a measurement, the angstrom (Å; $1 \text{ Å} = 10^{-8}$ cm), to note this, the resolving power of a light microscope being described as 2000–8000 Å. At this level, given the relatively long wavelength of light, one can see bacteria.

In 1933 Ruska succeeded in building an instrument of greater resolving power than the light microscope. By replacing light waves with the shorter wavelengths of electron beams he developed the earliest electron microscope that made it possible for the first time to see viruses.

Electron optics

In the modern electron microscope, an external high-voltage electron beam is concentrated by a condenser lens and directed toward the specimen. As the electrons pass through the specimen, they are scattered and thereby create an electronic image. A series of focusing lenses converts this electronic image into a visual one when projected onto a viewing screen. By varying the voltage of the electron beam and incorporating a number of recent instrumental improvements, it has become possible to magnify an object more than one million times while achieving a resolving power of three to five angstroms. This is equivalent to seeing structural details barely three to five atoms apart. At this level, one can begin to see minute groups of molecules, nucleic acids, and enzymes.

History has proven that what man can see he can eventually hope to duplicate, and it is as a means for adding important information to models of living systems at the molecular level and possibly providing the tools by which such systems can be manipulated that the electron

364

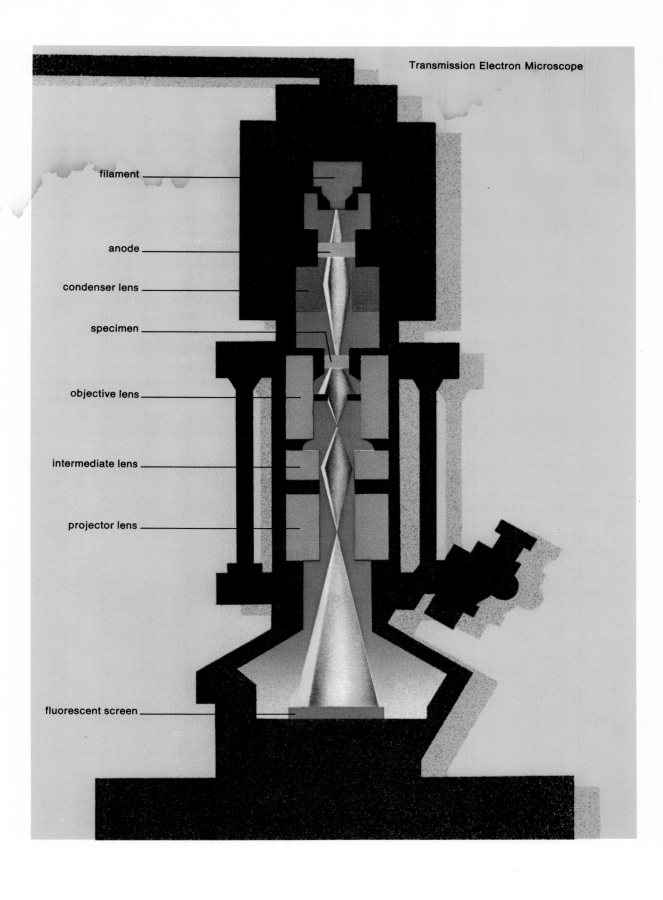

Transmission Electron Microscope

filament

anode

condenser lens

specimen

objective lens

intermediate lens

projector lens

fluorescent screen

microscope may well make its major contribution to medical science. However, this is by no means the limit of its usefulness. It can also be used in reverse, like a giant inverted telescope, to demagnify. In this respect, it might give rise to miniature computers with superconducting memories that rival the human brain in complexity and efficiency, or make possible the recording and storing of all of the volumes contained in the Library of Congress on a single page not much bigger than this one.

The problems that face the electron microscopist can be roughly divided into two areas: technical and interpretative. The former include the development of controlled laboratory facilities and adequate preparation techniques and instrumentation so that the maximum amount of information can be obtained from each specimen examined. The latter involve understanding and explaining what is seen.

In specially designed laboratories such as those at the University of Chicago's research institutes, scientists have created a virtually ideal environment in which to operate the extremely delicate microscopes. Special lighting, wiring, air conditioning and filters, and power supply systems are used routinely to ensure that the microscopes will be absolutely free from any random electromagnetic interference, instrument vibration, or external impurities that could disrupt the image. External impurities are a particular problem because even the smallest of them can cause difficulties in observation and interpretation, and thus the laboratories must be kept as dirt-free as possible.

Edge of a diamond knife (below) has a radius of only one-tenth of one-millionth of an inch. A diamond knife is used with an ultramicrotome (right) to slice specimens thin enough to be suitable for use in an electron microscope. These photographs and the subsequent micrographs were taken at the University of Chicago, courtesy of H. Fernandez-Moran.

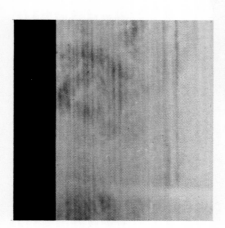

Preparing the specimen

Hand in hand with the special precautions described above go improved methods of specimen preparation and the development of new instruments to increase the microscope's resolving power. Beginning in the early 1950s much work centered on these problems.

In order for the electrons to pass through the specimen and thus create the image, the specimen must be extremely thin. The early microscopists used glass knives to cut slices only several hundred angstroms thick from small frozen tissue blocks of fixed or fresh material. While this was an important development in specimen preparation, some drawbacks remained. The glass cutting edge wore out relatively fast, leading to deterioration in the precision with which the thin sections could be cut. Oil and dirt from the rotor of the microtome, the instrument used to feed the specimen into the knife, also caused impurities within the sections.

In the late 1940s research scientist Humberto Fernandez-Moran realized that the sectioning problem might be solved by a device that would provide a precise, circular motion in a smoothly recurring flow system. This was the beginning of the development of the diamond knife and associated special ultramicrotome, which eventually became the most sophisticated and precise cutting tools available to the electron microscopist. Because the diamond is chemically inert and the hardest cutting substance known to man, its precise edge, which measures only a few thousandths of a micron (1 micron = 0.000039 in.),

Six-sided head, stalk, and tail structure of the bacterial virus T4 phage can be seen in bright-field (left) and dark-field (right) electron micrographs. Ultrathin carbon films were used to achieve high resolution while magnifying the specimen 400,000 times.

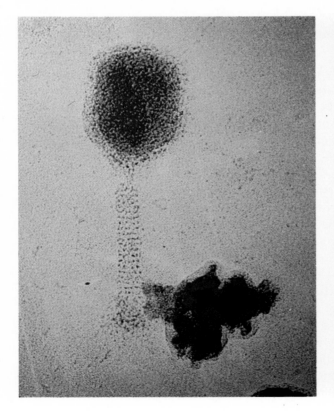
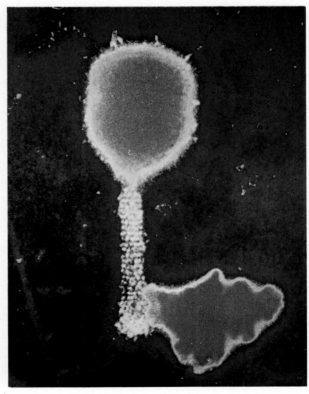

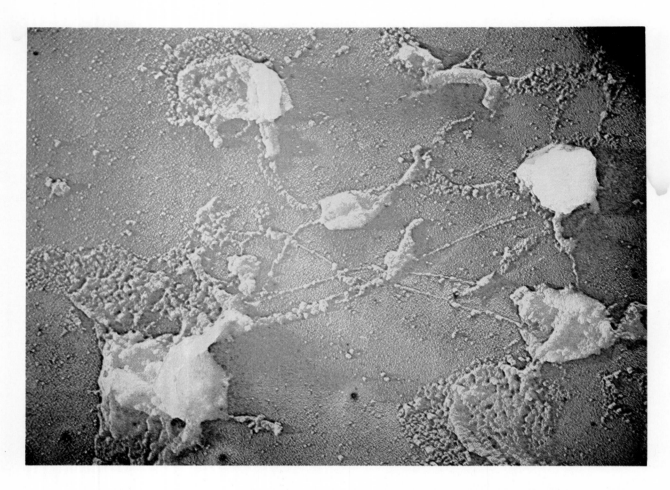

Platinum-shadowed portion of a chloroplast disrupted by osmosis and DNA strands associated with membranes are magnified 30,000 times in this high-resolution electron micrograph. Shadowing techniques enable researchers to see three-dimensional specimens more clearly. Opposite page, magnification of 1 million times achieved a micrograph of circular DNA.

does not readily wear out or contribute impurities to the specimens. In addition, the ultramicrotome can be operated at liquid helium temperatures (1.8°–4.2° above absolute zero, – 459.7° F), thereby keeping rearrangement of molecules within the heat-sensitive biological specimens to a minimum. By using this system researchers can routinely produce sections on the order of 10–100 Å thick and literally cut starch molecules into sugar. This type of molecular chemistry or "editing" may one day enable scientists to eliminate a host of genetic diseases in man. In order to protect the specimens and hold them after slicing, a special chamber was devised in which the thin sections could be sealed between layers of ultrathin graphite films. Electrons can pass freely through the thin graphite.

Low-temperature systems
Improvement of the electron microscope has played a major role in this research. As the electrons pass through the specimen during examination, they cause radiation damage. At the same time, the heat generated by the electric current as it passes through the circuits causes variations in the focus of the lenses and, consequently, difficulty in obtaining a true specimen image on the viewing screen.

368

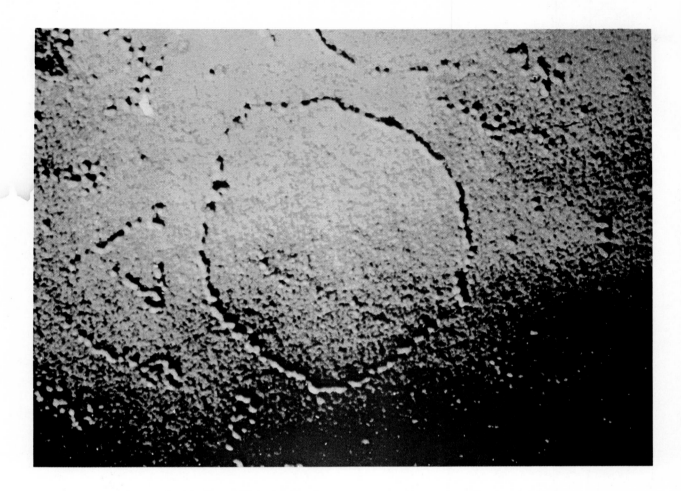

The introduction of special superconducting lenses in the 1960s was a major step forward. Designed to be operated at extremely low temperatures, they made it possible to put the current into a state of superconductivity. This means that the current can continue to flow even after the power has been turned off. Because it meets no electrical resistance under these circumstances and thus generates no heat, the current can sustain a very high, constant magnetic field, which stabilizes the image. These controlled magnetic fields focus the electron beam and thus act like the glass lenses of conventional light microscopes. Under these conditions, a selected specimen area can be viewed for extremely long periods with a significant decrease in specimen radiation damage.

Superconducting lenses become even more important for microscopes operating at high voltages. At such voltages, ranging from about 200,000 to 3,000,000 v, the wavelength of the electron becomes very short, allowing researchers to obtain resolutions on the order of one angstrom. Such resolutions allow the electron microscopist to approach his ultimate goal, a direct view of the structure of molecules. Therefore, in working with these instruments scientists must increase their efforts to reduce heat-generated interference and specimen

369

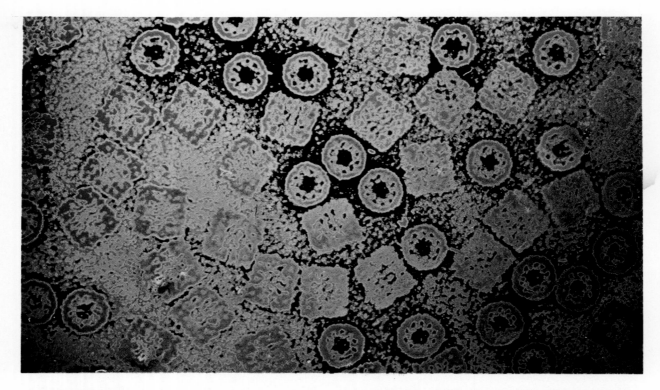

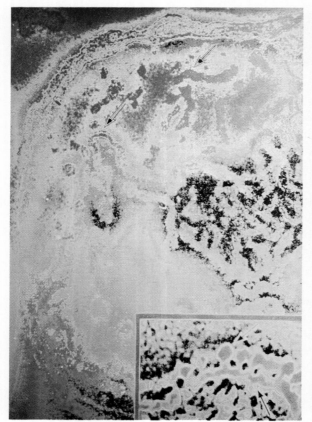

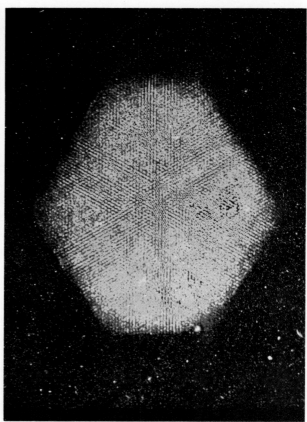

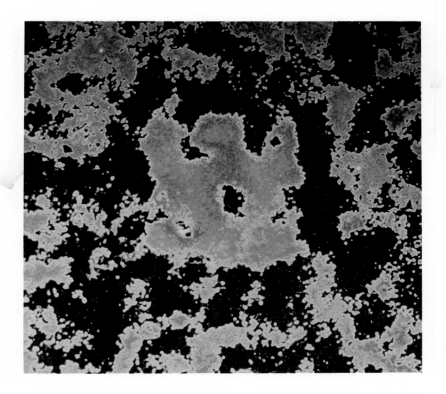

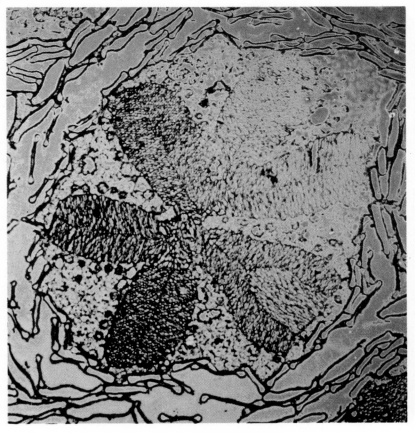

The structure of cells can be seen
in great detail through electron
microscopes. Hemocyanin molecules
from Helix pomatia (opposite page,
top) display circular and rectangular
forms when magnified 240,000 times.
Elementary particles of mitochondria
(opposite page, bottom left)
help produce the enzymes that lead
to the production of ATP, which
provides energy for the cell.
The large micrograph is magnified
68,000 times; the arrows
on it indicate the area of the insert,
which has been magnified 320,000
times. The periodicity of crystals
associated with mitochondria
(opposite page, bottom right)
is revealed by high-voltage (200 kv)
electron diffraction studies.
By achieving a magnification
of 2.5 million, researchers
clarified the structural details
of protein (top). Transverse section
of the compound eye of the South
American moth Erebus odora
is magnified 7,350 times (bottom).

radiation damage which generally increases as the electron beam becomes more intense. This can be done in several ways and with various combinations of methods.

Much effort has been concentrated on the low-temperature approach. The first systematic work was done with a closed-cycle superfluid-helium refrigeration system fully integrated with the high-voltage microscope to provide stable power in a nearly vibration-free environment for biological studies. This machine, really a giant refrigerator designed by Samuel C. Collins, operates at 1.8°–4.2° above absolute zero. At these very low temperatures, the lens system becomes superconducting, as described above. The extremely low temperatures also effectively freeze activity within the specimen, and, because the electron beam is more penetrating, thick specimens can be viewed in their entirety, without sectioning. For the biologist and the biomedical researcher this means that living systems, such as bacteria and viruses, can be observed under conditions more closely resembling those of their native state.

Observing DNA

By using the modified high-voltage electron microscope with superconducting lenses at liquid-helium temperatures, researchers can begin to make quantitative predictions about the behavior of biological matter. The most exciting area of this predictive research is in the "master tape" of life, the minute deoxyribonucleic acid (DNA) molecules that contain all of the programming information necessary to create all of the different systems in the human body. To give some idea of how small these master molecules of life are, each of the more than 50 trillion cells in a person's body, except sperm or ova, contains 46 chromosomes. Each of these chromosomes, in turn, carries more than 1,000 genes, and each gene contains vast numbers of double-helix strands of DNA each inscribed with the genetic code in atomic dimensions.

The recent application of dark-field carbon shadowing techniques has been particularly important in allowing scientists to see the geometry and contours of unstained DNA and related specimen molecules. Essentially, carbon shadowing involves evaporating a very thin (50 Å) carbon film onto the specimen in a controlled vacuum prior to electron microscope examination. The carbon atoms cling to the contours of the specimen and provide a reliable, finely detailed picture of its morphology in the form of measurable shadows, even after radiation damage and heat have caused the specimen itself to deteriorate during examination. By using a precisely controlled electron beam or ion, researchers can also pare away successive layers of carbon, perhaps opening the way for controlled molecular dissection of specimens during examination in the electron microscope.

Based in part on the types of information derived from X-ray and electron microscopy studies of genes and nucleic acids, Har Gobind Khorana of the Massachusetts Institute of Technology in 1973 succeeded in making a large synthetic gene. By radioactively tagging the gene with new homogeneous silicone emulsions developed at the University of Chicago, researchers can carry out a full series of ultra-high-resolution studies of its base-pairs (the complementary bases on each of the twin strands of a DNA molecule). This information can then serve as a reference file for comparative investigations of "living" genes and whole chromosomes. The end result of this process would be the first glimpse by biomedical researchers of the complex mechanisms by which human development is controlled.

Once this information can be used to determine a given genetic code, the ultrasharp cutting edge of the diamond knife may give scientists the ability to modify that code. This is particularly important because many scientists believe that the most effective control of genetic diseases, and maybe of cancer itself, can be accomplished by developing a means for interfering with the disease at its core, the complex process of transforming and translating genetic information that takes place in the invisible world of nucleic acids.

Color electron micrographs

Seeing into the atomic world is only part of the electron microscopist's problem; he must also interpret what he sees. Therefore, developing a means for recording as much information as possible from each specimen has been a major area of research. Until very recently, high-resolution black-and-white electron micrographs were the primary means

Gray scale/color scale comparison (opposite page) shows the advantage of using the new color translation process to record high-resolution electron micrographs. Shades of gray not distinguishable by the human eye become clearly distinct color differences, permitting scientists to see previously unobserved details of specimens. Black-and-white and color treatments of the same micrograph of ring patterns in single-crystal gold film (below) demonstrate the difference.

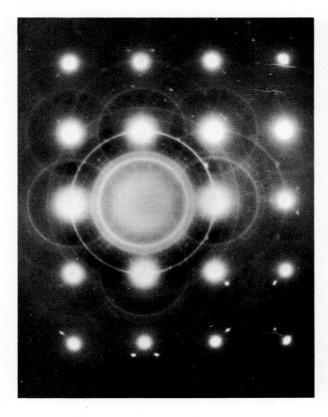
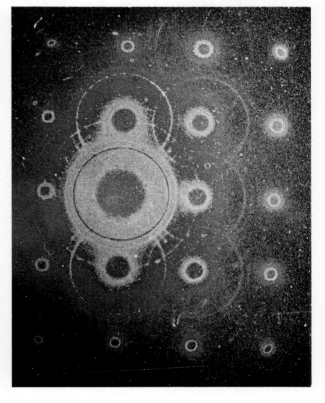

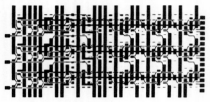

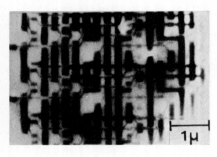

*Atomically thin, layered
ultrastructure of TaS$_2$ (top),
a superconducting compound
that may become useful in providing
memories for new kinds of computers,
is revealed in a 200-kv
high-resolution micrograph.
A cryotron array (middle)
is reduced electron optically
(bottom) with a demagnification
ratio of 1:10,000.*

of doing this. These were recorded on silver-halide-based films, which have a resolution limit of approximately 100 Å. In addition, since the human eye cannot readily distinguish between subtle shades of gray, much of the extremely fine structural detail that might have been available to researchers was lost.

A major breakthrough in this field occurred in the early 1960s. For example, by using new types of silver-free, high-resolution organic recording media and a color translation process based on the late-19th-century experiments of Gabriel Lippmann, Charles Hough at the University of Chicago developed a means for transforming the various shades of gray into clearly distinct color differences. As a result, microscopists could begin to see details of the molecular world with a clarity that no one had previously achieved.

By 1974 the best results had been achieved by coating clean glass plates with 1000-Å-thick sheets of evaporated aluminum. The aluminum surface was coated with a special emulsion containing the dichromate radical Cr_2O_7 and then placed in contact with the positive electron image plate, resulting in a product something like a photographic contact print. The aluminum plate was then exposed to high-intensity mercury lamps, developed, and dried, after which the colors became visible.

In another method for achieving color, based on the crosslinking of silicone, the electron micrograph is recorded directly on new, essentially grainless, electron-sensitive, organic films of silicone-based polymers similar to those first used in 1969–70 by Koichi Kanaya in Japan. The film emulsions used in the process record something like a topographical relief map. Blue areas indicate the thickest specimen regions, red areas show the thinnest, and the shades in between correspond to intermediate thickness gradations. Thus, the differences in thickness of biological specimens can be estimated directly from the measured variations in color and photographic density. By using a precisely controlled microbeam to irradiate the specimen through a full series of color recordings, researchers can obtain progressive visual records of specimen radiation damage, an approach that may prove valuable for those working in radiation research.

Since the electron microscope provides only a two-dimensional picture of a three-dimensional specimen, these color-recording processes are particularly important. It is through contrast, through shadows, that the third dimension can be calculated to complete a picture of a specimen. In high-resolution black-and-white electron micrographs the different shades of gray merge, but in the color recording processes each shade becomes a distinct hue that can be more precisely seen and measured.

Role of holography
By combining the new color micrographs with improved laser technology, researchers can also reconstruct three-dimensional color holograms of a wide variety of specimens recorded under special con-

374

ditions. Holography is a method of recording images on film without a lens, pioneered by Dennis Gabor in the late 1940s. A coherent illuminating beam (one with energy waves traveling in the same direction, at the same frequency, and in step with the stimulating radiation) is divided so that one part illuminates the specimen and a second part, the reference beam, goes directly to the film. The result is an interference pattern across the film, each part of which contains all the information needed to recreate the complete image in three dimensions when it is viewed using another coherent illuminating source.

Gabor's idea was to record holographic images of specimens with the electron beam and then reconstruct them at greater magnifications with visible light. Though the perfection of the laser has made this possible to some extent, not until scientists have succeeded in producing an electron beam comparable in coherence to the laser beam will they be able to view interatomic distances within specimens and see structures in three dimensions routinely. Such an achievement is considered likely by the late 1970s.

In 1974 holograms were proving useful in explaining and duplicating the most sophisticated ultraminiaturized information storage and retrieval system known to man, the memory portion of the human brain. Human memory banks appear to be highly repetitive, all of the data being stored in every portion of the brain just as all of the information needed to reconstruct a hologram is contained in every part of

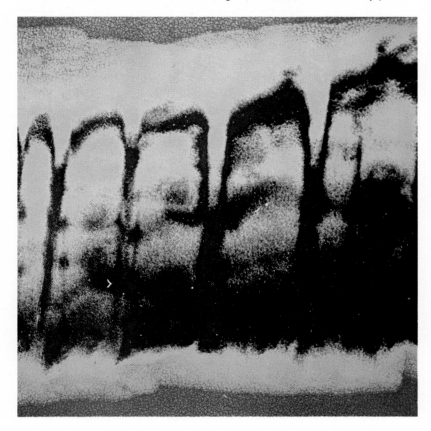

Pyroxene returned from the moon by the Apollo 11 astronauts demonstrates a regular band structure when magnified 300,000 times in an electron microscope. When combined with geophysical techniques and X-ray crystallography, this information may shed new light on the mysteries of the origin of the moon.

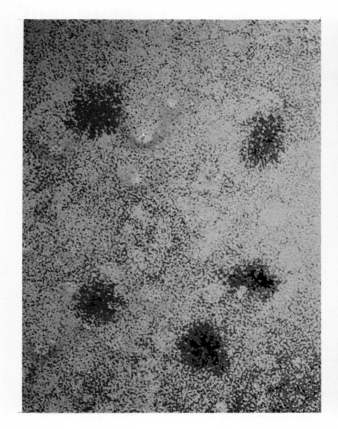

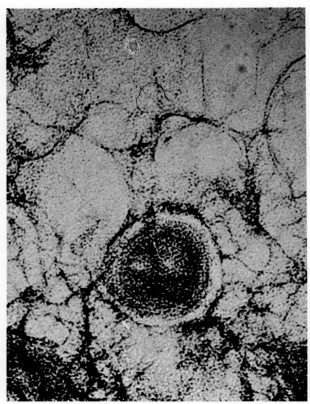

*Particles of the murine sarcoma
leukemia virus (above), which causes
cancer in mice, and the ultrastructure
of reverse transcriptase (right),
a nucleic enzyme
that might play a role
in causing viral cancers,
can be studied in detail
in high-resolution micrographs.
By this means electron microscopy
contributes potentially valuable
data to cancer research.*

the recorded interference pattern. The brain's retrieving mechanism might act something like reverse holography in that some type of "illuminating beam" or current is matched to the stored bit of information to produce a reference or memory.

In the near future scientists may be able to demonstrate this theoretical memory mechanism experimentally with visible light holography. An enormous quantity of microfilmed or electron-optically reduced data might be recorded on the new silver-free emulsions and passed in front of an illuminating beam that shines through a hologram of the item being sought. When the hologram matches the data on file, the reference beam flashes. If it could be shown that the brain uses a similar system for summoning thoughts, scientists would begin to understand the phenomenon of perception and could begin to experiment with the cells or cell clusters responsible for storing sensory information.

Integrated circuits

At several universities, researchers have already been able to use electron-optical techniques to develop printed circuits of micron and submicron sizes. The use of similar ultraminiaturized circuits for transmitting functional data recorded within the cellular and subcellular domains of living systems is within the scope of current technology. It is now possible to conceive of integrated circuits that begin

376

to approach the dimensions of macromolecular assemblies and can be incorporated into key junctional sites of living nerve membranes or be placed on red blood cells without disturbing the cells. Such submicroscopic prosthetic sensors, with biosynthetically produced protein coats, could form integral parts of the human nervous system and serve unique functions. For example, they could harness the natural electricity of the body to produce a direct operational link between the nervous system and man-made information-processing systems of similar complexity. Such a link might serve unique diagnostic functions and perhaps even help physicians to restore paralytic nerve damage at the macromolecular level.

Applied as a routine tool for diagnosis in hospitals, the electron microscope might become an integral part of the technology for performing and analyzing submicroscopic or macromolecular biopsies. It might eventually become possible to perform submicroscopic biopsies at key sites within a person's central nervous system. The damage that such a procedure would cause need be no more than the natural rate of nerve-cell decay taking place in the average adult. By using a system of vacuum locks a diagnostician could feed the minute tissue section directly into the electron microscope for nearly perturbation-free examination. A modification of this idea is being used in diagnosing certain liver ailments.

The future
The electron microscope will make its most important contributions in the next decade in the fields of biology and medicine. It will do this by allowing scientists to see how normal cells are transformed into cancerous ones and by identifying the genes responsible for hereditary diseases, thereby providing a better understanding of the basic units of life on which man depends.

Of course, the electron microscopist does not work in a vacuum; he correlates his results and the interpretations of what he sees with a host of data derived from biochemistry, physics, X-ray crystallography, and other branches of the physical and biological sciences. As a result, he can begin not only to see but also to interpret how life functions at the invisible level—the world of atoms and molecules.

The Tropical Forests: Endangered Ecosystems?
by Peter J. Fogg

A far cry from the teeming jungles of Tarzan films, tropical forests are a vast potential resource facing the danger of thoughtless exploitation.

For most people who have never seen a tropical forest, the words conjure up an image of steaming, impenetrable jungle, knee-deep in slime and undecayed vegetation and replete with hanging vines, biting insects, fierce animals, venomous snakes, and perpetual rain. To some extent, these elements exist in many tropical forests, but to find all of them fully developed in any single acre would be the exception rather than the rule. The image is more the product of fiction, the film industry, and travel writers than a reflection of reality.

Tropical forest is the natural vegetation that covers large areas of land lying mostly between the Tropics of Cancer and Capricorn. It occurs in three major zones: tropical America, with the greatest development in the Amazon basin; tropical Africa, with the greatest development in the Congo basin; and the Indo-Malayan areas between the Indian and Pacific oceans. Part of it is tropical rain forest, regarded by botanists as the most richly developed flora in the world. However, much of the area is actually covered by other, less luxuriant types of vegetation.

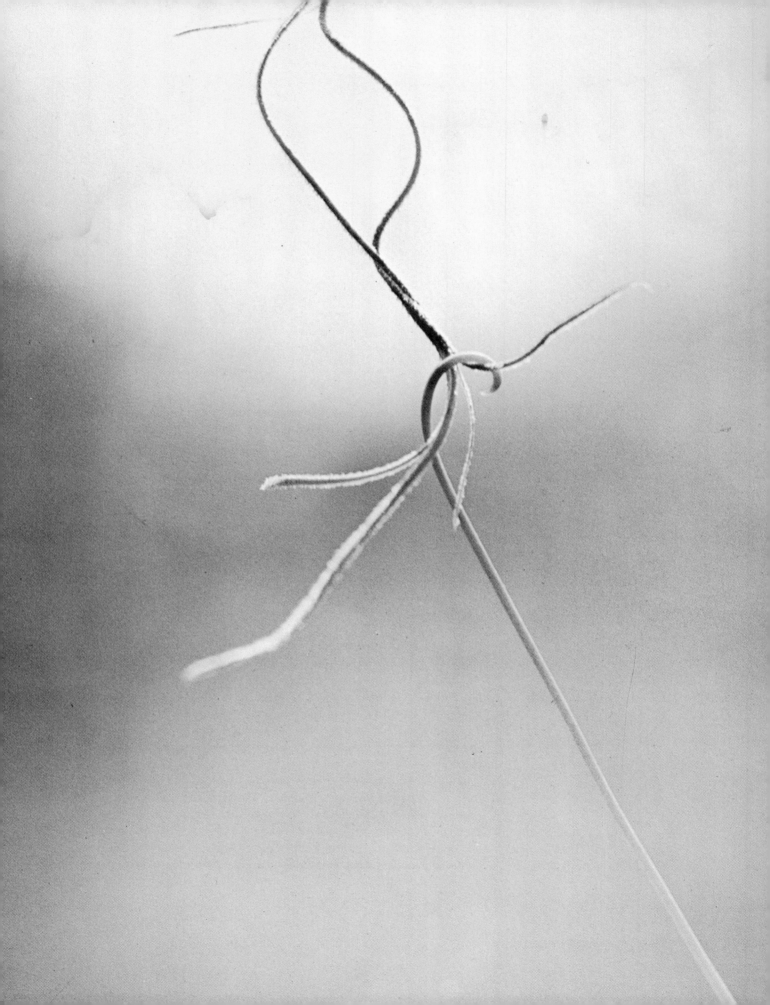

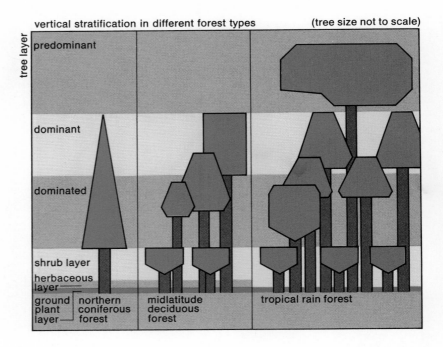

vertical stratification in different forest types   (tree size not to scale)

tree layer

predominant

dominant

dominated

shrub layer

herbaceous layer

ground plant layer — northern coniferous forest

midlatitude deciduous forest

tropical rain forest

The vast tropical forests are a largely untapped resource, but one that is coming under increasing pressure. The voracious appetite of modern industrialized economies for wood fiber is straining the capacity of temperate zone forests, and user countries are beginning to eye the tropical forests as a possible new source of wood products. The less developed countries, where most tropical forests are located, are attempting to expand their agricultural and industrial production, often at the expense of forest land. At the same time, many ecologists believe that tropical forests play a vital role in regulating the world's atmosphere and climate, and that any large-scale reduction in forested area might have disastrous worldwide effects. In order to assess the possible benefits and dangers of exploitation, it is necessary first to examine the characteristics of this unique environment.

Climate of the tropics

Of the various environmental components that determine the type of vegetation in a given area, climate must be considered the most important. It influences the plant cover directly, through such factors as temperature and humidity, and indirectly through the soils on which the plants depend. It has also had an important indirect effect on vegetation through the various restrictions it places on the activities of man.

In terms of climate, the tropics are often defined as those areas that have an average annual temperature of at least 20° C (68° F); there are, however, many other different but equally satisfactory definitions. The Tropics of Cancer and Capricorn roughly mark the extreme northern and southern spread of tropical climate, although even at the Equator cooler temperatures occur at higher elevations. Tropical climate tends

PETER J. FOGG is associate professor of forestry at Louisiana State University, Baton Rouge.

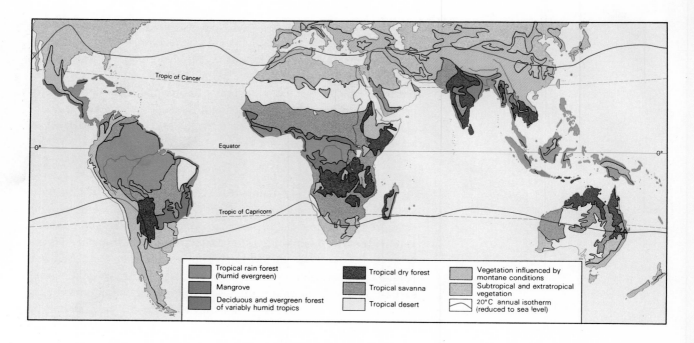

Tropical rain forest (humid evergreen)	Tropical dry forest	Vegetation influenced by montane conditions
Mangrove	Tropical savanna	Subtropical and extratropical vegetation
Deciduous and evergreen forest of variably humid tropics	Tropical desert	20°C annual isotherm (reduced to sea level)

to be uniform; with little variation in temperature from hour to hour, from day to day, or from season to season. This is partly because of the small range in length of daylight. At the Tropics of Cancer and Capricorn the longest day is only about two and a half hours longer than the shortest day, while at the Equator day length is nearly constant throughout the year. In many tropical areas the moderating influence of the oceans also contributes to the maintenance of uniform temperatures. While a high average annual temperature is characteristic of the tropics, higher extremes of temperature are experienced beyond the tropics in the major deserts of both the Northern and Southern hemispheres.

High rainfall is also considered characteristic of tropical areas. In fact, the debilitating effect of tropical climates on man is probably due to the sustained high humidities that accompany such rainfall rather than to the high temperatures. However, the pattern of rainfall varies widely in different localities. Rains may be abundant and evenly spread throughout the year, or they may be more moderate and more irregular. In extreme cases, one or two distinct dry seasons occur per year. Besides differing in seasonal distribution, rainfall also varies considerably in quantity, from as high as 500 in. annually to as low as 10 in. At one extreme the prevailing vegetation will be lush, evergreen forest, while at the other there will be tropical grassland or savanna, or even desert. Under intermediate conditions one finds vegetation types that are intermediate in character, such as deciduous forest or thorn forest. In the wetter portions of the tropics, the cloud cover associated with periods of rain helps to moderate the temperatures, so that the highest temperatures generally coincide with the drier periods of the year.

381

Thus, although one might characterize tropical climate as consistently hot, with high rainfall distributed quite evenly through the year and with concomitant high humidity, this description is true only in very general terms and there are many local variations. Divergences from this "typical" tropical climate are due primarily to differences in rainfall pattern, resulting from seasonal shifts in the trade winds, combined with the local influence of mountains and large bodies of water and with elevation above sea level.

Soils in the tropics

Seeing the abundant vegetation of the tropical forests, early observers concluded that the soils of the tropics were inexhaustibly fertile. This mistaken idea is now being replaced by a more reasonable assessment. In general, despite the often luxuriant growth, tropical soils are not as fertile as those in the temperate zones; they yield less and degenerate more rapidly. Many belong to a group of soils known as laterites, which occur where there are abundant rains and high temperatures. Such conditions cause the leaching of many important plant nutrients from the soil, leaving behind a high proportion of iron and aluminum compounds. Most of the fertility of these lateritic soils is retained by the organic matter present in the upper layers of the soil under forest cover. When the natural vegetation is removed, the exposed surface often becomes hard, reddish, and crusty. Once this condition is reached, the soil can no longer support the luxuriant forest that once covered it. Under certain conditions, loss of forest cover is followed immediately by considerable erosion, giving rise to deep gullying and reddish sedimentation in the river systems.

Not all tropical soils are of the lateritic type; those of relatively recent volcanic origin and those found in river floodplains and deltas are usually more fertile. This fertility tends to be maintained even under cultivation by the continued settling of airborne cinder dust from volcanic activity or the continued deposition of river-borne sediment.

Tropical rain forest

Tropical rain forest may be regarded as the ultimate in development of a plant community in the tropics. Such a forest can occur only where there are satisfactory environmental conditions: sustained high temperatures, abundant and regular (daily) rainfall, high humidity, moderately flat terrain at fairly low altitudes, and sufficient time without drastic interference from nature (flood, fire, cyclone) or man (harvesting, clearing, grazing). The particular species of plants found in tropical rain forest differ across the world. However, despite these local or regional differences, certain features are characteristic of all tropical rain forests. A person who is familiar with American rain forest will recognize its affinity with its African and Indo-Malayan counterparts, even though the species are different.

The large trees in a rain forest grow to a height of 80 to 120 ft, forming a more or less continuous layer or canopy. Through this layer

Exotic flora of the Venezuelan rain forests (opposite page) illustrate the variety of plant life found in this ecosystem. Epiphytic orchids like Ionopsis utricularioides *(top) live in treetops with their roots suspended in air, obtaining nutrients and water from the air and from rain. Intimidating thorns (bottom left) stud the trunk of a palm in the Amazonas Federal Territory. Fleshy fungi on a decaying log (bottom center) belong to the class Basidiomycetes. Such fungi perform an important function by speeding decomposition and the return of nutrients from dead matter to the soil. Bromeliad (bottom right) belongs to the same family as pineapple and Spanish moss. Rain forest bromeliads collect water in their rosettes, providing a habitat for small forms of animal life.*

Photos, Jacques Jangoux

Stilting is exhibited by trees in a rain forest near Victoria Falls in southern Africa. The "stilts" are actually roots that spring from the lower trunk, providing additional support for the tall but shallowly rooted tropical trees. Usually associated with mangrove forests and other swampy areas, stilt roots also appear in dry regions but disappear outside the tropics.

individual giant trees with rounded crowns may rise to 200 ft. Below the canopy layer are one to several layers of smaller trees, capable of surviving in the deep shade of the upper layers. Each layer of trees includes species that are different from those of the other layers, although small trees of the type present in the upper layers will sometimes be found occupying less prominent positions in the forest. These will eventually work their way to the top as the opportunity presents itself. Below the smallest trees there are usually a very few shrubs, while on the darkened forest floor will be found a quite sparse distribution of herbs, ferns, and, rarely, mosses.

The forest is seldom dominated by a single species of tree. Even the more abundant species may occur at the rate of only one or two per acre. More than 40 species may be present in a single acre, and a larger area might reveal many more than 100 species per square mile. There are occasional exceptions, however. For example, in the so-called mahogany gardens of Central America hardly any species besides mahogany is represented in the upper canopy.

In many cases, the bases of the tree trunks are flared out into buttresses, a feature rarely found outside the tropics. The bark is typically thin, smooth, and light-colored. The trees are evergreen, and their leaves or leaflets have a remarkably uniform, elliptical shape, with

384

smooth outlines and tips drawn out into narrow points, or "drip tips." These tips apparently hasten the draining of rainwater from the leaf surfaces, although the significance of this leaf shape is still a matter of some controversy. The flowers of the trees are chiefly pollinated by insects, and frequently the seeds are large, a feature that helps them reach the soil and helps the seedling trees to survive. Some of the larger trees have winged seeds that are carried a considerable distance by the wind.

There is a tremendous concentration of life in the canopy of the rain forest, where the supply of light is abundant. Water is plentiful throughout the forest, from top to bottom, but light is in relatively short supply beneath the canopy. The crowns of the larger trees are filled with plants that live their entire lives perched in the branches. These plants, known as epiphytes, have special ways of catching the water they need, and they manufacture most of their own food through photosynthesis, depending on the trees solely for support. Magnificent orchids, bromeliads, ferns, and mosses are characteristic epiphytes of the rain forest. Copious creepers or lianas reach all the way into the upper tree crowns, and so-called stranglers encircle the trunks and claw their way toward the sun, eventually killing the tree in the process.

The world's largest water lilies belong to the genus Victoria, *which occurs in the backwaters of Amazonas and the Guianas in South America. The giant pads with their veined, upturned edges may be up to six feet in diameter, and the fragrant flowers are from 7 to 18 in. wide.*

385

Among other noteworthy features of the rain forest is the presence of palm trees. These trees are found throughout the tropics and beyond, but their greatest development is in the humid tropics and they are found in the tropical rain forest in considerable numbers. Their trunks are often protected by murderous-looking thorns, and their bases are frequently supported by stilt-like roots that appear to prop up the trunks to prevent them from falling.

Tropical rain forest as an ecosystem

Although this account has emphasized the plant-related aspects of the tropical forest, and the preceding description may appear to suggest that the tropical rain forest is a stable, unchanging community of plants, it should be recognized that this forest is an ecosystem, in which the plants, animals, climate, and soils have arrived at a certain balance. This balance does not exclude the possibility of change. On the contrary, change occurs constantly, usually at a steady, almost imperceptible rate but sometimes suddenly and drastically. Locally, the death of a large tree and its collapse to the forest floor bring sudden change to its immediate vicinity. Regionally, a tropical cyclone may suddenly cause widespread destruction. Such violent changes are followed by a repairing process known as secondary succession, which continues until the damage is not easily discernible. But even in the absence of such drastic events, each organism in the community is growing, stagnating, or dying so that change is going on continuously.

As in all ecosystems, the plants and animals are closely interdependent. Some animals live their entire lives in the crowns of the trees and may be unable to survive in the absence of certain tree species. Others live exclusively in the small pools of water trapped in the leaves of an epiphyte. Certain trees need a particular species of insect to pollinate their flowers; if that insect is absent, seed will not be produced and the tree species will die out. The soil depends on the trees for protection from erosion and excessive heat as much as the trees depend on the soil for support, water, and nutrients. The trees depend on the climate to provide favorable growing conditions; yet the climate may be influenced by the presence or absence of forest. Thus, modification of any one of the four elements of an ecosystem — plants, animals, climate, soil — will result in a modification, however minute, of at least one other element. In addition, the tropical rain forest ecosystem has a certain interdependence with neighboring ecosystems. For instance, destruction of a portion of rain forest adjacent to an inland lake would result in changes within the ecosystem in the water.

In any discussion of the rain forest as an ecosystem, the effect of time should be mentioned. After some disturbance of the existing balance, secondary succession takes place over a period of time. The extent to which the damage has been disguised will depend on the length of time that has elapsed since the disturbance occurred. Getting back to the condition of balance is really a process of maturing.

Buttressing (opposite page, left) is another form of support for trees that is peculiar to the tropics. The triangular plates of thin, very hard wood may extend as much as six feet from the base of the trunk. (Opposite page, top right) Epiphytic moss hanging from a vine in a Venezuelan rain forest. (Center right) passionflower of the Guiana Highlands. (Bottom right) butterfly (Coelites euptychioides) of the rain forest of Borneo.

Jacques Jangoux

Jacques Jangoux

Jacques Jangoux

M.P.L. Fogden from Bruce Coleman Inc.

Tree ferns in a Malayan montane forest. Despite their height, these are true ferns, relatively primitive plants lacking flowers, fruits, and seeds and reproducing by means of spores. Once the dominant form of plant life on earth, ferns, while they occur throughout the world, are now most abundant in the humid tropics. Giant ferns such as these are almost entirely confined to the tropical regions.

Certain features characterize a mature ecosystem. For example, the growth of organisms is balanced by their decay, the dominant plants are mostly large and mature, the total volume of organic matter is large, nutrients are well conserved, and resistance to disturbance is high.

The boundaries of any ecosystem are not usually well defined; there is a carry-over effect from one ecosystem to another. Where conditions are not ideally suited to a tropical rain forest, some other ecosystem will be found. However, the change from one vegetation type to another may be so subtle and gradual that it would be hard to determine where one ends and the other begins. In some cases, of course, the changes may be abrupt, as at the shore of a lake, or there may be a patchwork of interwoven ecosystems resulting from variations in elevation, drainage, and soil. In the tropics, the rain forest is regarded as the standard against which other forest ecosystems are compared. The productivity (rate at which solar energy is converted into complex

388

foodstuff by the green plants for consumption by plants and animals of the community) of tropical rain forest is the greatest of any terrestrial vegetation type. It is this productivity that has given rise to the myth of the great fertility of tropical soils.

Other tropical forest types

In areas that have a distinct dry season, the forest usually consists partly of deciduous trees (those that lose their leaves for part of the year—in this case during the dry season) and partly of evergreen trees. This forest has only two layers of tree crowns, with the upper layer completely closed. The growth is less luxuriant than in the rain forest, and the tallest trees are not as high. Drier conditions result in a less productive type of forest consisting almost entirely of deciduous trees in two layers, with the upper layer somewhat broken or incomplete. In still drier areas one finds a single-layered forest that contains only medium- to small-sized trees, many of which bear thorns. This type of

Cloud forest in the Ruwenzori mountain range of Zaire, approximately 9,000 ft above sea level. Trees in these fog-shrouded montane forests are shorter than those of the typical rain forest, but their twisted trunks and branches, covered with a luxuriant growth of mosses, lichens, epiphytes, and climbing ferns, give a striking and distinctive appearance to this forest type.

389

Jacques Jangoux

M.P.L. Fogden from Bruce Coleman Inc.

forest is quite unproductive and often alternates with patches of grassland or savanna. Semidesert or desert occurs where the rainfall is very low.

Another series of forest types is found in areas where the rainfall is well distributed throughout the year but the quantity is smaller than in the rain forest, and where soil drainage helps to reduce the amount of available moisture. With increasing elevation above sea level, a series of types called montane forests occurs. None is as luxuriant as the true rain forest. One, the cloud forest, is so named because it is almost continually blanketed by clouds.

Low-lying areas subject to either seasonal or permanent flooding support different types of swamp forests. A very well-known type is the mangrove forest that grows throughout the tropics. Sometimes fire, resulting either from man's activities or from lightning, results in the establishment of other forest types. An example of this is provided by the fairly extensive pine forests that exist in certain parts of Central America.

Pressures on tropical forests

The so-called developed countries, most of which lie in the temperate zones, are using an increasing amount of wood fiber, much of it destined to go into newsprint and packaging. There is also a growing demand for sophisticated building materials made from wood fibers, as well as a continuing demand for high-quality, decorative hardwoods. All this has placed an increasing strain on temperate zone forests. Meanwhile, in the tropics there is a vast supply of standing forests, relatively unexploited, to which these nations are turning.

At the same time, most tropical countries are attempting to speed their economic development by increasing agricultural production and expanding industrialization. This inevitably means converting forest land to these uses. Much land is also being cleared or at least disturbed to accommodate urban development and modern road systems. In addition, much needed foreign exchange can be obtained from the export of forest products.

These pressures on the use of forest land are considerable, but they are not the first man has placed on the tropical forests. For centuries inhabitants of these lands have practiced shifting agriculture, in which the farmer fells a block of forest, burns it, grows crops until the fertility is lost—usually in a matter of a few years—then abandons the land and moves on to repeat the process in a new block of forest. Under this system, vast areas of forest land have been cleared, abandoned, and have regrown into poor secondary forest. Where land is in short supply the farmer may return to his original plot after as little as 10 to 15 years, reducing the fertility of the soil still further. Each time the land is farmed poorer crops are produced or the farmer must move on sooner. In extreme cases, severe erosion results in removal of the topsoil, and the land becomes practically useless for either agriculture or forests.

Aerial view taken along the St. Paul River in Liberia (opposite page, top) clearly shows the canopy, formed by the crowns of the highest trees, that is a typical feature of the fully developed tropical rain forest. Unless adequate steps are taken, the picture below, of a forest cleared for cultivation in Uganda, may represent the fate of the great tropical forest resource in the face of pressures from both developed and less developed economies.

391

Most successful, continuing systems of agriculture have been evolved in nontropical regions. Some experts believe that intensive cultivation of such areas as the Great Plains of the western U.S. and the newly plowed "virgin lands" of the U.S.S.R. will cause topsoil to be lost faster that it can be replenished, but serious consequences may not be apparent for several generations. In the tropics, however, the application of agricultural techniques that are reasonably successful in other places is usually doomed to imminent failure.

Clearing the forest removes the protective cover from the soil. The subsequent exposure hastens decomposition of the organic matter incorporated into the upper soil layers. Important plant nutrients are then released and are liable to be leached out from the soil by the high rainfall. Physical changes, especially in lateritic soils, accelerate the rate at which water runs off the surface, and erosion is speeded up.

Burning the felled vegetation releases a temporary supply of nutrients in the ashes. If a crop is established quickly, it will be able to utilize these nutrients, while at the same time creating a protective cover that reduces erosion. However, the process is one of rapidly diminishing returns. If the cleared land is permanently abandoned to forest regrowth, the secondary forest that establishes itself is less luxuriant, less productive, and further removed from the rain forest type than the original. This forest usually consists of shorter-lived trees that produce poorer-quality woods.

Apart from the effect of land clearing on conditions at the actual site of operations, many ecologists believe the loss of tree cover has broader environmental repercussions. In the cleared area, temperatures are higher in the daytime and lower at night and the relative humidity is reduced. This has an effect beyond the cleared area, particularly when a large amount of land is involved. Thus, widespread land clearing for agricultural or other purposes has a potential for modifying the climate of a broad area. Should land clearing in the tropics become excessive, the entire tropical environment might be affected.

Management of the forest resource

If only trees of selected species are felled and removed from the forest, the resulting gaps in the protective cover are much smaller than those that occur when land is cleared for agriculture. Partial protection is provided by the remaining trees, so that damage within the gaps is less intense, and the effect on the surrounding area is minimal. However, such selective cutting is feasible only when the value of the trees being harvested is sufficiently high and there are enough of them to yield a profit. In tropical forests, there are relatively few tree species that meet these criteria. Thus, although protection of the forest soil and environment is fairly well assured by this system, in many cases it is not practical from an economic standpoint.

This type of tree harvesting also gives rise to another problem: valuable trees are removed and species of lower value are left. Seedlings of the undesirable species may predominate in the gaps created by felling, and the composition of the forest may be steadily modified as the proportion of unusable trees rises. There are, of course, ways to counteract this by reintroducing desirable species, and foresters have devised various types of plantings to achieve this. Foresters are also working on various methods of enriching existing forests with desirable species and on regeneration techniques that will convert poorer forest types to more vigorous, productive ones.

Although selective cutting, combined with proper replanting techniques, will provide adequate protection for the tropical environment as a whole, it does little to meet the need for more agricultural land and for increased production of wood fiber. Balancing the needs of one segment of the population against those of another and one potential land-use pattern against another is a problem that is not specific to the tropics, but the way it is solved in the tropics may be of considerable consequence to other regions. The interdependence of man and his environment is only now being properly recognized, but the interdependence of the various environments of different regions of the world has yet to be widely appreciated. Solutions for tropical forests will affect the people throughout the world as well as throughout the tropics. The problem cannot be solved by experts in any one field. It has elements that are biological, climatological, sociological, and political — to name only a few. Solutions will be forthcoming only as a result of intensified research, coupled with tremendous cooperative efforts among specialists and laymen in all fields.

FOR ADDITIONAL READING:

Aubert de La Rüe, E., Bourlière, F., and Harroy, J.-P., *The Tropics* (Alfred A. Knopf, 1957).

Longman, K. A., and Jenik, I. J., *Tropical Forest and Its Environment* (Longman, 1973).

Richards, P. W., *The Tropical Rain Forest* (Cambridge University Press, 1952).

Bruno Barbey from Magnum

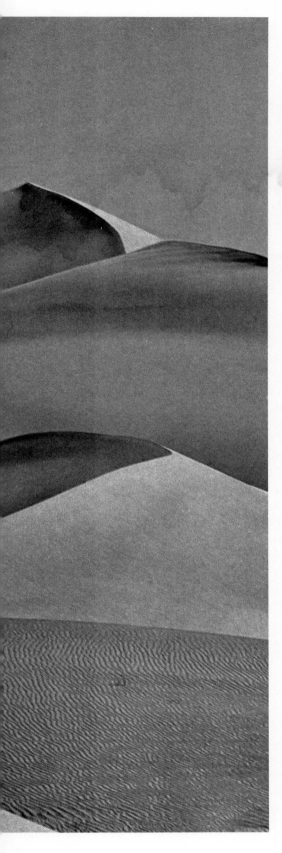

The Advancing Sahara
by Lawrence K. Lustig

As drought and famine continue to ravage sub-Saharan Africa, scientists ask whether this is a weather anomaly or the herald of a major shift in the earth's climates.

Embracing about 3.5 million sq mi of northern Africa between the Atlantic coast and the Red Sea, the Sahara is the world's greatest desert. It is also one of the most hostile environments on earth, one in which air temperatures exceeding 130° F have been recorded and ground temperatures commonly rise to 170° F or more. It is therefore most appropriate that the name Sahara was derived from the Arabic word *sahra*, meaning "wilderness" or "desert." Ever harsh and unforgiving toward the careless or unwary traveler who would trespass on its interior fastness, the Sahara has been the final resting place for thousands of men since ancient times. Even group effort has met with disaster upon occasion; whole caravans devoted to trade in slaves, ivory, gold, or salt vanished without a trace before trade routes became well established.

Periodic droughts of relatively recent vintage have produced tragedies of far greater scope. The latest of these culminated in 1973 in unprecedented devastation, affecting millions of people in more than a dozen countries. Famine stalked Mauritania, Senegal, Mali, Upper Volta, Niger, and Chad and extended eastward into Ethiopia and southward into northern Nigeria. Events have yet to run their course, and predictions of the ultimate magnitude of the disaster vary. It is clear, however, that catastrophe is the only truly suitable word for describing what is happening. An entire way of life, based on nomadism and cattle raising, appears destined for extinction throughout the affected region.

So disastrous an occurrence prompts reexamination of the nature of the Sahara generally and of its possible expansion over time. Specifically, whether the arid zone is in fact expanding and, if it is expanding, the reasons for such encroachment constitute the principal concerns of this article. Before considering the evidence bearing upon this question, however, it is necessary to examine the character and history of the Sahara.

Extent of the Sahara

The east-west extent of the Sahara is well defined. The Red Sea and the Atlantic Ocean serve as obvious boundaries, though it should be noted in passing that the Red Sea boundary is merely a geographer's convenience; there is no real physical or climatic basis for classifying the Arabian Desert as an entity distinct from the Sahara. The northern and southern boundaries of the Sahara are far more troublesome. They are gradational in character and thus debatable. Some authorities, for example, consider the Atlas Mountains to be a part of the Sahara, whereas others would place the desert's northern boundary along the southern margin of this range, the Mediterranean coast from Gabès, Tunisia, to the Sinai Peninsula being the logical eastward extension. Locating the southern boundary of the Sahara is the most perplexing question of all. A commonly drawn line extends from the mouth of the Senegal River eastward, through Timbuktu in Mali, Agadez in Niger, and Lake Chad and across the Sudan in the vicinity of Khartoum, where the waters of the Blue Nile and White Nile commingle.

The difficulty surrounding establishment of the southern boundary of the Sahara—or, indeed, the boundary of any desert—rests upon the fact that the limits of the arid environment are climatically defined. The commonly used basis is mean annual rainfall (the 10-in. or 250-mm isohyet), perhaps in combination with other factors. But nature will not cooperate by consistently distributing a precise amount of precipitation along a neat line. As the annual amount of precipitation decreases in any region, its variability generally increases, and this variability is, in fact, characteristic of the true Sahara. Boundary lines shown on a map are therefore somewhat illusory. It is partly in recognition of this fact that the semiarid, steppe-like southern borderlands of the Sahara are designated the Sahel, from an Arabic word meaning "boundary."

It should be noted that much of the debate over possible expansion of the desert is related to these problems. The lines delimiting arid and semiarid areas do not correspond exactly to conditions on the ground because they cannot. Demonstrating that the Sahara has expanded beyond some previously known "desert edge" may defy the surveyor's best efforts; hence, absolute proof of the desert's migration is difficult to obtain.

Geologic history

The Sahara occupies one of the older regions of the earth's crust, resting as it does on Precambrian basement rocks that are more than 570 million years old. In later ages, sedimentary deposits of various kinds were laid down upon this crystalline basement, forming the so-called Saharan Platform. The deposits include some marine sediments, indicating that the area was submerged beneath a shallow sea for long intervals, and continental sediments that suggest formation in freshwater lakes and streams. Certain rocks of Ordovician age are thought to be tillites, representing deposition by ice sheets and gla-

LAWRENCE K. LUSTIG served as a senior editor of Encyclopædia Britannica.

Ruined Roman aqueduct (top) once carried water from the Zahgouan Mountains in present-day Tunisia to the city of Carthage. Today the harsh desert claims even camels and vast areas are cracked and sterile.

Generations of moviegoers have visualized the Sahara as consisting entirely of vast seas of sand, or ergs, like the one shown on pages 394–395. In reality, the great desert contains a variety of landforms. The jagged peaks of the Ahaggar massif in Algeria (opposite page, top) form a part of the central Saharan ridge. High plateaus surround the ridge, characterized by bare rock plains, or hammadas, cut through by extinct watercourses (opposite page, bottom) and boulder-strewn plains, or regs (top). In contrast, dry lake beds and saline lakes, like the Chott Djerid in Tunisia (bottom), often plunge below sea level.

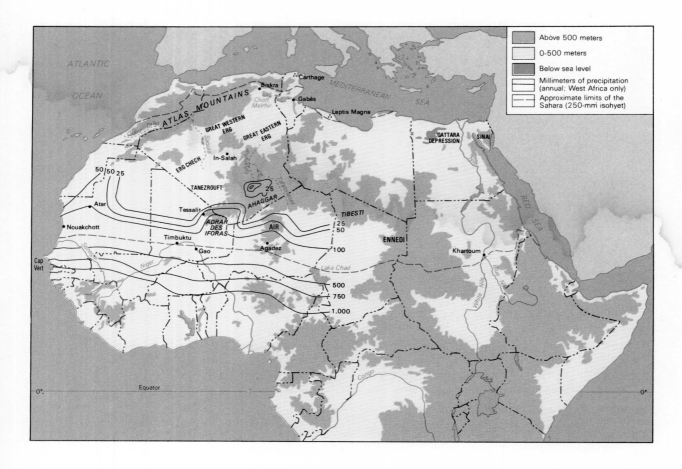

Extent of the Sahara as defined by the isohyet denoting 250 mm (10 in.) of precipitation annually, based on the 1930–60 average. Isohyets marking rainfall in the Sahel, shown for the same years, appear to have shifted southward since 1960. The irregularities in the isohyets indicate variations in relief, with precipitation increasing somewhat at higher altitudes.

ciers. Frigid climatic conditions are therefore indicated for this geologic period, particularly for central and southern Algeria where evidence for this ancient Saharan ice age has been examined in detail.

Rocks of Cretaceous age include the widely deposited Nubian sandstone, a porous sedimentary unit that serves as a principal aquifer (water-bearing stratum) in the eastern Sahara. Some beds within the Nubian series of strata have been interpreted as ancient sand dune deposits that became cemented and thus lithified — i.e., transformed into solid rock. They clearly suggest the presence of arid conditions during the time of their formation, but younger rocks of Tertiary age (from 2.5 million to 65 million years ago) again reflect more humid periods and the presence of lakes over large parts of the Sahara.

Volcanism also occurred during the Tertiary Period, as evidenced by basaltic plateaus and peaks in various places. Saharan sands and gravels, which mantle the whole sequence of rocks today and conclude the geologic history, are of Quaternary age. They were derived from preexisting rocks by wind and water erosion during the last 2.5 million years. The climatic history of the Sahara will be treated later, but this summary of the geologic record should make clear that the Saharan region has not always been arid. A wide variety of former environmental conditions is implied by the rocks that are present.

400

Physical features

All the world's deserts exhibit landforms of many kinds, but for purposes of this article the Saharan landscape may be said to consist essentially of ergs, regs, hammadas, and massifs. The ergs are vast areas of ever shifting sands, literally great sand seas. Together with other sandy regions, they blanket an area that has been estimated to comprise 10 to 17% of the entire Sahara. Prominent examples include the Great Western and Great Eastern ergs in Algeria and Libya and the Erg Chech to the south in the Tanezrouft or "land of thirst."

The regs are barren desert plains characterized by prominent, widely strewn pebbles, cobbles, and boulders, whereas hammadas are bare rock plains, generally dissected by gorges formed by extinct or rarely flowing drainage systems. Regs and hammadas occupy the plateaus that comprise much of the Saharan landscape, surrounding the central Saharan ridge on all sides. The latter is an east-west trending structural feature that includes the high massifs of Tibesti and the Ahaggar, many of the peaks of which attain elevations of 9,000 ft or so above sea level (the maximum elevation exceeds 11,000 ft). Smaller massifs of note are the Adrar des Iforas, Air, and Ennedi, all to the south. Because the average elevation of the Sahara is about 1,000 ft, the principal massifs stand like semiarid "islands" in a sea of extreme aridity (climatic conditions almost always ameliorate with altitude).

The Sahara also contains pronounced depressions, including a line of chotts, or dry lakes, south of the Atlas Mountains between Biskra, Alg., and Gabès. The Chott Melrhir, for example, is below sea level. A similar line at the western end of the Atlas is marked by the Oued Draa, a major ephemeral drainage system. Another example is the well-known Qattara Depression in Egypt, which descends to 435 ft below sea level. Finally, the Chad Basin, Niger Basin, and Nile Basin should be mentioned. The first two extend from the line of central Saharan massifs southward, into tropical Africa. The Nile Basin is especially noteworthy by reason of its contained river system — the marvelous Nile which traverses the whole of the burning eastern Sahara to discharge into the sea.

Climatic history

The climatic history of the Sahara has been deduced from abundant evidence of a highly varied nature. The geologic record certainly indicates that the Sahara has not always been the natural furnace it is today. Several kinds of nonarid environments existed in the distant past, even including glacial conditions. At another extreme, the presence of silicified tree trunks in Tertiary beds near In-Salah, Alg., signifies that a forest requiring relatively humid conditions must have existed there at that time. It is the Pleistocene record of 10,000 to 2.5 million years ago, however, that is replete with evidence of the lack of permanent aridity.

The Pleistocene was an epoch of global refrigeration that witnessed at least four major advances and retreats of ice sheets in mid-latitude

401

regions. In Europe, the Alps served as one locus of ice generation, with glaciers advancing from the higher parts of the mountains to the plains below, where they merged, coalesced, and spread across the landscape. Increased precipitation elsewhere was part of the climatic pattern associated with the growth of glaciers. In some instances, notably in North America, it is thought that these pluvial intervals or moist periods were synchronous with the ice ages. This may not have been true for the Sahara, relative to ice advances in Europe, but there is no question that large parts of the Sahara were both wetter and cooler during certain periods of the Pleistocene than they are today.

Consider, for example, the fact that one of the world's greatest groundwater reservoirs exists beneath the Sahara. Wherever this water has been dated by radiocarbon techniques, an age of about 20,000 to 35,000 years has been obtained. This means that the groundwater originated as rainfall in highland areas during that time interval and subsequently migrated downward and toward basin centers. In general, groundwater cannot be forming today beneath the central Sahara because evaporation far exceeds the scanty precipitation that occurs. The mere presence of this fossil water thus constitutes one kind of evidence that somewhat wetter and cooler climatic conditions once prevailed.

Another line of hydrologic evidence is provided by Lake Chad, which is known to be the impoverished offspring of a Pleistocene lake approximately 120,000 sq mi in extent. This ancient body of water was comparable in size to the modern Caspian Sea, implying the existence of a much different hydrologic regimen for its maintenance. Large lakes also occurred in the Niger Basin, the Algerian chotts, and elsewhere, and what are now sand-choked drainage systems, such as the Saoura, Igharghar, and Tafassasset of the central Sahara, flowed frequently if not continuously. Moist climatic conditions would have been required for the sustenance of all these lakes and streams.

Red soils of a type common to humid regions today are present in the Sahara. Together with certain kinds of rounded bedrock hills that are thought to reflect tropical weathering conditions, these red soils are also a testament to an absence of aridity in the past. Limestone rubble along the Mediterranean coast is widely regarded as the result of frost action, that is, the repeated freezing and thawing of moisture in cracks and crevices which serves to break up solid rock. And the presence of fossil pollen indicates that forests of cedar and oak thrived about 20,000 years ago in Tibesti and other highland areas. Even more impressive, perhaps, are the faunas with existing tropical relatives. These include crocodiles found in waterholes at Ennedi and Tibesti which could not have reached these localities under present conditions, and large game animals that roamed the Atlas Mountains until they were eliminated by hunting in historic times.

In a similar vein, there is much archaeological proof in the form of cave paintings and rock carvings that prehistoric man hunted elephant, rhinoceros, lion, buffalo, and other large species in presently arid parts of the Sahara. Artifact collections from many sites in the southern Sahara, for example, include large spearpoints as well as arrowheads. The inescapable conclusion is that grasslands and other vegetation must have existed to sustain the requisite herds of grazing animals. This does not necessarily mean that the entire Sahara was a pleasant, well-watered park, as is sometimes implied, but rather that grassland conditions must have prevailed over tracts perhaps hundreds of miles wide in the northern and southern fringe areas and around the "islands" created by the central Saharan massifs.

Still more recent wet or pluvial phases also have been detected. The last of these periods occurred between approximately 2,000 and 6,000 years ago. Lakes, springs, and permanent streams again were sustained, and the original inhabitants of the Sahel region were able to practice agriculture and cattle raising.

The subsequent period of Roman occupation in North Africa is well documented. Among the ruins of obvious significance are the remains of the great ancient aqueduct that extended about 35 mi from the Zaghouan Mountains (in present-day Tunisia) to Carthage, and the amphitheater at Leptis Magna, Libya, which could accommodate thousands of persons. These and other Roman monuments bear silent witness to a substantial civilization, now vanished. The population of those times required food, and the practice of agriculture inevitably required water. The ruins, therefore, have been cited as evidence of moist conditions during the early Christian era (agriculture survived the Vandal invasion of the 5th century A.D.) and of subsequent Saharan expansion. On the other hand, it has also been argued that the many cisterns, aqueducts, and bridges built during the Roman period may reflect attempts to adjust to and conserve what were actually relatively meager water supplies.

In any case, this overview of the climatic history of the Sahara may be summarized by saying that climatic conditions have varied dramatically through time, and there is no doubt that expansion and contraction of the arid Sahara have occurred in the distant and not too distant past. This being true, the possibility of desert encroachment in this century cannot be dismissed out of hand; the evidence is worthy of examination.

Modern encroachment of the Sahara

The advance of the Sahara appears inexorable to those inhabitants of the Sahel whose very lives depend upon maintenance of the status quo. Farming villages in Mauritania, for example, have survived since the Middle Ages by employing agricultural practices that require about 400 mm of rain a year. Precipitation values today are about 200 mm in this part of West Africa, and complaints of falling levels in water wells are common in Libya and other countries in the northern Sahara as well.

Aside from declining water levels and remembrances that the "old days" were better, substantial physical evidence of desert encroach-

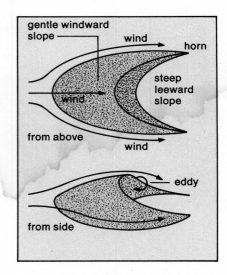

Barchans, or crescent-shaped dunes, found in parts of the Saharan ergs are formed as the prevailing wind blows sand from the rear slope to the front. In this way they move slowly forward. In areas where the wind shifts, one horn of the barchan may become elongated, giving rise to seifs, or ridges of sand, that may stretch up to 50 mi in length.

403

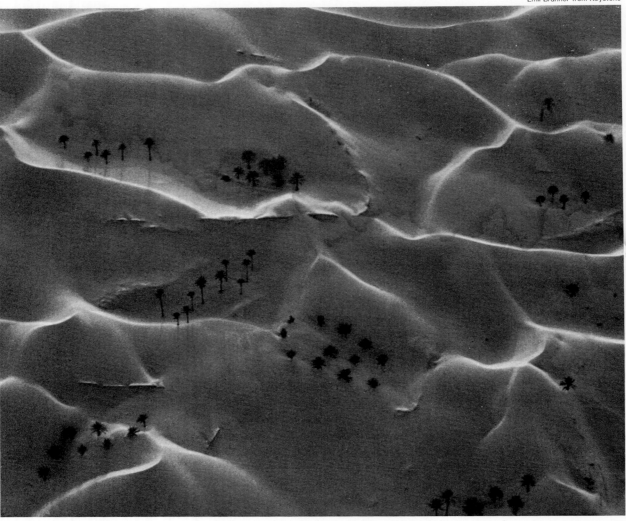

Aerial photograph highlights the drama of a shifting dune as it inexorably overtakes a desert oasis. In some places only the tops of palm trees are visible above the sand.

ment exists in scores of villages and oases of Tunisia and Algeria, where preventive measures have failed to stem the advance of dunes and drift sand. Whole communities have been abandoned, and the former sparse vegetation cover has been eliminated entirely. The erosion of the soil by wind and occasional flood has been intensified accordingly, and the onset of gullying and badland formation has coincided with a human index of deterioration—the migration of nomads and farmers to the few urban population centers of the Sahara where they swell the ranks of the poor and hungry.

Arguments have raged for many years as to whether loss of vegetative cover, advance of dune sand, and falling water levels in wells truly point to desert advance induced by climatic change, or whether man has actually created deserts through overgrazing and poor tillage practices in these fragile, marginal, semidesert areas. In this instance, however, there is considerable evidence that climatic deterioration is to blame, even if man has exacerbated natural events.

404

Meteorological hypothesis

The meteorological factors that have produced the Sahara as it exists today are well known. The two principal ones are radiation from the sun and the rotation of the earth; in combination, they are responsible for the general circulation of the atmosphere and for much of the existing weather pattern on earth. Solar radiation is most intense in the equatorial region, heating air masses that tend to rise, cool, condense, and release their moisture in the tropical zone. The dry air masses then subside near latitudes 30° N and 30° S, creating two great subtropical high-pressure belts—which coincide with a number of the world's great deserts. Under the impetus of the earth's rotation and the Coriolis effect (which, as a result of the earth's rotation, deflects moving objects to the right in the Northern Hemisphere and to the left in the Southern Hemisphere), these belts of descending air are responsible for the easterly trade winds that blow toward the Equator. In the Sahara these dry easterly winds are called the harmattan. Their occurrence leads to many severe dust and sandstorms.

Precipitation in deserts is extremely irregular in both space and time and, indeed, the more intense the aridity the more irregular the rainfall. In temperate regions, 30 to 40 in. of annual precipitation may be relatively uniformly distributed over 10 or 12 months of the year. In the central Sahara, however, the annual total may be 3 or 4 in. and it may fall in one or two intense storms.

On the other hand, the southern margin of the Sahara—the semiarid Sahel—is frequented by the so-called African monsoon. Each year the Sahel relies upon the same type of compensatory rains that visit India at the eastern end of the summer monsoon belt. The rains have not always fallen to the best advantage of the inhabitants, of course. In some years they have arrived early or late, and their duration also has varied. Accordingly, it is not surprising that a number of severe droughts have struck the Sahel in this century, notably those of 1910–14, 1939–42, and the disastrous trend of the 1960s during which summer rainfall across the great reach between Mauritania and India has been practically halved. Drought also has persisted in recent years in the Cape Verde Islands off Africa's west coast, and it is noteworthy that they are located within the same latitudes.

Much more scientific research needs to be undertaken before long-term climatic fluctuations can be predicted with absolute confidence, but some of the work that has been done is highly suggestive. For example, a detailed analysis of the available precipitation data in Africa and India was undertaken in 1973 by Derek Winstanley, a British meteorologist. The method used involved construction of 5-year and 20-year running mean values of the precipitation at eight stations, six in the Sahel (Atar and Nouakchott in Mauritania, Tessalit and Gao in Mali, Agadez, and Khartoum) and two in India. The construction of a running mean is a common statistical technique that serves to smooth or average out the fluctuations in a series of data and thus to reveal any basic underlying trends that may exist. Winstanley found that

405

summer monsoon rainfall has been gradually decreasing in the southern Sahara since the late 1920s. The more precipitous decline of the 1960s (more than 40% at most stations) therefore appears to be a continuation of the long-term climatic trend.

If the climate is indeed changing, the next question is why. The answer is suggested by a second phenomenon. The winter-spring rainfall in the Mediterranean region of North Africa and in the Middle East also has changed during this century. Comparable analysis of precipitation data in this region shows that the basic trend is negatively correlated with that in the Sahel to a statistically significant degree. In other words, an inverse relationship exists—summer rainfall is decreasing in the southern Sahara while winter rainfall is increasing in the northern border region. This would appear to indicate a basic change in the global circulation pattern at high altitudes and a concomitant southward shift of associated weather systems: the belt of summer monsoons is shifting southward toward the Equator and the southern margin of the Sahara is advancing for this reason. Considering the shift of the northern margin, however, the meteorological portent seems to be for a southward shift of the Sahara as a whole rather than for expansion of the desert per se.

The magnitude of the southward shift can be calculated from knowledge of the rainfall gradient across the Sahel region in a north-south direction and the average decline in rainfall at the observation stations cited above. The gradient is approximately 1 mm of annual rainfall per 1.4 km, and the rainfall at Atar, Nouakchott, Gao, Tessalit, Agadez, and Khartoum decreased by 65 mm between 1960 and 1970. During this ten-year period, isohyets marking precipitation across the Sahel have therefore shifted southward by 65 x 1.4 km, or about 9 km (5.5 mi) per year. The hue and cry over expansion of the Sahara thus appears quite warranted in the Sahel region. Climatic forecasting is still an inexact science, but extrapolation of existing trends suggests that the southern margin of the Sahara will be 100–200 km farther south by the year 2030, depending upon the assumptions made. At the same time, the northern margin also will retreat southward if winter-spring rainfall trends in the Mediterranean region continue to hold.

Prognosis and overview

If this reading of the situation is correct, it would appear that there is no hope for the Sahel region, insofar as nomadism and cattle raising are concerned. The long-term prediction must be in favor of continually decreasing monsoon rainfall and persistent drought, although a wet year or a series of relatively wet years may well occur from time to time. From a long-range viewpoint, however, the tendency for Saharan expansion in this zone must be regarded as a single swing on the pendulum of time. Climatic change, after all, brought about the boundary shift toward the center of the Sahara in Pleistocene time, estimated to have been 200–300 km, and the climatic history discussed earlier is replete with evidence for similar or greater fluctuations. Pulsation of deserts about some mean, steady-state configuration may well be the rule rather than the exception in the long-term scheme of things.

All this is of scant interest to the victims of climatic events, but it is, perhaps, best to quench false hopes. Assuming that Saharan expansion is linked to the worldwide atmospheric circulation and attendant climatic changes, there is no possibility that the works of man can defeat those of nature. The cause is too general to be affected by the erection of shelterbelts of planted trees across the Sahara, or attempts at weather modification. Clearly, if desert expansion is an effect of climatic change and not the cause, true solutions must be sought in the political arena. The affected nomads must be permitted to migrate southward across geopolitical boundaries, to follow the life-sustaining summer monsoons that no longer reach them.

FOR ADDITIONAL READING:

Campbell, I. A., "Ephemeral Towns on the Desert Fringe," *Geographical Magazine* (June 1973, pp. 669–675).

Dresch, Jean, "Drought Over Africa," *UNESCO Courier* (August-September 1973, pp. 44–47).

Lamb, H. H., *Climate: Present, Past and Future,* Vol. 1 (Methuen, 1972).

Winstanley, Derek, "Rainfall Patterns and General Atmospheric Circulation," *Nature* (Sept. 28, 1973, pp. 190–194).

The Productivity of Research Scientists
by Derek de Solla Price

The brilliant young scientist burned out by 25 is a well-known figure of popular folklore, but a systematic study of how researchers actually work yields some different—and surprising results.

World civilization is increasingly dependent upon the skill with which we manage our advanced technologies. Behind all the engineering for war and for peace, the provision of food and comfort, and our competences in medicine and in agriculture lie the steadily increasing scientific understandings that have made these things possible. Clearly each nation must weigh very carefully its investment in science and in scientific research. As the burden of supporting research has grown from a relatively trivial expense to a significant item of government spending in all advanced nations, the problem of making some sort of cost-benefit analysis of science has become acute.

Although it is not particularly difficult to measure the investment in science in terms of input—money spent on highly trained manpower and instrumentation—it is a very tough problem to estimate the output. The obvious thing would be to evaluate science by its effect upon society, but any approach of this sort measures only the actual production of those technologies that societies decide to buy rather than the increase in understanding and know-how that made all such production and innovation possible. Getting behind this to make estimates of the production of new knowledge in science leads to many interesting paradoxes that require unusual caution if one is to avoid naive misunderstandings that may lead to costly policy errors.

The measurement of productivity

The basic position taken by most of those who attempt to measure the production of new knowledge by research scientists is that it may be done by counting suitable items in the special literature of learned journals, research reports, monographs, and similar communications that are the normal end products of such scientists' work. The prime paradox is that, although there are a host of objections indicating that

409

any such head counts would be grossly unreliable and even irrelevant, the measures when used give remarkably good agreement with all other indicators and theories. If reasonably conducted, such counts of papers appear to constitute an objective indicator that behaves rather more regularly in international comparisons than the monetary measures used in econometric analysis. Once proper precautions are taken, and due caveats entered about what is being measured and to what accuracy, it is possible to measure the research activity and productivity of individuals, teams, research institutions and programs, even whole nations, and from these measurements, basic balance sheets for science policy may be drawn up.

Perhaps the most important caveat is that any count of publications is an indicator of only part of the activity of a fraction of the population of scientists. Many people who have been trained in science have never been brought up to the research front where new knowledge is created, and even among those who have there is a large group who thereafter spend all their energies behind the front. They may teach already available knowledge; they may apply such knowledge to the delivery of services in medicine or agriculture or in geology, for example; or they may apply this knowledge to the practical end of making new products or operating old ones in industry or in government. Many may spend all or part of their time performing managerial or administrative functions, or transfer to some completely different occupation that has no connection with their scientific training.

For those who are working at the research front, publication is not just an indicator but, in a very strong sense, the end product of their creative effort. Some scientists, especially those in private industry, may produce discoveries that are not immediately published but, rather, are translated directly by the industry into new products or into licenses and patents for the production of such products; this takes place frequently, for example, in the field of pharmaceuticals or the design of new models of airplanes, missiles, or computer programs. For the majority of scientists, however, the connection between research and product is not direct. Their function is to add to the world consensus of knowledge, be it in what is normally termed basic science or in the so-called applied sciences, such as metallurgy and electronics, where the same system of publication exists.

In this system, whatever a scientist may discover—however great or small his contribution—becomes effective only through being published, judged, incorporated somehow into the stock of knowledge, and used by his peers. The act of discovery is incomplete until it is sanctioned by the acquiescence and acceptance of other workers in the field. This gives rise to the interesting paradox that private property in scientific discovery is gained only by open publication: the more open and complete the publication, the more certain the knowledge and the claims (if they are made) of the discoverer. The ideas and the data in such a publication may be wholly or partly true or they may be false, but only insofar as they are effective in producing more contri-

DEREK DE SOLLA PRICE is Avalon professor of the history of science at Yale University.

410

butions do they add to the transnational corpus of scientific knowledge. In a McLuhanesque sense, the publication *is* the research, it *is* the innovation, and as such it may be counted and measured.

Unlike the other creative efforts of mankind, science acts as though it were lying there waiting to be discovered step by step. All the world's scientists are playing the same game, pitting their wits against the same set of problems as they gradually fall into place like the pieces in a jigsaw puzzle. Action at any front gives rise to fresh clues that result in more action, so that researchers tend to hunt in packs, competing with each other for the next hot clue.

Quantity and quality

Any crude head count of publications would run a high risk of being absurd if it ignored the obvious fact that such publications differ enormously in their size and quality. In addition to the mere physical length —ranging from a single-line correction note to a multivolume encyclopaedic reference work—publications may differ in the extent of their substantive content. A far-reaching new generalization in mathematical physics, even if short in terms of pages, may be much more extensive in content and represent more man-years of labor than a long but routine experimental report on the properties of one of a very large class of organic chemical compounds. Even publications of similar length in the same field may differ widely in quality and, therefore, in importance. Fortunately, it is possible to throw light on this complex of variables by using citations to measure the actual impact of a publication on the corpus of knowledge.

For more than a decade, an important tool for indexing and sorting the huge body of international scientific literature has been the *Science Citation Index,* published by the Institute for Scientific Information in Philadelphia. This index takes all the references given in all the papers published each quarter year in more than 2,000 of the world's most important scientific journals and sorts them alphabetically by author and title. Thus one can look up any previously pub-

411

lished piece of work and see who is using it and in what papers it has been cited.

Incidental to its main indexing purpose, the *Science Citation Index* enables one to make what is, in effect, an evaluation of previous papers by finding out how often they are invoked in the course of new work. Clearly, one must be cautious about the interpretation of such citation counts, particularly if they refer to individual researchers or publications, but measures of larger groups, programs, institutions, departments, and nations seem to be reasonably valid and consistently in agreement not only with subjective and intuitive peer judgments but with most of the other indicators that have been proposed.

The distribution of quantity and quality

The most striking result of all attempts to measure quantity and quality in any population of scientific researchers is the discovery of a "millionaires and peasants" effect. Scientific productivity seems to be distributed more unequally than money or athletic ability or any other human capacity. The distribution is like an enormously steep pyramid in which the largest numbers of people are at the lowest levels while at the top there exist a few individuals of such great attainment that their combined contribution equals that generated by all the lower masses. From one point of view the situation is highly elitist. From the opposite aspect, it may equally well suggest José Ortega y Gasset's hypothesis that no other activity requires so much work by so many people of commonplace activity and competence.

Technically, the distribution both for quantity and for quality as measured by citations follows mathematical laws that can be represented with fair accuracy by the hyperbolic, lognormal, or negative binomial distributions—it does not seem to matter much which particular representation is used. Probably the best results are from variants of Alfred J. Lotka's Law of Scientific Productivity, which states that the number of scientists with just n publications is proportional to $1/n^2$. Similar results are obtained if one considers other probability situations, such as the relative frequency of use of all the words in the dictionary, but in the case of science the effect seems at least as pronounced. As an example of this, in one of the great bibliographies of old astronomical writings, the most productive 1% of the authors produced one quarter of the published papers, while it took the least productive 66% of the authors to produce another quarter. Many other collections of scientific writings show the same pattern. Hence, one can affirm that, whether or not it is openly manifested, science is one of the most competitive activities of mankind.

In general, then, about half of the total output of papers will come from the small elite of highly productive authors whose number is equal to approximately the square root of the entire group. For example, in a population of 10,000 scientific authors, the top 100 (1% of those writing) will contribute about half the papers, and the other 99% of the authors will produce the other half. Strikingly, this result, which

holds true for both quality and quantity in scientific writings, appears to be valid for all scientific fields, in all countries, and it has been equally verified for 17th-century collections and for the modern period. This high degree of competitiveness appears to be deeply and permanently built into the social structure of the scientific community.

The mechanism governing this competitiveness seems to be that, even after one has reached the research front, there is a considerable chance of falling away again and ceasing to produce. Each subsequent success in publication reduces the chance of failure, however, giving rise to what has been called the Matthew Principle: "Unto every one that hath shall be given . . . but from him that hath not shall be taken away even that which he hath." At any rate, success breeds success, and failure breeds failure, so that the chance of further publication increases steadily with the number of items already published.

There also appears to be a clear but not complete correlation between quality of work and probability of continuation. The first paper offered for publication must be of considerable value in order to be accepted, but acceptance gives the worker the motivation and reinforcement to continue. The same sort of result is true for institutions and for subfields of activity as well as for individuals. Each success makes further successes more probable. For the research scientist the effect of this law is that at each stage in his career, before the ages bordering on retirement, there is approximately an even chance that he will be able to double the number of his publications. There is a 50% chance that the person with a single paper will publish a second, and a similar 50% chance that the person with a bibliography of 100 will reach 200. It may be that there are actually twice as many would-be researchers with only half a paper in them as there are workers who break through the barrier of getting out their first paper—probably in the course of their doctoral research.

Productivity and age

Because there is a very large but decreasing chance that any given researcher will discontinue publication, the group of workers that reaches the front during a particular year will decline steadily in total output as time goes on. Gradually, one after another, they will drop away from the research front. Thus the yearly output of the group as a whole will decline, even though any given individual within it may produce at a steady rate throughout his entire professional lifetime. We need, therefore, to distinguish this effect of research-front mortality from any differences in the actual rates of productivity at different ages among those that remain at the front.

Age variation in productivity has been measured for several populations of researchers, but in dealing with large groups it is difficult to disentangle the two effects noted above. More directly, there is the evidence of the life records of individuals known to have had long and productive careers. The typical case history can be divided into three distinct periods: (1) the formative period, with a rapid rise to a

414

maximum rate of productivity, in which the groundwork of the author's reputation is laid; (2) maturity, during which the production rate oscillates around a stable level or perhaps several different levels corresponding to career circumstances in various well-defined periods; (3) retirement, in which there is usually a gradual diminution of activity.

The formative period usually seems to last a little less than ten years, culminating in maximum activity around 30 years of age. For mathematics the age is notoriously low, for physics a little higher (the superstition is that if you haven't made it by 35, quit), and for chemistry higher still. For such subjects as medicine, where the apprenticeship is long and where maturity and wisdom rather than native ability are at a premium, the age is highest, though it is perhaps in the most nonscientific of the humanities that the maximum occurs latest.

In many—but by no means all—subjects, there seems to be some small drop in productivity once reputation and position have been established, but from then on productivity remains stable for long periods and changes usually occur in conjunction with major shifts in life-style. Anecdotal wisdom to the contrary, there seems to be no clear evidence that a general decline with age occurs among those who continue to produce. The eventual age at which the retirement decline takes place appears to vary a great deal from case to case, field to field, and country to country. The average working lifetime of those researchers who continue to produce steadily seems to be about 40 years from first work to last, including a span of 25–30 years of the mature rate(s).

The popular impression that there is a large and continuing drop in productivity after the maximum of the formative period seems to have arisen from the rapid exponential growth of the scientific research population coupled with the equally rapid research-front mortality. Ever since the 17th century, when scientific learned societies were first organized and the mechanism of scientific journals and research papers was invented, the numbers of scientists and their publications have been rising at a compound interest rate of some 6–7% every year. This extraordinarily rapid and steady growth implies a doubling in numbers about every 10 to 12 years, though there have been periods of relative quiescence; the exact rate of development has varied slightly from country to country; and now, in the most developed countries, there seems to be a general slowing as the system of higher education expands to cover virtually the entire available population.

Though there was no sudden burgeoning of science after World War II, the startling consequences of this steady, regular growth seem very immediate. At any given time half of all the papers ever published have appeared during the last decade, so that science seems perpetually new. If the productive lifetime of a scientist is some 40 years, it must span at least three periods during which the scientific research population has doubled, so that for each person who died before he appeared there are seven now alive—one from his first period, two from the second, and four from the third. Consequently, seven scientists out of every eight who ever worked (87.5%; say 90%) are alive right now. More than that, four out of the seven living scientists—more than half—are in their first decade of work; they are young and still in their formative years. Thus the age distribution of scientists is heavily stacked toward the young, and even if discoveries are distributed evenly among all scientists, the young will make most of them.

Because of this, any individual who escapes the high research mortality of the formative years and achieves his reputation and institutional position will find his work in his own hunting pack steadily and continuously diluted by growing numbers of newly emergent researchers. Even if the historian considers all researchers as a group and makes an age distribution of discoveries of any constant quality, this effect will still obtain. Such distributions tend to show that, even though the individual may work at a constant rate and level of quality during his productive lifetime, the relative number of discoveries made by researchers at any given age must approximately halve for every decade after the peak of the initial formative period.

Ever since Harvey Lehman first published this result in 1953 and established it clearly for many fields and countries, it has been a source of perennial discussion, and contradictions between the different types of data have been demonstrated repeatedly. The absolute productivity of an individual researcher remains rather stable once he has survived the high mortality that attends the beginning of a scientific career; his productive lifetime may be relatively long and may be enhanced with each success through the working of the Matthew

416

Principle. However, the productivity of the individual relative to the community peaks out in the mid-30s and declines rapidly thereafter.

Perhaps because of this, it is commonly found that the most outstanding researchers execute a leapfrog tactic in their campaigns. It is typical of the scientific genius that he is outstanding not merely in one subfield but in several, often quite widely separated in substantive content. As soon as the formative peak has been attained in one area, the researcher makes a new sortie in a fresh one, leapfrogging over the intervening territory and making a new start instead of continuing in an aging subfield. For example, though Nobel Prizes are customarily given (usually a dozen years or so after the fact) for work done in the 30s—age 33 for physics, 37 for chemistry, 41 for medicine—there are many prizewinners who could have received their awards for any one of a string of almost unrelated pieces of work done at either earlier or later ages. No doubt this phenomenon also occurs in some researchers below the genius level.

Production rates, transience, and collaboration

Not only must one be wary of the difference between absolute and relative productivity, but also one must take care to separate the effects of research mortality and work done behind the research front from rates of steady publication by those who are more or less continuously active. For example, in several investigations of whether—as has often been supposed—women have lower productivity than men,

it has been found that the absolute productivity of women may not be significantly less but the chance that their publication will be interrupted is much higher because of social differences that affect career possibilities. Similarly, a finding that Soviet scientists are only half as productive as those in the U.S. may be due to the fact that the Soviet figures are based upon a much larger proportion of scientific workers whose activities are behind the research front and not in publication at all.

Clearly, any attempt to link manpower figures with publication data must be based upon a definition of manpower that is universally and satisfactorily valid. Preferably, it should be related to the number of people who have been brought to the research front to the point of publishing at least once. Fortunately, since there is universal competition in solving the same scientific jigsaw puzzles, the standard of what constitutes a publishable paper in any particular subfield seems to be much the same everywhere and appears not to have changed radically with time. Therefore, one may reasonably take the numbers of people who publish in any of the world's acknowledged scientific journals in any period or in a lifetime as the measures of manpower, and base productivity figures on them.

The immediate result of such a technique is to reveal the fact that a rather large number of those publishing in any given period have never been heard of before and will never be heard of again in any sort of scientific publication. They are transients who reach the research front, publish one or two papers, and then go to serve other functions in society. The existence of these transients brings down the average

rate of productivity from the fairly high levels found for those who continue to publish. For such persons who continue a rate of approximately four items per year seems about average—Bernice Eiduson reports 4.41; Anne Roe 4.2, 3.4, and 3.7 for various groups; Price 4.37 —though there are strong suspicions that the average rate varies a great deal from field to field, corresponding to a large number of less important papers in chemistry and biochemistry and the taxonomic sciences and a small number of more weighty papers in mathematics, physics, and astronomy.

If one includes the transients with those who publish steadily—for example, by counting from the Source Index of the *Science Citation Index* annual—the rate is about halved, being some two items per year on the average for all fields. It must not be supposed, however, that the total number of papers is found simply by multiplying the number of authors by the number of titles attributed to them. Many papers have more than a single author, and indeed the practice of collaborative authorship has been increasing rapidly in all fields and countries. From being a rare phenomenon in the 19th century, it has become so common that the average is now two authors per paper, and the number is still rising. Papers with more than a dozen authors caused some concern at first but are now commonplace and acceptable. There are good reasons to suppose that this tendency to collaborate is a consequence of the growing institutionalization of scientific research and the increase in economic support for it. One result of massive support of research is that it becomes possible and even necessary for leading researchers to purchase the labors of their col-

419

leagues and assistants. The names of such workers appear on the papers as a form of fiscal accounting for the funds spent on the product. Thus the degree of support of any particular field or institution may be judged from the average number of authors per paper.

Perhaps one should also correct publication rates by computing fractional authorships in which the unit of the paper is divided into equal shares for the participating authors. Though this seems important, it actually makes surprisingly little difference. With an average of about two authors for each paper, those who continue to publish have approximately two primary and two secondary authorships each year. If the transients are included, the count from an annual index is an average of one primary and one secondary authorship per author per year. Therefore, with an overall productivity of about two authorships per author per year, the number of source authors publishing in a given year is about the same as the number of papers published in that year.

Such average figures, however, say little about the small core of highly productive authors on whom so much of the total research effort rests. Even including the transient researchers and those in mid-career, the five-year cumulation of the *Science Citation Index* lists 5.8 papers per cited author. At a publication rate of about half a primary authorship per person per year, this corresponds to more than a decade of production for each person. For the full working life of those who continue to publish, bibliographies of several hundred published and citable items are not at all uncommon. Among the most eminent, Lord Kelvin published more than 650 items, many of them

420

still very heavily cited, while among less-known researchers the world record seems to be held by Theodore Dru Alison Cockerell (1866–1948). This Colorado naturalist, an entomologist specializing in bees, has a total bibliography of 3,904 items published over 67 working years; he peaked at more than two publications per week and continued to publish more than one item per week for most of his life. Lord Kelvin's rate was a steady 8.5 items per year for the first 25 years and 15 per year for the next 30 years of his active period.

Though quantity and quality are quite separate measures, one should not succumb to the strong temptation to reject quantitative measurements as irrelevant in judging a worker's real contribution. Many scientists, especially the best, have a marked resistance to quantitative methods because of real fear that the entire system will be endangered if anything but insider peer judgments is admitted. It seems clear, however, that quantity and quality in research production distribute themselves over the same type of hyperbolic curve, so any sort of quality evaluation and weighting by objective or subjective means would leave all the stated conclusions about the statistics of productivity substantially unchanged. Only some of the names of the individuals concerned would be different.

For one thing, a good correlation exists between quantity and quality because of the workings of the Matthew Principle; qualitative success leads to further publication and lack of success tends to terminate publication. For another, the deviant cases seem to be rare. There are few who publish a single magnificent piece of research or even a sparse few golden pieces and leave it at that. The best researchers tend to have many publications, and only rarely will these be trivial and uncited. There are perhaps more instances of prolific authors of junk, especially when they capture some means of unrefereed publication. The well-known Dean's Principle of "publish or perish" had its roots in the fact that the one clear way of knowing that a teacher's mind was full was to see that it constantly ran over. The fiscal workings of this principle, as has often been noted, tend to result in the regular publication of papers that are frequently undistinguished and account only for time and money spent rather than new knowledge gained. However, citation weighting is an excellent diagnostic for such cases.

For Science and For Sweden: A Nobel Prize Judge and His World

By William K. Stuckey

Although not known by name to most of the world, the judges who award the Nobel Prizes are scholars of considerable influence. Their decisions have helped shape some of the most important events of the 20th century.

For almost two decades, the center of the secret and formal world of Professor Erik Rudberg was a genteel, red-walled building resembling a cross between a small castle and a library. It sits on the crest of a small hill in the northwest suburbs of Stockholm, its black-tiled roof capped by a sculptured stone crown in weathered green. The style of both the building and Professor Rudberg is benignly monarchical, suggestive of the elitist Sweden of the past. Yet the activities of Professor Rudberg and his colleagues bring great international prestige and power to the prosperous, egalitarian, permissive Social Democratic Sweden of today. As leading members of the Royal Swedish Academy of Sciences, headquartered in this elegant hilltop structure, Rudberg and 14 other Swedes select the winners of three of the six Nobel Prizes each year. They are the Nobel judges.

Since the prizes were founded in 1901, Nobel judges such as Rudberg have helped to shape some of the most important events of the 20th century. Prize awards, for example, have given encouragement and recognition to men who developed the great artificial marvels of the era—lasers, sophisticated communications systems components, penicillin, artificial fibers and fertilizers, Green Revolution strains of food grains, perceptive medical diagnostic and treatment machines, unbelievably sensitive scientific measuring devices, and the most terrible weapons in the history of man. The decisions of the Nobel judges have helped the names of Einstein, Madame Curie, Marconi, Fermi, Planck, Dirac, Townes, Röntgen, and Bardeen to be enshrined in textbooks, newspapers, and histories almost everywhere. Although the judges discharge their prize-selection duties as representatives of the private Nobel Foundation, the international scientific intelligence they gather through the "Nobel network" is of potential use to the public and industrial sectors of Sweden as well. The quality of their decisions also has long reinforced Sweden's reputation for objectivity and neutrality.

Until his retirement in January 1973, Professor Rudberg was the permanent secretary and principal executive officer of the Royal Swedish Academy of Sciences. Spiritually the 200 or so Swedish scientists of the academy are the inheritors of the great scientific traditions of the botanist Linnaeus, of the chemist Berzelius, and of all the other dead masters of the Age of Enlightenment in Scandinavian natural philosophy. Their official duties: to advise the Swedish government upon request, to operate libraries and small research stations, and to publish learned journals and almanacs. However, the unofficial duties assigned them in the will of the Swedish industrialist Alfred Nobel—to select winners of the physics and chemistry prizes (and, since 1969, the economics prize)—lift them above the other science academies of the world in importance. Outstanding academy scientists are selected to serve on the three Nobel Prize committees, and for the 11 years preceding his retirement Professor Rudberg was chairman of the physics prize committee.

Of all the six prizes, perhaps physics imparts the most prestige to its winners. Among other things, physics was the first prize mentioned in the Nobel will of 1896. Also, it is the most basic and ambitious of the sciences, and, aside from pure mathematics (for which there is no Nobel Prize), physics requires the most exacting efforts in logical thought. Physics seeks the function, character, and relationships of all forms of matter and force in the universe, from atomic nuclei and electromagnetic waves to massive quasars and strange "black holes" on the boundaries of the cosmos. Almost to a man, physicists agree that the Nobel Prize is the highest honor in their field. Fewer than 100, out of the tens of thousands of physicists who have lived in this century, have won it. And those who have won it often have risen to high posts of military, economic, educational, or governmental leadership in their own countries. Much of what they have come to be known for

WILLIAM K. STUCKEY *is a science writer and former editor of* MIT Reports on Research.

they owe to the sets of values, the sense of history, and the character of people such as Erik Rudberg.

Probably the best time of the year to have met Professor Rudberg before his retirement was in late summer. That is when I first met him, in the summer of 1971.

The sun is such an infrequent visitor to Sweden that when it finally appears, even those with urgent work to do are compelled to leave their desks, bathe in it, walk in it, smile at it. Rudberg is among them. As the permanent secretary of the Swedish science academy, he was assigned the largest single plot in a manicured garden near the academy building—a garden which in earlier times was used to train gardeners for the Swedish aristocracy.

Rudberg used the garden primarily for recreation, however. He would walk among the neat rows of beans, cabbages, and greens to smell the earth, taste something green, and clear his mind in the sun, a mind which in late summer was occupied with completing his most important task of the year, the writing of a book-length document with the uninspiring title "Committee Reports and Recommendations." In it, Rudberg presented reports on 20 to 30 outstanding physicists from the major science-oriented nations of the world. In the recommenda-

Erik Rudberg as he appeared in 1966. At that time he was permanent secretary and chairman of the physics committee of the Royal Swedish Academy of Sciences.

425

tions, he would state that no more than three should win the Nobel Prize for Physics for that year.

Although Rudberg is in his early 70s, he was so relaxed in these moments spent in the garden that he looked two decades younger. He is tall, walks with an easy gait, and when he smiles his nose wrinkles like that of a child. His face is long, thin, and kind and drops from a large head covered in combed-to-the-side gray hair with some last shades of reddish-brown still showing through. He is a man who first strikes you as being the product of a calm and war-free life.

The Rudberg character traits resemble in varying degrees those of other Nobel judges I met in Stockholm. Their character, their Sweden, is one which believes in the Lutheran work ethic and a modest, understated life-style; in a struggle toward objectivity no matter how painful and self-effacing that struggle may be; in being so secretive in conducting its affairs as to be almost invisible; in revealing the real substance of its views only to other members of the elite, rather than sharing it with everyone as the Social Democratic holy cry of "equality" would suggest; and in faithful adherence to rules, traditions, and customs regardless of outside pressures for change. The judge's world is humane, honest, apparently apolitical, committee-prone, not overly imaginative, and based on a rock-bound conservatism. It is also the world of the "network."

Ulf von Euler shared the Nobel Prize for Physiology or Medicine in 1970, and in 1974 was president of the Nobel Foundation. His father, Hans von Euler-Chelpin, shared the chemistry prize in 1929 and was an early colleague of Rudberg.

The "Nobel network"

The "Nobel network" is one of the world's most exclusive clubs, and it has influenced Rudberg's life for half a century. In essence, the Nobel network *is* Erik Rudberg's life.

The network is an international pyramid with the Nobel judges, such as Rudberg, occupying the top; all living Nobel laureates in the space not far below the top; and perhaps a thousand or more senior university professors and directors of scientific installations in Scandinavia, West Germany, the U.S., the U.K., France, and the Soviet Union (plus a handful of academics from "emerging" nations such as Israel, India, and Japan) fleshing out its base. The purpose of the network is the collection of scientific intelligence. Its members scour the earth for the brilliant discovery, the breathtaking bit of research that "opens a whole new field of activity," the laboratory or theoretical answer to a long-standing scientific question. Very discreetly, most indirectly, they will make inquiries about promising men whose work may be of Nobel caliber. They customarily feed such information to the judges in Stockholm by filling out forms which place a candidate in nomination for the prize, or, less formally, by presenting or attending international seminars on research which a Nobel judge might also, very discreetly, be attending.

No scientist may nominate himself for the prize. Only a member of the network can do that, and even then only by invitation from the judges (except for former prizewinners, who may nominate whomever they choose).

There is a similar but separate network for each of the prize committees. In addition to the networks radiating from the science academy, there is a network of senior life scientists and physicians that reports to the physiology/medicine committee based at Stockholm's Caroline Institute; of literary scholars, critics, and officials of various pen clubs reporting to the literary prize committee of the Swedish Academy, headquartered in Stockholm's charmingly medieval Old Town; and of senior diplomats, politicians, and social scientists who propose candidates for the peace prize to a committee appointed by the Norwegian parliament, the Storting.

Rudberg enters the network

Rudberg was not born into the network, as some judges and active networkers have been, but he nevertheless came to its attention at an early age. He was born in 1901, the first year the Nobel Prizes were awarded, into the family of a Stockholm high school physics teacher. Young Rudberg had a natural bent for the techniques and concepts of physics, and by the age of 12 was helping his father design classroom physics experiments for high school students.

It was soon obvious that Rudberg's interests and intelligence would lead him into a university career, something only a small number of Swedish youths could expect in that elitist and pre-Social Democratic day of the Swedish monarchy. (Only in the past two decades has the idea of higher mass education and open admissions to universities been accepted in Sweden.) He was active physically and spent a summer working in railroad construction in the north of Sweden. An engineering career interested him—outdoor activity, science-based—until he learned that most engineers of his day spent their days in offices. He chose a university diet of physics and chemistry instead—and stumbled into the network when he was just 20.

Today Rudberg is not quite sure of the exact position which Hans von Euler-Chelpin, an early teacher of his at the University of Stockholm, held in the network. Such things are simply not discussed. Yet a part of the network he was, as indicated by the fact that von Euler nominated some winning candidates and also shared the 1929 Nobel Prize for Chemistry.

Von Euler illustrated two facets of network life: a number of people who are invited by the Nobel judges to nominate prize candidates eventually win the prize themselves; and, on occasion, separate members of the same family of network members will win it. Von Euler's son, Ulf von Euler of the Caroline Institute, won his own prize in physiology/medicine in 1970 and is now the president of the Nobel Foundation. Perhaps the most famous father-son team of Nobel laureates was Sir William Henry Bragg and Sir Lawrence Bragg, who shared the 1915 prize for developing X-ray crystallography. Also, a current senior member of the Nobel physics committee, Kai Siegbahn of the University of Uppsala, is the son of 1924 physics prizewinner Karl Manne Siegbahn.

Rudberg's brightness impressed the elder von Euler at the University of Stockholm, and soon the two were doing research in physical chemistry together and publishing the results in learned journals.

"I merely started out attending his lectures," Rudberg recalled in 1973. "Von Euler was continually seeking out students who should be encouraged, and asked his assistants to watch for them. When I came to his attention, he gave me a problem to work on to see how I would do. I did all right, and we began a full professional collaboration. He encouraged me very much and helped me to get into Göttingen."

Göttingen was the holy place of physics before World War II. This small German university town was the home of David Hilbert, one of the century's most brilliant mathematicians, and also was an academic stopping station for such physicists as J. Robert Oppenheimer of the U.S., and squads of future Nobel laureates such as Paul Dirac, Werner Heisenberg, and the 1925 physics laureate with whom Rudberg studied at Göttingen, James Franck. All of these men were in the Nobel network. The 1920s have been termed the "golden age of physics," and it was at Göttingen that the gold shone most brightly.

During the 1920s and 1930s, Rudberg worked with several networkers. He conducted research for Svante Arrhenius in Stockholm (chemistry, 1903) and with O. W. Richardson in the University of London's King's College (physics, 1928). By the time Rudberg became an assistant professor of physics at MIT in the 1930s, he had worked also in the three principal network countries, Great Britain, Germany, and the U.S. (Sweden is also a network country, of course, but the painfully objective Swedes think long and hard before giving one of their countrymen a prize.)

Rudberg's own research won the respect of physics professionals in all the network countries. His specialty was using electrons to bombard various metals, useful both in determining the characteristics of electrons and of the metal targets. Yet, as far as he knows, "I was never nominated for the prize." But he obviously was being groomed, albeit slowly, for something perhaps more important, a post as director of a key segment of the network.

Rudberg returned to Sweden in 1936 to conduct his first industrial research, for the Swedish General Electric Company, and dropped out of academic life until 1939, when he became chairman of the physics department at Chalmers University of Technology in Sweden's second largest city, Göteborg. While Sweden stayed neutral in World War II — it has not been in a war since Napoleonic times — Rudberg handled some research projects for the Swedish military.

"It was modest work, dealing with things like instruments for submarine detection, military uses of magnetism, and finding the right kind of paint for our submarines," he recalled.

After the war Rudberg was appointed director of the Swedish Institute for Metal Research — "it was almost too industrial for me, and I wanted to return to more basic research" — and continued to serve as secretary of the Swedish National Committee on Physics. This brought him back into the Nobel network, for the president of that committee was Swedish laureate Karl Manne Siegbahn. But in the meantime he had made important Swedish industrial contacts, and industry too plays a role, if somewhat indirect, in Nobel Prize matters. Of the five board members of the Nobel Foundation, two are bankers.

Rudberg's first chance to move into the network's inner circle came in 1947, when a vacancy appeared in the physics section of the Royal Swedish Academy of Sciences. He was nominated for it, since everyone in the academy knew him by then. However, another candidate received more votes than Rudberg, and he had to wait until another vacancy occurred in 1954. Finally, at age 53, he was considered seasoned enough for membership.

Also, Rudberg was perfect for the role of a judge: an attentive listener to any scientist who might have a tidbit of scientific intelligence for the network, yet tough enough to resist a lobbying effort on behalf of an overly ambitious scientist. He had remained conversant with many areas of physics, was a "natural committeeman" who preferred the seeking of consensus to table-pounding pushing of his own views, and, of course, was discreet. Another quality which made him obvious judge material was that he was Swedish by birth, yet international in training. Although Nobel statutes do not exclude non-Swedes from being judges, the science academy has never appointed a foreigner to such a position.

In his first year as a science academy member, Rudberg was invited by the physics committee to be a "joint member." That meant that he would work on the committee for one year only, as opposed to the four-year terms of full members. Joint membership is often used as a probationary period to see if a scientist has the makings of a full-fledged Nobel judge. It is not known what Rudberg worked on during that year as a joint member, although his former MIT student, William Shockley, emerged as a co-winner of the physics prize two years later.

Rudberg's committee life moved on secretly but apparently quite fruitfully, since, in 1959, he was named permanent secretary of the academy and four years after that became chairman of the physics committee. He had arrived at the center of the network, at the age of 62.

The Nobel judges at work

Rudberg has seen many judges at work. In commenting on the qualities a good Nobel judge should possess, he observes "He doesn't have to be the most brilliant scientist in the world as long as he has good international contacts and knows his own field thoroughly."

Such a statement perfectly describes Rudberg himself. As to "good international contacts," he must have more than any physicist in Sweden, since he has served as president of the principal European physical society and worked for decades with the International Scientific Union.

A judge should also be a persuasive, informed advocate, who through sheer force of knowledge and conviction can persuade other judges to support a controversial prize candidate. Such a man was Stig Lundqvist of Sweden's Chalmers University of Technology and Göteborg University. Lundqvist was named a joint member of the physics committee in 1972 and had custody of the very difficult case of John Bardeen of the University of Illinois. The problem was that Bardeen had already won a physics prize, sharing the 1956 prize with William Shockley and Walter Brattain for the discovery of the transistor. And no man had ever been awarded two physics prizes, as Lundqvist was proposing.

Bardeen's part in developing the first accurate and comprehensive theory of superconductivity, explaining why all electrical resistance disappears in certain metals when they are cooled to temperatures near absolute zero, was easily as spectacular and significant as was his transistor work, Lundqvist argued before the committee. The committee's conservatives resisted, deeply disturbed by the thought of breaking a tradition. There was also considerable support for another candidate, Brian Josephson of the University of Cambridge. Lundqvist finally won, however, in part because he pointed out that Josephson's work owed much to Bardeen's superconductivity theory. The committee finally ended its resistance. Bardeen and two co-workers, Leon Cooper and John Schrieffer, received the 1972 physics prize. And Stig Lundqvist, who was no more than a temporary joint member

of the committee in 1972, became its chairman, succeeding Rudberg, early in 1973.

Yet resistance to pressure on behalf of a candidate is still one of the highest virtues a judge can possess. An easily persuaded person would probably not be elected to the academy in the first place, Rudberg stated, much less to its physics section or the physics prize committee.

Nothing, the true judge believes, must be done to compromise the world's belief in the fairness, neutrality, and correctness of the prize. Every judge tells this to the outsider. To test it facetiously, I asked Rudberg if he thought the physics committee would award the prize to a man who was unquestionably brilliant and accomplished in physics, but who also was a heroin addict with a bad credit record and unspeakable sexual perversions, and a celebrant of Black Masses. His reply: "Of course they would give the prize to him if they were convinced of the value of his scientific work—just to make a point about the objectivity of the prize."

But is such eccentricity, if it can be called that, allowed on the part of the judges? Are good judges the celebrants of Black Masses? Of course not, Rudberg replied, although their tastes and behavior are not homogeneously boy scout (as they sometimes appear to outsiders).

"We are not required to be alike or to like each other," said Rudberg. "There definitely have been judges whom I would not associate with socially. Some might have an absurd family life. Others might have had tastes I didn't approve of, or have been totally dull or nega-

Crown Prince Carl Gustaf of Sweden (left, now King Carl XVI Gustaf) presents the 1972 Nobel Prize for Physics to John Bardeen of the University of Illinois. Bardeen became the first person to win two physics prizes, receiving his first award in 1956.

431

tive regarding things I was rather fond of. But most of us have gotten along well enough and share the same general professional goals."

Perhaps the most important quality a judge should have is patience. He should be able to restrain his own personal enthusiasm for a prize candidate long enough, for decades if necessary, until he is certain that his man's research is of enough importance. The average prize candidate is considered by the committee for at least 10 years, and sometimes 25 or 50.

Nevertheless, such patience pays off for the reputation of the prizes. The judges will admit they have been wrong on occasion, but they have not been wrong many times in the seven-decade history of the prizes. Few physicists will quarrel with the list of winners.

At the same time, almost every physicist claims to know of a good man who should have won, but did not, or a field of physics which ought to have received Nobel recognition, but has not. None of the great developments in computers, space research, or rocketry has been laureled by the committee, for example. Tradition has limited the number of co-winners of a prize each year to three, and such developments were wrought by teams of many more than three.

Rudberg admits that such flaws exist. "On looking back at the winners, I can say that occasionally a man got it who was excellent, but who was less intellectually sharp—less insightful into nature—than a contending candidate who lost. So the winner was lucky, that's all.

"We can't measure excellence all that accurately," he continued. "There are no generally accepted measurements which apply equally to all areas of science. In a foot race, you know without a doubt who won. In science, it is a matter of opinion as to who is best."

As to the more neglected fields of natural science such as geology and astronomy, Rudberg suggested that the outside world be patient. Such fields will produce their Nobel laureates "in time."

And that is the way Rudberg talks about secret affairs to an outsider —occasionally candid, protective to the institutions which form his world, mature enough to recognize that all of his decisions will not be right, a calm, pleasant, charming gentleman, and a skilled operator of one of the most effective public relations machines for intellectualism of all time.

Index

Index entries to feature and review articles in this and previous editions of the *Yearbook of Science and the Future* are set in boldface type, *e.g.,* **Astronomy**. Entries to other subjects are set in lightface type, *e.g.,* Radiation. Additional information on any of these subjects is identified with a subheading and indented under the entry heading. The numbers following headings and subheadings indicate the year (boldface) of the edition and the page number (lightface) on which the information appears.

> **Astronomy 75**–181; **74**–408, 103; **73**–184
>> aurora polaris **74**–144
>> Center for Short-Lived Phenomena **75**–151
>> honors **75**–247; **74**–249; **73**–251
>> Orbiting Solar Observatory **73**–182
>> planetary research **75**–178
>> SAS-A Program **74**–396
>> spectrograph **73**–175
>> topography of Venus **73**–223

All entry headings, whether consisting of a single word or more, are treated for the purpose of alphabetization as single complete headings and are alphabetized letter by letter up to the punctuation. The abbreviation "il." indicates an illustration.

A

AAA (American Anthropological Association) **75**–191
Abelson, Philip **75**–321
Aborigines **73**–141
Abortion **74**–271
Abscisic acid **75**–199
Abu Simbel ils. **74**–122, 123
Academy of Sciences (U.S.S.R.) **74**–179, 413; **73**–179
Accelerator, Particle: *see* Particle accelerator
Accelerographs **74**–427
Accidents and safety
  alcoholism **73**–277
  biological rhythms **74**–83
  building codes for earthquake design **74**–424
  coal mining **73**–249
  hazards of virus experimentation **74**–301
  metric system for signs **73**–153
  nuclear power **74**–239
  passenger buses **74**–335
  traffic regulation **73**–383
  V/STOL aircraft **73**–330
Acetone **75**–285
Acetylene **75**–211
α-acetyl-methadol (drug) **73**–290
ACGP (Arctic Circumpolar Geophysical Program) **75**–226
Acoustics (Sound) (phys.) **74**–370
  echo techniques **75**–340
  motion picture sound recording **75**–307
  physics of stringed instruments **74**–129
Acqua alta (High water) **75**–348
Acronycine **75**–208
ACS: *see* American Chemical Society
ACTH: *see* Adrenocorticotropic hormone
Actin **75**–295
Action potential (bot.) **73**–202
Activated charcoal
  pest control **75**–160
Acupuncture 74–50; **73**–268
  Chinese policies **73**–45
  dental research **74**–275
  evaluation **75**–273
  psychic phenomena **73**–68
  surgery il. **73**–46
ADA: *see* American Dental Association
Adamenko, Victor (physic.) **73**–81
Adams, Roger (chem.) **73**–307
ADAS (Air data acquisition system) **75**–329
Additives
  food **75**–244; **73**–243
Adenosine diphosphate (ADP) **73**–293
  chlorophyll fluorescence **75**–293
Adenosine 5′-monophosphate (5′-AMP,

5′-adenylic acid) **74**–380; **73**–293
Adenosine 3′, 5′-monophosphate: *see* Cyclic AMP
Adenosine triphosphate (ATP) **74**–382; **73**–293, 302
  southern corn leaf blight **75**–197
Adenylyl (Adenylate) cyclase **75**–291, **74**–382
3′, 5′-adenylic acid: *see* Cyclic AMP
5′-adenylic acid: *see* Adenosine 5′-monophosphate
ADP: *see* Adenosine diphosphate
Adrenaline (Epinephrine) **74**–381
Adrenocorticotropic hormone (ACTH) **74**–88, 384; **73**–214
Adriamycin **75**–208
Advanced Research Projects Agency (ARPA) **74**–220; **73**–220
AEC: *see* Atomic Energy Commission
Aegilops (Goat grass) **75**–62
Aeronautical satellite (Aerosat) system **74**–174
Aeronomy **74**–193
Afo-A-Kom (statue) **75**–161
Agency for International Development, U.S. **75**–243
Agnes (tropical storm) **74**–229
AG/PACK (AGricultural Personal Alerting Card Kits) (U.K.) **74**–257
Agricultural Research Service (ARS) **75**–159; **74**–159
Agriculture 75–159; **74**–159; **73**–160
  applied chemistry research **75**–203
  archaeological finds **75**–162
  botanical research **75**–197
  Chinese policies **73**–44
  cloning **75**–34
  diet requirements **73**–131
  drought **75**–227
  earth resources surveys **75**–172; **74**–172; **73**–174
  effects on tropical forests **75**–391, il. 390
  energy crisis **75**–238
  food supplies **75**–243
    need for pest control **75**–114
  Origins of Agriculture, The **75**–56
  plant-animal interactions **73**–200
  rice paddies il. **73**–42
  Sahara Desert **75**–403
  Tasaday tribe **73**–50
  veterinary medicine **75**–335
  waste materials
    fuel production **75**–285
Agriculture, U.S. Department of (USDA) **75**–159; **73**–241
  food research **75**–203, 244; **73**–134
  insect pest control **75**–160
  VEE vaccination program **73**–336
AGU (American Geophysical Union) **75**–259

Airbus **75**–328
Air conditioning **74**–169
  energy conservation **75**–166
  geothermal energy **75**–112
  solar energy **75**–78
    solar furnace ils. **75**–77, 80; **74**–246
Air-cushion vehicle **75**–335
Air data acquisition system (ADAS) **75**–329
Airfoil **75**–95, il. 94
Airlock module **75**–176; **73**–177
Air pollution: *see under* Pollution
Airports **75**–329
  Dallas-Fort Worth ils. **75**–330, 331
  Montreal's jetport **74**–169
  noise problems **74**–372
Airship **75**–330
Airtrans (automated airport transportation system) **75**–329
Akuri (people, S.Am.) **75**–155, il. 154
ALA-dehydrase (Delta-aminolevulinic acid dehydrase) **75**–370
Albumin
  genetic research **75**–297
Alcohol **73**–275
  bacterial production of fuel **75**–285
  yeast production **75**–244
Alfalfa **75**–159, 203
Alfano, R. R. **75**–293
Alfvén, Hannes **73**–252, 256
Algae **73**–338
  role in evolution **75**–223
Algebraic geometry **75**–263; **74**–262
Algebraic K-theory **74**–262
Algol (star) **75**–21, 188
Allergy **74**–293
  lymphocytes **75**–133
  rheumatic diseases **75**–277
Allograft **75**–135, il. 142
Alphanumeric character (computer technology) **74**–232
Alpha particles **75**–313; **73**–316
Alpha waves (brain) **73**–78
  monitoring ils. **73**–79, 80
ALSEP: *see* Apollo lunar surface experiments package
Altered states of consciousness (ASCs) **73**–77
Altimeter
  Skylab **75**–175, 262
Aluminum **75**–220
  electron micrography **75**–374
"Alvin" (submarine) **74**–260
AMA: *see* American Medical Association
Amchitka Island, Alaska **73**–225
American Academy of Family Practice **75**–272
American Anthropological Association (AAA) **75**–191
American Association for the Advancement of Science **75**–321

environmental issues **75**–238
  Mead, Margaret **75**–190
American Astronomical Society **74**–415
American Cancer Society **75**–269; **73**–275
American Chemical Society (ACS) **75**–203, 321; **73**–326
American Dental Association (ADA) **75**–281; **74**–275; **73**–282
American Geophysical Union (AGU) **75**–259
American Hospital Association **75**–272; **74**–111
American Medical Association (AMA) **75**–271; **73**–274
  nutrition **73**–133
American Physical Society **75**–316; **74**–316; **73**–325
American Psychological Association **75**–194; **73**–197, 328
American Science and Engineering, Inc. **74**–396
American Society for Neurochemistry **74**–324
American Society of Civil Engineers (ASCE) **73**–326
American Society of Heating, Refrigerating, and Air-Conditioning Engineers **75**–166
American Telephone and Telegraph Company (AT & T) **75**–215; **74**–218
  mobile communications **73**–216, 231
  satellite communications **75**–170
American Veterinary Medical Association (AVMA) **75**–337; **74**–336
Amino acids
  applied chemistry **75**–203
  coenzyme-binding sites **75**–291
  high-lysine corn **73**–131
  lunar rocks **74**–338
  protein structure **74**–38; **73**–297
  protein synthesis **74**–291
  residues **74**–296
  structural chemistry **75**–213
5-amino-3-chloropyrazinecarbox-aldehyde oxime **73**–211
Ammonia
  geothermal energy technology **75**–107
  recycling by plants **73**–199
  solvated clusters **73**–207
  water disinfection **75**–203
5′-AMP: *see* Adenosine 5′-monophosphate
Amphetamine (chem.) **74**–271; **73**–272
  research on rats **74**–89
Amphioxus **75**–343
Amplitude analysis (phys.) **73**–314
Amtrak **75**–333; **74**–332; **73**–333
  turbine-powered train il. **75**–334
Analog (chem.) **74**–210
Analog communication **75**–214
  optical fibers **75**–50

433

Acknowledgments

4   Illustration by Peter Lloyd
6   Photographs by (top left) Nelson Merrifield from FPG; (top right)
      Louis Goldman from Rapho Guillumette; (center left and bottom left)
      Jacques Jangoux; (center right) courtesy, Bell Laboratories;
      (bottom right) courtesy, H. Fernandez-Moran, University of Chicago
10–11 Photograph by Lee Boltin
32–33 Illustration by Peter Lloyd
35, 38,
40–42 Illustrations by Tim Clark
44–45 Photograph courtesy, Bell Laboratories
46–47, 49, 51,
53, 54 Illustrations by Dave Beckes
56–57 Illustration by Dennis Magdić
61, 65 Illustrations by John Craig
72–73 Illustration by Jan Wills
74–75 Illustration by Ben Kozak
78–79 Illustrations by Dave Beckes
86–87 Illustration by Ben Kozak
94–95 Illustrations by Dave Beckes
98–99 Illustration by Ben Kozak
106, 112 Illustrations by Dave Beckes
115   Illustration by Kerig Pope
125   Illustration by Dave Beckes and Zorica Dabich
128   Illustration by Dennis Magdić
128   Photograph by Manfred Kage from Peter Arnold
129   Bettmann Archive
131, 134, 139–
142   Illustrations by Dennis Magdić
144–145 Illustration by John Craig
158   Photographs (from top to bottom) courtesy, IBM Corp.; David Hughes
      from Bruce Coleman Inc.; courtesy, Smithsonian Institution;
      courtesy, Fermi National Accelerator Laboratory
167, 194, 200, 202, 204–212, 214, 240, 289, 292–294,
296, 338 Illustrations by Dave Beckes
344–345 Photograph by Fritz Henle from Photo Researchers
362–363 Photograph by Dan Morrill—EB Inc., courtesy, University of Chicago
365   Illustration by Tim Clark
379   Photograph by Shelly Grossman from Woodfin Camp
380, 403 Illustrations by Dave Beckes
408–420 Illustrations by John Everds